RANKING OF MULTIVARIATE POPULATIONS

A Permutation Approach with Applications

RANKING OF MULTIVARIATE POPULATIONS

A Permutation Approach with Applications

Livio Corain
University of Padova
Vicenza, Italy

Rosa Arboretti
University of Padova
Vicenza, Italy

Stefano Bonnini
University of Ferrara
Ferrara, Italy

with contributions from
Luigi Salmaso and Eleonora Carrozzo

CRC Press is an imprint of the
Taylor & Francis Group, an **informa** business
A CHAPMAN & HALL BOOK

CRC Press
Taylor & Francis Group
6000 Broken Sound Parkway NW, Suite 300
Boca Raton, FL 33487-2742

First issued in paperback 2022

ISBN-13: 978-1-466-50554-4 (hbk)
ISBN-13: 978-1-03-234005-0 (pbk)
DOI: 10.1201/b19673

Publisher's Note

The publisher has gone to great lengths to ensure the quality of this reprint but points out that some imperfections in the original copies may be apparent.

Library of Congress Cataloging-in-Publication Data

Names: Corain, Livio, author. | Arboretti, Rosa, author. | Bonnini, Stefano, author.
Title: Ranking of multivariate populations : a permutation approach with applications / by Livio Corain, Rosa Arboretti, and Stefano Bonnini.
Description: Boca Raton : CRC Press/Taylor & Francis Group, 2016. | Includes bibliographical references and index.
Identifiers: LCCN 2015042960 | ISBN 9781466505544 (alk. paper)
Subjects: LCSH: Ranking and selection (Statistics) | Order statistics. | Sequential analysis. | Multivariate analysis.
Classification: LCC QA278.75 .C67 2016 | DDC 519.5/35--dc23
LC record available at http://lccn.loc.gov/2015042960

Visit the Taylor & Francis Web site at
http://www.taylorandfrancis.com

and the CRC Press Web site at
http://www.crcpress.com

Contents

Preface

The purpose of this book is to introduce a novel permutation-based nonparametric approach for the problem of ranking several multivariate populations using data collected from both experimental and observation studies and set within some of the most useful designs widely applied by research and industry investigations such as MANOVA, that is, multivariate independent samples and multivariate randomized complete block (MRCB) design, that is, multivariate dependent samples also known as repeated measures.

This topic is not only of theoretical interest but also has a practical relevance, especially to business and industrial research, where a reliable global ranking in terms of performance of all investigated products/prototypes is a very natural goal. In fact, the need to define an appropriate ranking of items (products, services, teaching courses, degree programmes, and so on) is very common in both experimental and observational studies within the areas of business and industrial research. In the field of New Product Development the research aim is often focussed on evaluating the product/service performances from a multivariate standpoint, that is, in connection with more than one aspect (dimension) and/or under several conditions (strata).

From the methodological standpoint, the problem of ranking several multivariate populations is dealt with by referring to the relative stochastic ordering of each population when compared with all other populations. Basically by pairwise counting how many populations are stochastically larger than any specific population we extend into a nonparametric multivariate framework the traditional definition of ranking. Usually in the literature, depending on the assumptions made about the random errors, the distribution of ranking parameter estimators can be derived in a parametric or nonparametric way. However, as demonstrated by many simulation studies and applications to complex real case studies, the parametric approach presents a number of drawbacks and limitations. Conversely, thanks to its robustness and flexibility, the permutation and combination approach appears to be more reliable and powerful.

This book has a double configuration: on one hand it is part of the research field of so-called methods of ranking and selection, and it integrates that theory from the perspective of a new nonparametric approach. On the other hand, the scheme of the book is very application and problem-solving oriented, as has been suggested by numerous case studies that are solved by reference to a customizable software according to practitioner's needs.

The book is directed mainly to postgraduate students (i.e. master's and PhD courses) in the areas of statistics and applied sciences (engineering, chemistry, materials, economics, biomedicine, etc.). Considering that this is an application-oriented book it can also be used by practitioners with a basic knowledge of statistics who are interested in the same applications described in the book or in similar problems. The presence of customizable Web-based software tools useful to solve the application problems makes the book very useful from a practical standpoint. Practitioners interested in the software can perform some application examples proposed in the book. For further requests, online assistance and consulting is provided at: info.globalranking@gest.unipd.it.

The book has two sections. Section I introduces and motivates the topic of ranking of multivariate populations by presenting the main theoretical ideas and an in-depth literature review. After presenting our Web-based NPC Global Ranking app we developed as companion software for the book, in Section II we present a large number of real case studies, by classifying them into four specific research areas: new product development in industry, perceived quality of the indoor environment, customer satisfaction in services and finally cytological and histological analysis by image processing.

We recognize the contribution of Professor Luigi Salmaso, who is actually leading our research group and who supervised all parts of the book with special contribution to the theoretical aspects and support in the development of the software. As for the software tools, we acknowledge Eleonora Carrozzo, who helped develop the R software code. The contribution of three authors to the book has to be considered equally likely with respect to all chapters provided that some chapters already extend some published papers listed in the references.

Authors share the full responsibility for any errors/ambiguities, as well as for the ideas expressed throughout the book. Although we have tried to detect and correct errors and eliminate ambiguities, there may well be others that escaped our scrutiny. We take responsibility for any remaining ones, and we would appreciate being informed of them.

The NPC Global Ranking software was co-authored by Luigi Salmaso and Livio Corain, who acknowledge also all valuable colleagues involved in this project, particularly Davide Ferro. This software can be customized according to the necessity of the user. For any request please contact: info.globalranking@gest.unipd.it.

Livio Corain
Rosa Arboretti
Stefano Bonnini

Authors and Contributors

Livio Corain, PhD, is assistant professor of statistics at the University of Padova, Italy. His main research areas concern nonparametric methods for multivariate hypothesis testing and multivariate ranking, statistical methods for quality and improvement of products and services and applied statistics to engineering and biomedical studies.

Rosa Arboretti, PhD, is assistant professor of statistics at the University of Padova, Italy. Her main research areas concern nonparametric methods for multivariate hypothesis testing and composite indicators, statistical methods for conjoint analysis and applied statistics to engineering and biomedical studies.

Stefano Bonnini, PhD, is assistant professor of statistics at the University of Ferrara, Italy. His main research areas concern nonparametric statistics and statistics applied to economic and social sciences, health sciences, and engineering.

Luigi Salmaso, PhD, is full professor in statistics at the University of Padova, Italy and he is leading the research group on nonparametric methods in statistics applied to real data. His main research areas concern nonparametric methods for multivariate hypothesis testing, ranking and composite indicators, statistical methods for quality and applications to real case studies in engineering and biostatistics.

Eleonora Carrozzo is a PhD student in management engineering at the University of Padova, Italy. Her research work is mainly concerned with nonparametric methods for construction of composite indicators and ranking of multivariate populations with application in customer satisfaction surveys.

Section I

Theory and Methods

1

Introduction and Motivation

The necessity and usefulness of defining an appropriate ranking of several populations of interest, that is, processes, products, services, teaching courses, degree programs and so on are very commonly recognized within many areas of applied research such as chemistry, material sciences, engineering, business, education, biomedicine and so forth. The idea of ranking in fact occurs more or less explicitly whenever the goal in a study is to determine an ordering among several input conditions/treatments with respect to one or more outputs of interest when there might be a 'natural ordering'. We remark that the 'natural ordering' should be directed to the way in which the response is interpreted and not to any kind of a priori knowledge on ordering of populations that in the framework of the methodology we propose is not assumed at all.

This happens very often in the context of management and engineering problems where the populations can be products, services, processes and so on and the inputs are, for example, the managerial practices or the technological devices that are considered in relation with some suitable outputs such as any performance measure. Given the frequency with which these situations occur in specific managerial and engineering areas, we can consider the typical issues of operations management, quality control and marketing, where numerous different theories and diverse methods have been developed in the literature surrounding the problem of ranking. At the same time, the ranking problem is a typical interdisciplinary subject; for example, in the development process of a new product where managerial practices, engineering issues and statistical techniques are jointly involved to achieve high-quality and potentially successful products.

Often the populations of interest are multivariate in nature, meaning that many of their aspects can be simultaneously observed on the same unit/subject. For example, in many technological experiments the treatments under evaluation provide an output of tens (see examples in Chapter 5) or even hundreds of univariate responses (e.g., think of the myriad of automated measurements that are performed on a silicon wafer during the manufacturing process in the microelectronics industry). Similarly, but in a completely different context, a survey-based observational research provides for each unit/subject (respondent) a long list of answers, as in the case of evaluating and monitoring of customer satisfaction (see examples in Chapter 7).

From a statistical point of view, when the response variable of interest is multivariate in nature, the inferential problem may become quite difficult to cope with, owing to the large dimensionality of the parametric space. Moreover, when the goal is to compare several multivariate populations, a further element of difficulty is related to the nature of the response variable. If we consider a continuous response, provided that the underlying distributional and sampling assumptions are met and the degrees of freedom are large enough, then inference on populations can be performed using classical methods (e.g., such as Hotelling T^2). But when the response variables are ordered categorically the difficulties of the traditional methods based on contingency tables may become insurmountable. Nonparametric inference based on the Nonparametric Combination (NPC) of several dependent permutation test statistics (see Pesarin and Salmaso, 2010a), as we see in Chapters 2 and 3, allows us to overcome most of these limitations, without the necessity of assuming any specified random distribution. The main advantages of using the permutation and combination approach to classify and rank several multivariate populations is that it is the only testing method that allows us to derive multivariate directional p-values that can be calculated when the number of response variables is much larger than the number of replications (so-called *finite-sample consistency* of combined permutation tests; see Section 2.4.2). It is worth noting that in this situation, which can be quite common in many real applications, all traditional parametric and nonparametric testing procedures are not appropriate at all.

To better illustrate the goal behind the ranking of multivariate populations and the related concepts such as stochastic dominance and ordering within a multivariate setting, let us consider three bivariate normal populations Π_1, Π_2 and Π_3 represented by the random variables $\mathbf{Y}_j \sim N(\boldsymbol{\mu}_j, I)$ for $j = 1, 2, 3$, where $\mathbf{Y}_1$ is dominated by $\mathbf{Y}_2$ and $\mathbf{Y}_3$ with respect to both univariate components, that is, Y_1 and Y_2, while $\mathbf{Y}_2$ dominates $\mathbf{Y}_3$ for the second component and the opposite holds for the first component (Figure 1.1).

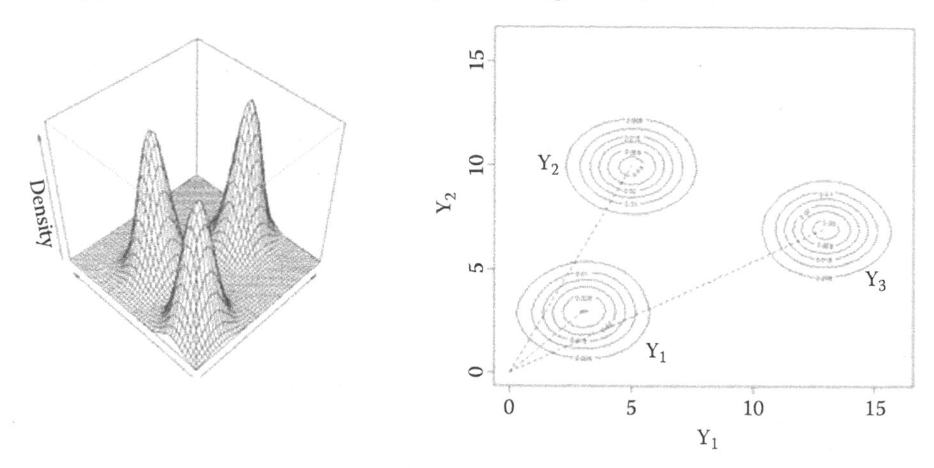

FIGURE 1.1
Density and contour plot of three spherical bivariate normal populations.

As quite often happens in many real-world situations, we assume that all populations can strictly take positive real values and the rule 'the larger the better' holds so that the origin may represent the minimum reference value. As suggested by the ranking and selection literature (Gupta and Panchapakesan, 2002), let us choose as the multivariate ranking parameter the Mahalanobis distance from the origin $D_j = \boldsymbol{\mu}_j^T I^{-1} \boldsymbol{\mu}_j = \sum_{k=1}^{2} \mu_{kj}^2$ and let $\hat{D}_j$ be the related sampling estimators, $j = 1, 2, 3$; as we are referring to spherical normal distributions the Mahalanobis distance is equal to the Euclidean distance (dashed lines in Figure 1.1); therefore because $D_1 < D_2 < D_3$ within this metric the true underlying population ranking can be defined as (3, 2, 1). Accordingly, we can reformulate the multivariate ranking problem into a univariate dominance problem focussed on the sampling estimators of D_j; in particular in this case the relation $\hat{D}_1 \overset{d}{<} \hat{D}_2 \overset{d}{<} \hat{D}_3$ takes place. Note that the distribution of $\hat{D}_j$ does depend on the multivariate characteristics of the related population distribution.

As can be deduced from the previous example, it is worth noting that possible opposing dominance of several univariate components from two or more given populations do not affect the possibility of defining and inferring the possible stochastic dominance and multivariate ordering among those populations. In fact, as in Dudewicz and Taneja (1978), the multivariate ranking and selection literature highlights that once a suitable scalar function of the unknown parameters has been chosen, this permits a complete ordering of the populations and the related inferences are based on a suitably chosen statistic that has a univariate distribution (Gupta and Panchapakesan, 1987). However, it is worth noting that when trying to perform parametric pairwise hypothesis testing on the Mahalanobis distances (via Hotelling-type statistics) with the goal of inferring the ordering that can be supported by sampling data, several complications are encountered, as pointed out by Santos and Ferreira (2012); in particular, the joint distribution for all pairs of mean vectors is unknown even under assumption normality. At any rate, several bootstrap and permutation solutions do exist; see, for example, Santos and Ferreira (2012), Minhajuddin et al. (2007) and Finos et al. (2007) for directional alternatives.

When the multivariate population distributions are not specified, that is, considering the ranking problem from a nonparametric point of view, we should refer to a more general and possibly metric-free distance measures. Similarly to what has been proposed by several authors within the nonparametric ranking and selection framework (Govindarajulu and Gore, 1971), in this book we consider as a multivariate ranking parameter the relative stochastic ordering of each population when compared with all other populations. This parameter can be estimated by a function of the distribution function *F*, specifically by pairwise counting how many populations are stochastically larger than any specific population. Note that this composite indicator can be viewed as a nonmetric 'distance measure' among multivariate distributions. In this connection the combination methodology (Pesarin and Salmaso, 2010a) appears to be a useful tool because of its ability to reduce the dimensionality so as to compare and rank the populations under investigation.

Informally speaking, the underlying idea behind the permutation approach for ranking of multivariate populations we propose in this book is quite simple: given two multivariate random variables $\mathbf{Y}_j$ and $\mathbf{Y}_h$, if $\mathbf{Y}_j$ dominates $\mathbf{Y}_h$ then the significance level function related to the combined test statistic T''_ϕ suitable for testing the null hypothesis of equality in distribution against the alternative $\mathbf{Y}_j \overset{d}{\gtrless} \mathbf{Y}_h$ will be stochastically larger under the alternative than under the null hypothesis of equality. Moreover, the significance level function under the true alternative will also dominate that one under the false directional alternative (see Expression 1.5 in Section 1.2). As pointed out by Pesarin (2001), the multivariate combined permutation p-values have the property to be *positive upper orthant dependent* (Lehmann, 1966), so that the probability to reject correctly the false multivariate null hypothesis is larger than the product of rejecting of each individual marginal component.

Actually, beyond the fact that we are referring here to a multivariate setting, using the pairwise p-values as tools for ranking univariate populations is not an entirely new idea in the literature. In fact, from Tukey's underlining representation of pairwise comparison results according to the increasing values of their estimated means (Hsu and Peruggia, 1994), one can argue which population can be overall considered the best, the second best and so forth. In this regard Bratcher and Hamilton (2005) proposed a Bayesian subset selection approach to ranking normal means via all pairwise comparisons and compared their model with Tukey's method and the Benjamini–Hochberg procedure (1995).

After presenting some guideline examples to help readers understand in practice the precise meaning of the problem of ranking of multivariate populations, this chapter introduces the formalization and the general solution of the multivariate ranking problem. As will be shown, we intend the ranking problem to be seen as a nonstandard data-driven classification problem, which can be viewed as similar to a sort of a special case of a post hoc multivariate multiple comparison procedure. In this view, the classification procedure is an empirical process that uses inferential tools, in particular permutation test statistics, with distance indicators and signals playing useful roles in ranking the populations.

As the problem of ranking is still addressed in the literature with respect to many other different points of view, in this chapter we briefly review the basic procedures proposed in the literature, classifying them within the main reference field where they have been developed, that is, statistics and operations research. Finally, because our proposed method has little relevance to the usual approaches proposed in the literature around the concept of ranking we close by illustrating the specificities and advantages of the permutation approach for multivariate ranking problems.

1.1 Some Guideline Examples

Assume that there are several (more than two) populations of interest to be compared with each other to establish if they are all equal, that is, if there is just a single population, or if they are different, in which case we want to classify those populations from the 'best' to the 'worst' according to a given prespecified criterion. From a general point of view and depending on the specific real context, both experimental and observational studies can be taken into account so that one sample is drawn from each population to classify and rank those populations, where by the term 'ranking' we mean a meaningful criterion that allows us to rank populations from the 'best' to the 'worst'. Also assume that the response we observe on each unit/subject is multivariate in nature, where each univariate component can be either continuous or discrete or binary or ordered categorical (we even admit the mixed situation). Finally, we assume that for each univariate component a unique criterion is defined such that it is possible to establish a natural preference direction, as for instance, 'the greater the better', 'the smaller the better' or 'the closer to the target the better'.

1.1.1 Experimental Example

In the first example we consider an experimental case study with very few replications and where the response is continuous in all univariate components and the criterion 'the higher the better' holds for all components (Bonnini et al., 2009). This is actually a simplified version of the case study presented in Section 5.1.

The solid dots in Figure 1.2 represent the hypothetical true unknown/unobservable population means, in this case product performances and through this experiment, where the four observed replications are represented by empty dots, we wish to establish if the products have the same performance or if they are different, in which case we want to classify them from the 'best' to the 'worst'. From the simple descriptive inspection of observed data, we expect that the ranking analysis will suggest that the three products are different.

Because the response has a natural direction, that is, the higher the response the better is the product, we also expect that the best product will likely be P1, which looks better than products P2 and P3, which in turn are likely not different and both at the second ranking position. Note that the underlying true ranking can be inferred from the solid lines representing the distances of the true means from the point of absolute theoretical maximum (top right). In fact, P1 is the product much closer to the maximal point.

It is worth noting that because the populations are multivariate in nature the ranking analysis and the classification process into a global ranking should be properly taken into account for the distribution of the multivariate response variable.

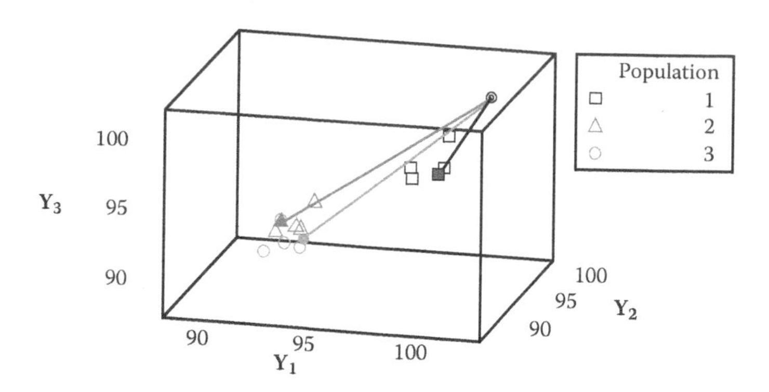

FIGURE 1.2
Graphical three-dimensional representation of an experiment.

1.1.2 Observational Example

A further element of difficulty of the traditional multivariate inferential methods is related to the nature of the response variable. If we consider a continuous response as in the previous example, provided that all assumptions were reasonably met, then we could properly use classical likelihood based inferential methods, but when some assumptions were met with uncertainty or even when the response variables are ordered categorically the difficulties of the traditional methods based on contingency tables may become insurmountable. Inference based on the NPC method, however, as we shall see in Chapters 2 and 3, allows us to relax most of the stringent assumptions of traditional parametric and nonparametric methods, providing a flexible solution that, from an inferential viewpoint, is exact for whatever sample size also in the case of ordered categorical responses.

To illustrate an example of ranking of multivariate populations in the case of an ordered categorical response variable, where observed data are not rigorously obtained by a random sampling procedure, let us refer to an observational survey-based investigation in the field of indoor environmental quality evaluations where a sample of pupils from three primary schools have been enrolled. The goal of the study is to rank the three educational buildings where the pupils are taught in terms of subjective well-being of indoor environmental quality, related to microclimatic conditions and other building-related factors. This is actually a simplified version of the real case study presented in Chapter 6, Section 6.1. For the sake of simplicity let us consider only two dimensions of the perceived environmental quality, for example an overall score on the winter and summer thermal discomfort, using a Likert scale 1–4 (here the rule 'the lower the better' holds). In Table 1.1 the observed frequencies in the three educational building are reported.

To classify the three schools from the best to the worst in terms of of perceived indoor environmental well-being it is clear that we have to consider the comparisons of the bivariate distributions of the two environmental

TABLE 1.1

Contingency Table of the Winter and Summer Thermal Discomfort Perceived in Three Educational Buildings

Educational Building	Thermal Winter Discomfort	Thermal Summer Discomfort			
		1	2	3	4
X	1	–	7	7	7
	2	2	33	77	16
	3	1	6	12	10
	4	–	1	1	2
Y	1	–	6	4	12
	2	3	23	38	14
	3	–	7	15	12
	4	–	–	2	3
Z	1	–	–	3	13
	2	3	5	30	32
	3	–	3	23	20
	4	–	–	2	9

quality evaluations. Roughly speaking, if the majority of the assessments of a given classroom will stay at the bottom right cells of the two-way contingency table it is more likely that the school will be classified in a low rank position. Note that, in contrast to what happens in the case of numeric/continuous responses, a simple global indicator measuring a sort of distance from the ideal worst situation (all frequencies in the bottom right cell [4, 4]) is not here an easily obtainable solution.

1.2 Formalization of the Problem and General Solution

Following Arboretti Giancristofaro et al. (2014) and extending the approach previously proposed by Corain and Salmaso (2007), let us assume that data were drawn from each of C multivariate populations $\Pi_1,\ldots,\Pi_C$ (i.e., items/groups/treatments), $C > 2$, by means of a sampling procedure, so as to make an inference on their possible equality and in case of rejection of this hypothesis to classify those populations to obtain a relative ranking from the 'best' to the 'worst' according to a prespecified meaningful criterion. We use the term *relative ranking* because we want to underline that it is not an absolute ranking but a kind of ordering that is referred only to the C populations at hand.

With reference to the so-called one-way MANOVA layout (for more complex designs we refer readers to Chapter 3, Section 3.1.1), let us formalize the problem within a nonparametric framework: let **Y** be the p-dimensional response variable represented as a p-vector of the observed data from population Π and let us assume, without loss of generality, that large values of each univariate aspect Y correspond to a better marginal performance, so that when comparing two populations the possible marginal stochastic superiority should result in a high ranking position. In other words, we are assuming the criterion 'the larger the better'. The term 'large values' has a clear meaning in the case of continuous responses, while in the case of binary or ordered categorical responses, this should be considered in terms of 'large proportion' and of 'large frequencies of high-score categories' respectively. The marginal univariate components of **Y** are not restricted to belonging to the same type; in other words, we can consider also the situation of mixed variables (some continuous/binary and some others ordered categorically).

We recall that our goal is to classify and rank $\Pi_1,\ldots,\Pi_C$ multivariate populations with respect to p marginal variables where C samples are available, from each population, of n_j independent replicates represented by the random variables $\mathbf{Y}_1,\ldots,\mathbf{Y}_C$, $j = 1,\ldots,C$. In other words, we are looking for an estimate $\hat{r}(\Pi_j)$ of the rank $r(\Pi_j)$, that is, the relative stochastic ordering of each population when compared among all other populations, that is, more formally

$$r_j = r(\Pi_j) = 1 + \sum_{j \neq h} I\left(\mathbf{Y}_j \overset{d}{<} \mathbf{Y}_h\right) = 1 + \left\{\# \mathbf{Y}_j \overset{d}{<} \mathbf{Y}_h,\ h = 1,\ldots,C,\ j \neq h\right\},\ j = 1,\ldots,C, \tag{1.1}$$

where $I(\cdot)$ is the indicator function and # means the number of times. This definition extends to a nonparametric multivariate framework the traditional definition of ranking and hence it is consistent with the ranking problem literature; see Gupta and Panchapakesan (2002) and Hall and Miller (2009). Note that the definition in Expression 1.1 of the rank of one given multivariate population within a set of several multivariate populations is derived by using the concept of stochastic superiority and by simple pairwise counting of the number of populations that are stochastically larger than that specific population. Let us consider an alternative definition of population ranking,

$$r_j = 1 + \left\{\# \left(C - \sum_{j \neq h} I\left(\mathbf{Y}_j \overset{d}{>} \mathbf{Y}_h\right)\right) > \left(C - \sum_{j' \neq h} I\left(\mathbf{Y}_{j'} \overset{d}{>} \mathbf{Y}_h\right)\right),\ j' = 1,\ldots,C,\ j' \neq j\right\},\ j = 1,\ldots,C. \tag{1.2}$$

The alternative definition in Expression 1.2 is derived by using the concept of stochastic superiority and by simple pairwise counting of the number of populations that are stochastically smaller than a given population. Note that even if this second ranking definition in Expression 1.2 works as an upward ranking procedure from the worst to the best (while the previous ranking definition in Expression 1.1 works from best to worst in a downward fashion), both definitions do provide exactly the same ranking. This is because starting from either the first or last position and then moving to the lower or higher positions obviously must necessarily provide the same ordering. The equivalence of definitions in Expressions 1.1 and 1.2 is illustrated by the following three examples labeled as *a, b, c*, in the case of C = 3 populations where $r_a = \{2, 3, 1\}$, $r_b = \{2, 1, 2\}$, $r_c = \{1, 3, 1\}$ and the matrices represent the pairwise results from the indicator function, that is, $I(\mathbf{Y}_j \overset{d}{>} \mathbf{Y}_h)$, $j, h = 1, 2, 3, j \neq h$, applied to all the $C \times (C - 1) = 6$ directional pairs, and where R is the rank operator.

Example	(a)	(b)	(c)
Definition	$\begin{bmatrix} - & 1 & 0 \\ 0 & - & 0 \\ 1 & 1 & - \end{bmatrix}$	$\begin{bmatrix} - & 0 & 0 \\ 1 & - & 1 \\ 0 & 0 & - \end{bmatrix}$	$\begin{bmatrix} - & 1 & 0 \\ 0 & - & 0 \\ 0 & 1 & - \end{bmatrix}$
(1.1)	$r_a = \{1+0+1, 1+1+1, 1+0+0\}$	$r_b = \{1+1+0, 1+0+0, 1+1+0\}$	$r_c = \{1+0+0, 1+1+1, 1+0+0\}$
r	$= \{2, 3, 1\} =$	$= \{2, 1, 2\} =$	$= \{1, 3, 1\} =$
(1.2)	$= R\{3-1-0, 3-0-0, 3-1-1\}$	$= R\{3-0-0, 3-1-1, 3-0-0\}$	$= R\{3-1-0, 3-0-0, 3-1-0\}$

Note that definition 1.1 actually provides the ranks by summing up the 1's and 0's along the columns, where we can find the stochastic inferiorities of each population when compared to all other ones, while definition 1.2 provides the ranks by summing up the 1's and 0's along the rows, where we can find the stochastic superiorities.

It is worth underlying that in definitions 1.1 and 1.2 no one piece of a priori knowledge or assumption on the true ordering is considered at all because r_j is simply obtained by counting how many populations are stochastically larger or smaller than the jth population, without any kind of restriction or 'a priori' information. Accordingly, for the existence of the multivariate ranking $r = \{r_1, \ldots r_j, \ldots, r_C\}$ it is sufficient that the expressions $\mathbf{Y}_j \overset{d}{>} \mathbf{Y}_h$ and $\mathbf{Y}_j \overset{d}{<} \mathbf{Y}_h$ are consistently defined, that is, the notions of multivariate stochastic inferiority/superiority are properly defined. In other words, the key point in our approach is that the ranking parameter $r = r(\Pi)$ can be properly well defined so that we can achieve a general and unambiguous definition of *multivariate stochastic ordering*. It is also worth noting that the subsequent empirical estimation process of population ranks will be performed via suitable multivariate directional pairwise comparisons and it could be affected by the so-called intransitivity problem (Dayton, 2003), that is, a possible inconsistency arising from pairwise results. For example, assuming that in the case of three populations the true ranking is $r = \{2, 3, 1\}$ but inferential results support only the largest stochastic difference, that is, $\mathbf{Y}_2 \overset{d}{<} \mathbf{Y}_3$ and two stochastic equalities $\mathbf{Y}_1 \overset{d}{=} \mathbf{Y}_2$, $\mathbf{Y}_1 \overset{d}{=} \mathbf{Y}_3$, the estimated ranks from Expressions 1.1 and 1.2 will be $\{1 + 0 + 0 = 1, 1 + 0 + 1 = 2, 1 + 0 + 0 = 1\} = \{1, 2, 1\}$ and $R\ \{3 - 0 - 0 = 3, 3 - 0 - 0 = 3, 3 - 1 - 0 = 2\} = \{2, 2, 1\}$, which are different one another and moreover one of them, that is, the first one, is clearly inconsistent. In fact, from the logical viewpoint, the last two comparisons suggesting $\mathbf{Y}_1 \overset{d}{=} \mathbf{Y}_2$ and $\mathbf{Y}_1 \overset{d}{=} \mathbf{Y}_3$ should imply $\mathbf{Y}_2 \overset{d}{=} \mathbf{Y}_3$ but on the contrary empirical data do support that $\mathbf{Y}_2 \overset{d}{<} \mathbf{Y}_3$. We return to this important issue later on (see Section 1.2.1).

Let us now formally define *multivariate stochastic ordering*. Following Shaked and Shanthikumar (2007) the random variable $\mathbf{Y}_j$ is said to be strictly greater than $\mathbf{Y}_h$ in multivariate stochastic order, denoted $\mathbf{Y}_j \overset{d}{\geq} \mathbf{Y}_h$, if $\Pr(\mathbf{Y}_j \in U) \geq \Pr(\mathbf{Y}_h \in U)$ for all upper set $U \in \mathbb{R}^p$, i.e., the set of p-dimensional real numbers. A set $U \in \mathbb{R}^p$ is called an upper set if $\boldsymbol{u} \in U$ implies that $\boldsymbol{v} \in U$ whenever $u_k \leq v_k$, $k = 1,\ldots, p$. It follows that if $\mathbf{Y}_2 \overset{d}{=} \mathbf{Y}_3$ are multivariate random variables then $\mathbf{Y}_j \overset{d}{\geq} \mathbf{Y}_h$ if and only if for all upper sets $U \in \mathbb{R}^p$

$$\sum_{t \in U} p_j(t) \geq \sum_{t \in U} p_h(t),$$

where for $p_j(t) = \Pr(Y_{j1} = t_1,\ldots, Y_{j1} = t_p)$ and $p_h(t)$ is similarly defined.

This notion of multivariate stochastic ordering generalizes the univariate one and simply indicates that one random vector is more likely than another to take on larger values. The relationship between multivariate and univariate stochastic ordering can be established by Theorem 1.1 (Davidov and Peddada, 2011).

Theorem 1.1

Suppose that $\mathbf{Y}_j \overset{d}{\geq} \mathbf{Y}_h$, *then* $\mathbf{Y}_j \overset{d}{>} \mathbf{Y}_h$ *if and only if* $Y_{jk} \overset{d}{>} Y_{hk}$ *for some* k, $k = 1,\ldots,p$.

Hence we can state that if $\mathbf{Y}_j$ is strictly stochastically greater than $\mathbf{Y}_h$ then a strict ordering between $\mathbf{Y}_j$ and $\mathbf{Y}_h$ holds provided that at least one of the marginal distributions is strictly ordered. This is very relevant in view of the general solution we are going to present because we will prove a possible multivariate stochastically superior through a combined analysis of the univariate ordering of marginal distributions. Note that, in general, the univariate stochastic ordering of marginal distributions does not imply the multivariate stochastic ordering of the joint distribution.

As pointed out by Davidov and Peddada (2011), the validity of the theorem does not depend on the type of the random variable that can be continuous or ordered categorical or binary/continuous in each marginal component so that the mixed response variable is also allowed. The relationship between the ordering of multivariate distributions and their marginals was noted also by Klingenberg et al. (2009) and Finos et al. (2007, 2008).

A more suitable and effective definition of multivariate stochastic ordering with respect to the problem at hand can be provided as follows: the two p-variate random variables $\mathbf{Y}_j$ and $\mathbf{Y}_h$ with distribution functions $\boldsymbol{F}_j$ and $\boldsymbol{F}_h$ and kth marginal distributions F_{jk} and F_{hk}, $k = 1,\ldots,p$, are said to be stochastically ordered, written $\mathbf{Y}_j \overset{d}{\gtrsim} \mathbf{Y}_h$, if and only if the relation between expectations $\mathbb{E}[f(\mathbf{Y}_j)] \geq \mathbb{E}[f(\mathbf{Y}_h)]$ holds for all increasing functions f: $\mathbb{R}^p \rightarrow \mathbb{R}$ such that the expectation exists. From Corollary 3 in Bacelli and Makowski (1989), it follows that $\mathbf{Y}_j \overset{d}{\gtrsim} \mathbf{Y}_h$ holds if and only if $F_{jk}(y) \geq F_{hk}(y)\ \forall\ y \in \mathbb{R}$, $k = 1,\ldots,p$, with at least one strict inequality holding at some $y \in \mathbb{R}$ for at least one k. It is worth noting that both definitions of stochastic ordering, that is, the last one and that provided by Shaked and Shanthikumar (2007), are actually equivalent, as can be found in Scarsini and Shaked (1990).

Under the hypothesis of distributional equality of the C populations, all true ranks would necessarily be equal to 1, hence they would be in a full ex aequo situation, that is

$$r\left(\Pi_j \middle| H_0\right) = \left\{1 + \#\mathbf{Y}_j \overset{d}{<} \mathbf{Y}_h,\ h = 1,\ldots,C, j \neq h\right\} = 1,\ \forall j.$$

This situation of equal ranking where all populations belong to just one ranking class may be formally represented in a testing-like framework where the hypotheses of interest are

$$\begin{cases} H_0 : \mathbf{Y}_1 \overset{d}{=} \mathbf{Y}_2 \overset{d}{=} \ldots \overset{d}{=} \mathbf{Y}_C \\ H_1 : \exists \mathbf{Y}_j \overset{d}{\neq} \mathbf{Y}_h, j, h = 1,\ldots,C, j \neq h. \end{cases} \tag{1.3}$$

In the case of rejection of the global multivariate hypothesis H_0, that is, when data are evidence of the fact that at least one population behaves differently from the others, it is of interest to perform inferences on pairwise comparisons between populations, that is

$$\begin{cases} H_{0(jh)} : \mathbf{Y}_j \overset{d}{=} \mathbf{Y}_h \\ H_{1(jh)} : \mathbf{Y}_j \overset{d}{\neq} \mathbf{Y}_h, j, h = 1,\ldots,C, j \neq h. \end{cases} \tag{1.4}$$

Note that a rejection of at least one hypothesis $H_{0(jh)}$ implies that we are not in an equal ranking situation, that is, at least one multivariate population has a greater ranking position than some others, more formally

$$\exists\, r(\Pi_j) \neq r(\Pi_h),\ j,h = 1,\ldots,C,\ j \neq h.$$

Note that, as usual in the framework of the *C*-sample inference, the rejection of the global null hypothesis is not informative on the specific alternative that has caused the rejection so that post hoc analysis is needed to look for which alternative is more likely. In this connection, to make an inference on which marginal variable(s) that inequality is mostly due to, it is useful to consider the inferences on univariate pairwise comparisons between populations, defined as

$$\begin{cases} H_{0k(jh)} : Y_{jk} \overset{d}{=} Y_{hk} \\ H_{1k(jh)} : \left(Y_{jk} \overset{d}{<} Y_{hk}\right) \bigcup \left(Y_{jk} \overset{d}{>} Y_{hk}\right) \end{cases},\ j,h = 1,\ldots,C,\ j \neq h,\ k = 1,\ldots,p, \tag{1.5}$$

because when $Y_{jk} \overset{d}{\neq} Y_{hk}$ is true, then one and only one alternative between $Y_{jk} \overset{d}{<} Y_{hk}$ and $Y_{jk} \overset{d}{>} Y_{hk}$ is true, that is, they cannot be jointly true.

Looking at the univariate alternative hypothesis $H_{1k(jh)}$, note that we are mostly interested in deciding whether a population is either greater or smaller than another one (not only establishing if they are different). In this connection, we can take into account separately the directional type alternatives, namely those that are suitable for testing both one-sided alternatives (see Bertoluzzo et al., 2013; Pesarin and Salmaso, 2010a, p. 163). In this connection, also Expression 1.4 can be reformulated as

$$\begin{cases} H_{0(jh)} : \mathbf{Y}_j \overset{d}{=} \mathbf{Y}_h \\ H_{1(jh)} : \left(\mathbf{Y}_j \overset{d}{<} \mathbf{Y}_h\right) \bigcup \left(\mathbf{Y}_j \overset{d}{>} \mathbf{Y}_h\right),\ j,h = 1,\ldots,C,\ j \neq h. \end{cases} \tag{1.6}$$

Let $p^{+}_{k(j,h)}$ and $p^{-}_{k(j,h)}$ be the permutation-based marginal directional *p*-value statistics related to the stochastic inferiority or superiority alternatives $H^{+}_{1k(jh)} : Y_{jk} \overset{d}{>} Y_{hk}$ and $H^{-}_{1k(jh)} : Y_{jk} \overset{d}{<} Y_{hk}$, respectively. Because by definition

$p^-_{k(j,h)} = 1 - p^+_{k(j,h)} = p^+_{k(h,j)}$, note that all one-sided inferential results related to the hypotheses in Expression 1.6 can be represented as follows:

$$P^+ = \begin{bmatrix} - & p^+_{1(1,2)} & p^+_{1(1,3)} & \cdots & p^+_{1(1,C)} \\ p^+_{1(2,1)} & - & p^+_{1(2,3)} & \cdots & p^+_{1(2,C)} \\ \cdots & \cdots & - & \cdots & \cdots \\ p^+_{1(C-1,1)} & p^+_{1(C-1,2)} & \cdots & - & p^+_{1(C-1,C)} \\ p^+_{1(C,1)} & p^+_{1(C,2)} & \cdots & p^+_{1(C,C-1)} & - \end{bmatrix}, \ldots, \begin{bmatrix} - & p^+_{p(1,2)} & p^+_{p(1,3)} & \cdots & p^+_{p(1,C)} \\ p^+_{p(2,1)} & - & p^+_{p(2,3)} & \cdots & p^+_{p(2,C)} \\ \cdots & \cdots & - & \cdots & \cdots \\ p^+_{p(C-1,1)} & p^+_{p(C-1,2)} & \cdots & - & p^+_{p(C-1,C)} \\ p^+_{p(C,1)} & p^+_{p(C,2)} & \cdots & p^+_{p(C,C-1)} & - \end{bmatrix}. \quad (1.7)$$

Finally, let be $p^+_{\bullet(j,h)}$ the directional p-value statistics calculated via nonparametric combination methodology (see Chapter 2) and related to the multivariate stochastic superiority alternatives $H^+_{1(jh)} : \mathbf{Y}_j \overset{d}{>} \mathbf{Y}_h$ in Expression 1.6. All the C × (C − 1) $p^+_{\bullet(j,h)}$ can be represented as follows:

$$P^+_\bullet = \begin{bmatrix} - & p^+_{\bullet(1,2)} & p^+_{\bullet(1,3)} & \cdots & p^+_{\bullet(1,C)} \\ p^+_{\bullet(2,1)} & - & p^+_{\bullet(2,3)} & \cdots & p^+_{\bullet(2,C)} \\ \cdots & \cdots & - & \cdots & \cdots \\ p^+_{\bullet(C-1,1)} & p^+_{\bullet(C-1,2)} & \cdots & - & p^+_{\bullet(C-1,C)} \\ p^+_{\bullet(C,1)} & p^+_{\bullet(C,2)} & \cdots & p^+_{\bullet(C,C-1)} & - \end{bmatrix}. \quad (1.8)$$

Note that p-value statistics in Expression 1.8 indicates if there is a significant global dominance between each pairs of populations and, if so, the global direction in which this dominance can actually exist. It is worth noting that, in contrast to what happens for the marginal directional

p-value statistics, the constraint to sum to 1 does not hold in this case, that is, $p^{+}_{\bullet(j,h)} \neq 1 - p^{-}_{\bullet(j,h)}$.

Now, let α be the chosen significance α – level and let S the $C \times C$ matrix which transforms the adjusted (by multiplicity) p-values $p^{+}_{\bullet(j,h)adj}$ into 0-or-1 scores where each element $s_{(j,h)}$ takes the value of 0 if $p^{+}_{\bullet(j,h)adj} > \alpha/2$, otherwise it takes 1 if $p^{+}_{\bullet(j,h)adj} \leq \alpha/2$, that is

$$S = \begin{bmatrix} - & s_{(1,2)} & s_{(1,3)} & \cdots & s_{(1,C)} \\ s_{(2,1)} & - & s_{(2,3)} & \cdots & s_{(2,C)} \\ s_{(3,1)} & s_{(3,2)} & - & \cdots & s_{(3,C)} \\ \cdots & \cdots & \cdots & - & \cdots \\ s_{(C,1)} & s_{(C,2)} & \cdots & s_{(C,C-1)} & - \end{bmatrix}. \tag{1.9}$$

Note that by using $\alpha/2$ as cut-off for the decision rule on the possible directional significances we are just applying the Bonferroni correction. In practice, S is nothing more than a synthetic representation of results from all multivariate directional pairwise comparisons suitable for estimating the possible pairwise dominances. If we consider either the sum of the $s_{(j,h)}$ 0-or-1 scores along the hth column or the jth row, then we are respectively counting the number of populations that, at the chosen significance α – level, are considered to be stochastically larger or smaller than the hth or the jth population. That is, we are defining an estimate $\hat{r}(\Pi_h)$ and $\hat{r}(\Pi_j)$ of the rank $r(\Pi_h)$ or $r(\Pi_j)$, that is, the relative stochastic ordering of each population when compared with all other populations by referring to the ranking definitions in Expression 1.1 or 1.2, that is, more formally

$$\hat{r}_h^D = 1 + \sum_{j=1}^{C} s_{(j,h)},\ h = 1,\ldots,C, \tag{1.10}$$

$$\hat{r}_j^U = 1 + \left\{ \# \left(C - \sum_{h=1}^{C} s_{(j,h)} \right) > \left(C - \sum_{h=1}^{C} s_{(j',h)} \right),\ j' = 1,\ldots,C,\ j' \neq j \right\},\ j = 1,\ldots,C, \tag{1.11}$$

where D and U represent downward and upward rank estimates, respectively. According to the ranking definitions in Expressions 1.1 and 1.2, we note that the ranking estimators defined in Expressions 1.10 and 1.11 are

TABLE 1.2

Marginal Directional Permutation p-Values for the First Guideline Example (Three-Product Experiment)

	Y_1			Y_2			Y_3		
	P1	P2	P3	P1	P2	P3	P1	P2	P3
$P^+ =$ P1	–	**.0021**	**.0012**	–	**.0018**	**.0011**	–	**.0007**	**.0045**
P2	.9979	–	.6710	.9982	–	.3418	.9993	–	.5519
P3	.9988	.3290	–	.9989	.6582	–	.9955	.4481	–

Note: Significant p-values at the 5% level are highlighted in bold.

derived by counting, on the basis of empirical evidence, how many populations are significantly stochastically larger/smaller than the hth/jth population at the chosen significance α – level. The two estimated rankings $\hat{r}^D$ and $\hat{r}^U$ of the true rank r are intentionally denoted with a different notation to highlight that sometimes they could provide different rank estimates for the same population because of the intransitivity issue as detailed in Section 1.2.1.

We remark that, as we outlined in our literature review, the use of a pairwise matrix as to derive a ranking is quite common, especially in the algorithmic ranking literature.

To make the proposed procedure clearer, let us apply it to the two guideline examples presented in the previous section. Tables 1.2 through 1.5 represent P^+, $P^+_{\bullet}$, $P^+_{\bullet adj}$ and S, that is, the directional univariate marginal and the directional multivariate unadjusted and adjusted (using the Bonferroni–Holm–Shaffer method) permutation p-values, and finally the score matrix.

From the S matrixes, it is very easy to estimate the rankings and classify the three products (first example) and the three educational buildings (second example) as follows.

- Global ranking of the three products (the rule 'the higher the better' holds): P1 = 1, P2 = 2, P3 = 2.
- Global ranking of the three educational buildings (the rule 'the lower the better' holds): building X = 1, building Y = 1, building Z = 3.

Note that in both cases the significant p-values (α = 5%) are actually arranged in blocks.

1.2.1 The Intransitivity Issue and the Revised Ranking Estimator

Focussing on the type II error, it is well known in the context of pairwise comparisons such as those we considered in Expression 1.4 the results may be affected by the so-called intransitivity problem (Dayton, 2003), that is, the possible inconsistency arising from pairwise results. As pointed out in

TABLE 1.3

Multivariate Directional p-Values and Score Matrix ($\alpha = 5\%$) for the First Guideline Example (Three-Product Experiment)

	P1	P2	P3		P1	P2	P3		P1	P3	P2
$P^{+}_{\bullet}$ = **P1**	–	**.0001**	**.0002**	$P^{+}_{\bullet adj}$ = **P1**	–	**.0003**	**.0002**	S = **P1**	–	1	1
P2	.9984	–	.8110	**P2**	.9984	–	.8110	**P3**	0	–	0
P3	.9841	.7510	–	**P3**	.9841	1.000	–	**P2**	0	0	–

Note: Significant p-values at the 5% level are highlighted in bold.

TABLE 1.4

Marginal Directional Permutation p-Values for the Second Guideline Example (Perceived Discomfort in Three Educational Buildings)

	Thermal Winter Discomfort			Thermal Summer Discomfort		
	X	Y	Z	X	Y	Z
P^{+} = X	–	.8272	.9994	–	.8031	.9992
Y	**.1728**	–	.9873	**.1959**	–	.9996
Z	**.0006**	**.0117**	–	**.0008**	**.0004**	–

Note: Significant p-values at the 5% level are highlighted in bold.

Shaffer (1986), when comparing C populations and the global null hypothesis is false, not all possible combination of true and false pairwise hypotheses are possible because there are some constraints related to the transitivity property that must be true from the logical perspective. For example, in case of $C = 3$, 4 or 5 populations a number of 2 out of 3 (there are three pairwise comparisons when $C = 3$), 4 and 5 out of 6 (for $C = 4$), and 7, 8, 9 out of 10 (for $C = 5$) cannot be simultaneously true because this would imply that all null pairwise hypotheses must be jointly true (Shaffer, 1986).

However, it is well known that actual pairwise results do not respect these kinds of logical constraints and the intransitivity issue may pose serious problems in the ranking estimation process as proved by the following example. Assume that we are comparing three populations $\{\Pi_1, \Pi_2, \Pi_3\}$ and the true ranking is $r = \{2, 3, 1\}$ and suppose that no one type III error occurs so that $s_{(1,3)}$, $s_{(2,1)}$ and $s_{(2,3)}$ cannot take the value 1 and must be equal to 0 (otherwise we would conclude in a direction opposite the true one). Among the remaining eight possible configurations ($8 = 2^3$, as $s_{(1,2)}$, $s_{(3,1)}$ and $s_{(3,2)}$ can take one out of the two values 0 or 1), assume that the number of the occurred type II errors (highlighted in bold) are respectively 2,1 and 0 so that the 0-or-1 score matrix S appears as one the following three configurations:

$$\text{(a) } S_a = \begin{bmatrix} - & \mathbf{0} & 0 \\ 0 & - & 0 \\ \mathbf{0} & 1 & - \end{bmatrix} \quad \text{(b) } S_b = \begin{bmatrix} - & 1 & 0 \\ 0 & - & 0 \\ \mathbf{0} & 1 & - \end{bmatrix} \quad \text{(c) } S_c = \begin{bmatrix} - & 1 & 0 \\ 0 & - & 0 \\ 1 & 1 & - \end{bmatrix}$$

When considering Expressions 1.10 and 1.11 to apply the ranking estimators to the three matrices S_a, S_b and S_c we obtain

$$\text{(a) } \hat{r}_a^D = \{1,2,1\},\ \hat{r}_a^U = \{2,2,1\} \quad \text{(b) } \hat{r}_b^D = \{1,3,1\},\ \hat{r}_b^U = \{1,3,1\}$$
$$\text{(c) } \hat{r}_c^D = \{2,3,1\},\ \hat{r}_c^U = \{2,3,1\}$$

TABLE 1.5

Multivariate Directional p-Values for the Second Guideline Example (Perceived Discomfort in Three Educational Buildings)

	X	Y	Z		X	Y	Z		X	Y	Z
$P^{+}_{\bullet}$ = **X**	–	.5553	.9382	$P^{+}_{\bullet adj}$ = **X**	–	1.000	.9382	S = **X**	–	0	0
Y	.3718	–	.8111	**Y**	.3718	–	.8111	**Y**	0	–	0
Z	**.0003**	**.0017**	–	**Z**	**.0009**	**.0017**	–	**Z**	1	1	–

Note: Significant p-values at the 5% level are highlighted in bold.

Note that only configuration (c) where no type II errors occurred leads to the fully correct inference on both the global ranking r and partial ranks r_j, while with configuration (a), where two type II errors occurred ($s_{(1,2)} = s_{(3,1)} = 0$), both of the estimated rankings $\hat{r}_a^D$ and $\hat{r}_a^U$ do not match the true one and only one and two individual rank estimates agree with the true one (for population Π_3). Finally, in case of configuration (b) one type II error occurred ($s_{(3,1)} = 0$) and both of the estimated rankings $\hat{r}_a^D$ and $\hat{r}_a^U$ do not match the true ranking; more specifically, one out of three individual estimates ranks does not agree with the true rank (for population Π_1). The intransitivity problem occurs at configuration (a) and in general whenever in the comparisons of three populations only one comparison out of three is significant. Intransitivity simply means that the logical ordering relations among the items do not hold empirically, providing an inconsistent estimated ranking. At any rate, under the hypothesis of homogeneity of all populations we pointed out that an observed intransitivity provides a false signal of an underlying stochastic ordering that actually does not exist. Conversely, under the hypothesis of nonhomogeneity, the intransitivity occurrence should be viewed as a signal related to possible type II errors. It is also worth noting that the intransitivity issue is related to the interpretation of results and obviously cannot suggest which types of inferential errors actually have occurred. In fact, we could get the same matrix S_a even under the hypothesis of homogeneity of all populations.

In case $\hat{r}_j^U$ does not match $\hat{r}_j^D$ for each $j = 1,\ldots,C$, then at least one intransitivity occurrence occurred and a revised ranking estimator $\bar{r}_j$ (suitable to overcome the intransitivity issue) can be defined as

$$\bar{r}_j = 1 + \left\{\#\left(\hat{r}_j^U + \hat{r}_j^D\right)/2 > \left(\hat{r}_h^U + \hat{r}_h^D\right)/2,\ h = 1,\ldots,C,\ j \neq h\right\},\ j = 1,\ldots,C. \quad (1.12)$$

It is worth noting that the revised ranking estimator (Expression 1.12) can be viewed as the average rank from the ranks derived from two types of counting: the significant observed stochastic inferiorities (i.e., $\hat{r}_j^U$) and the significant observed stochastic superiorities (i.e., $\hat{r}_j^D$).

When applying the revised ranking estimator $\bar{r}_j$ to the three matrices S_a, S_b and S_c we obtain

$$\text{(a) } \bar{r}_a = 2,3,1 \quad \text{(b) } \bar{r}_b = \{1,3,1\} \quad \text{(c) } \bar{r}_c = \{2,3,1\}.$$

The usefulness of the revised version in Expression 1.12 for the ranking estimator is self-evident: with configuration (a) the intransitivity occurrence is solved in favor of the true solution whereas in the case of configurations (b) and (c) the previous ranking estimates do not change. Note also that the $\bar{r}$ estimator can mitigate the effect of type II errors, as demonstrated in the case of configuration (a) where two type II errors occurred but all estimated ranks match the true ranks.

1.2.2 Properties of the Ranking Estimator

As our conclusion on the population ordering is inferential in nature, the ranking estimator $\bar{r} = \{\bar{r}_1, \bar{r}_2, \ldots, \bar{r}_C\}$ of the true ranking $r = \{r_1, r_2, \ldots, r_C\}$ related to the C multivariate populations $\{\Pi_1, \Pi_2, \ldots, \Pi_C\}$ is obviously affected by both type I and II errors. Moreover, our estimator is also affected by a type III error (Shaffer, 2002) because possible wrong decisions could be related to some false directions on the alternatives. In general, a type III error can occur when one accepts a specific directional alternative when in fact the opposite alternative is true; in our case, when a given population takes a false-higher/false-lower rank than the rank of a dominated/dominating population this means that a type III error occurred.

Let Correct Global Ranking (CGR) and Correct Individual Ranking (CIR_j) be the events that occur when $\bar{r} \equiv r$ and $\bar{r}_j \equiv r_j$ that is, when the ranking estimator (Expression 1.12) jointly and individually correctly estimates the ranks for all and the jth individual population respectively.

It can be proved that the ranking estimator defined in Expression 1.12 satisfies the following properties:

P1. $\Pr\{CGR|\text{homogeneity}\} = \Pr\{\bar{r}_j = r_j = 1, \forall j|\text{homogeneity}\} = 1 - \alpha$; $\Pr\{CIR_j|\text{homogeneity}\} = \Pr\{\bar{r}_j = r_j = 1|\text{homogeneity}\} \geq 1 - \alpha_j^*$, $j = 1, \ldots, C$;

P2. if $\mathbf{Y}_j \overset{d}{>} \mathbf{Y}_h$, then $\Pr\{\bar{r}_j < \bar{r}_h|\text{nonhomogeneity}\} > \alpha^*$, $j,h = 1, \ldots, C, j \neq h$;

P3. $\lim_{n\to\infty} \Pr\{CGR|\text{nonhomogeneity}\} = \lim_{n\to\infty} \Pr\{CIR|\text{nonhomogeneity}\} = 1$; where α and α_j^* are respectively the chosen significance α – level and the resulting adjusted individual α – level in all the pairwise comparisons involving the jth population, as it is formalized in Expressions 1.2 to 1.7, $n = \min_j n_j$ (sample size for the jth population) and homogeneity or nonhomogeneity refer to the situation where we are assuming that the null hypothesis of equality of all populations is true or false, respectively.

The proof of property P1 is trivial, because it simply states that the multivariate combined permutation testing procedure is able to control both familywise and individual type I errors (see Pesarin and Salmaso, 2010a), while properties P2 and P3 derive from unbiasedness and consistency of multivariate combined permutation tests (Pesarin and Salmaso, 2010a, pp. 83–104, 137–156). In fact, P2 means that the probability of rejecting a false pairwise null hypothesis is greater than the adjusted individual α – level, which means in turn that the combined permutation test is an unbiased test; moreover, property P3 means that the combined permutation tests are consistent so that as the sample sizes increase the probability that the estimated ranking matches the true one, that is, to reject either one and all false null hypotheses, converges to 1.

Let us consider the following theorems.

Theorem 1.2

(estimated ranking under homogeneity). *Assume that homogeneity holds, that is* $r_j = 1\ \forall j = 1,\ldots,C$. *It follows that* $Pr\{\overline{r}_j = r_j = 1, \forall j|homogeneity\} = 1 - \alpha$ *and* $Pr\{\overline{r}_j = r_j = 1|homogeneity\} \geq 1 - \alpha_j^*, j = 1,\ldots,C$ *where* α *and* α_j^* *are respectively the chosen significance* α *– level and the resulting adjusted (by multiplicity) individual* α *– level in all the pairwise comparisons involving the jth population.*

Proof

The proof is trivial because by definition of $\overline{r}_j$ the events $\{\overline{r}_j = 1, \forall j|\text{homogeneity}\}$ and $\{\overline{r}_j = 1|\text{homogeneity}\}$ occur when under the global null hypothesis H_0 (see Expression 1.3), no one pairwise false rejection and no one pairwise false rejection involving the *j*th population are observed. From the properties of combined permutation tests (see Theorem 1.6), it follows that

$$\Pr\{\overline{r}_j = r_j = 1, \forall j|\text{homogeneity}\} = \Pr\{\text{no one rejection}|H_0\} = \Pr\{s_{(j,h)} = 0\ \forall j, h = 1,\ldots,C, j \neq h|H_0\} = 1 - \alpha,$$
$$\Pr\{\overline{r}_j = r_j = 1|\text{homogeneity}\} = \Pr\{\text{no one rejection}|H_0\} = \Pr\{s_{(j,h)} = 0\ \forall h = 1,\ldots,C|H_0\} \geq 1 - \alpha_j^*, j = 1,\ldots,C, j \neq h,$$

where $s_{(j,h)}$ are the 0-or-1 scores in Expression 1.9, α is the chosen significance α – level and $\alpha_j^* < \alpha$ is the familywise type I error related to the sub-hypotheses $\{H_{0(jh)}, j = 1,\ldots,C, j \neq h\}$. ■

Note that under the assumption of homogeneity the estimated ranking necessarily cannot be an unbiased estimator of the true ranking. In fact, because under the global null hypothesis the true ranks are equal to 1 for all populations, the estimated ranks will obviously differ from the true values each time any type I error occurs. So that under the assumption of homogeneity $\overline{r}_j$ is a negatively biased estimator of r_j; more specifically, the bias for the *j*th population is given by

$$\begin{aligned}\text{bias}\left(\overline{r}_j\middle|\text{homogeneity}\right) &= \mathbb{E}\left[\overline{r}_j\middle|\text{homogeneity}\right] - 1 \\ &= (1\times(1-\alpha^*)) + \sum_{j=2}^{C} j \times \Pr\left\{\overline{r}_j = j\middle|\text{homogeneity}\right\} - 1.\end{aligned}$$

At any rate, although under the assumption of homogeneity a negative bias of $\overline{r}_j$ does exist, it can be definitely controlled by the chosen significance

α-level. Moreover, we may presume that it should be possible to take relatively small values and aim at evaluating the actual size of this negative bias via Monte Carlo simulation studies. As far as the variance of the ranking estimator is concerned, this is actually quite a complicated issue because the estimator is basically related to a sum of dependent Bernoulli random variables (the $s_{(j,h)}$ 0-or-1 scores in Expression 1.9) the probabilities and dependence structure of which rely on the true multivariate distributions that are assumed to be unknown within our framework.

Theorem 1.3

(estimated dominance under nonhomogeneity). *Assume that nonhomogeneity holds, that is* $\exists\, r_j = r_h,\ j, h = 1,\ldots,C,\ j \neq h$. *If* $\mathbf{Y}_j \overset{d}{>} \mathbf{Y}_h$, *then* $Pr\{\bar{r}_j > \bar{r}_h \mid nonhomogeneity\} > \alpha^*, j, h = 1,\ldots,C, j \neq h$ *where* α^* *is the resulting adjusted (by multiplicity) individual* α – *level in the pairwise comparisons procedure.*

Proof

First let us note that by definition of $\bar{r}_j$ the event $\{\bar{r}_j > \bar{r}_h \mid \text{nonhomogeneity}\}$ occurs when either $\left\{\hat{r}_j^D < \hat{r}_h^D\right\}$ or $\left\{\hat{r}_j^U < \hat{r}_h^U\right\}$, that is, either when the sum along the rows or along the columns of the 0-or-1 $s_{(j,h)}$ scores in Expression 1.8 are larger for the jth population than for the hth population. More formally, $\{\bar{r}_j > \bar{r}_h \mid \text{nonhomogeneity}\} = \{(\hat{r}_j^D < \hat{r}_h^D) \cup (\hat{r}_j^U < \hat{r}_h^U)\}$. Because the two subevents are not independent, then $\Pr\{(\hat{r}_j^D < \hat{r}_h^D) \cup (\hat{r}_j^U < \hat{r}_h^U)\} > \Pr\{\hat{r}_j^D < \hat{r}_h^D\} + \Pr\{\hat{r}_j^U < \hat{r}_h^U\}$, so that let us focus on $\Pr\{\hat{r}_j^D < \hat{r}_h^D\}$, which can be rewritten as

$$\Pr\left\{\hat{r}_h^D > \hat{r}_j^D\right\} = \Pr\left\{\sum_{j'=1}^{C} s_{(j',h)} > \sum_{j'=1}^{C} s_{(j',j)}\right\} = \Pr\left\{\sum_{j'=1}^{C} (s_{(j',h)} - s_{(j'-j)}) > 0\right\}$$

$$> \Pr\{s_{(1,h)} > s_{(1,j)}\} + \ldots + \Pr\{s_{(j,h)} > s_{(h,j)}\} + \ldots + \Pr\{s_{(C,h)} > s_{(C,j)}\}$$

$$> \Pr\{s_{(j,h)} > s_{(h,j)}\}.$$

If $\mathbf{Y}_j \overset{d}{>} \mathbf{Y}_h$, because the combined permutation test is an unbiased test, it can proved (see Theorem 1.6) that

$$\Pr\{s_{(j,h)} > s_{(h,j)}\} = \Pr\left\{\text{reject } H_{0(jh)} \middle| H_{1(jh)}^{+} : \mathbf{Y}_j \overset{d}{>} \mathbf{Y}_h\right\} > \alpha^*,$$

where α^* is the resulting adjusted by multiplicity individual α – level in the pairwise comparisons procedure. It follows that $\Pr\{\bar{r}_j > \bar{r}_h | \text{nonhomogeneity}\} > \alpha^*$. ■

Note that under the assumption of nonhomogeneity, the estimated $\bar{r}_j$ is also a biased estimate of r_j; more specifically, the bias is given by

$$\begin{aligned}\text{bias}\left(\bar{r}_j\right) &= \mathbb{E}\left[\bar{r}_j \middle| \text{homogeneity}\right] - r_j \\ &= \left[r_j \times \Pr\left\{\text{CIR}_j \middle| \text{nonhomogeneity}\right\}\right. \\ &\quad \left. + \sum_{j' \neq j} r_{j'} \times \Pr\left\{\bar{r}_j = j' \middle| \text{nonhomogeneity}\right\}\right] - r_j.\end{aligned}$$

At any rate, although under the assumption of nonhomogeneity a bias does exist, the $\bar{r}$ estimator is asymptotically unbiased thanks to the consistency of combined permutation tests (see Theorem 1.8). In fact, because $\lim_{n \to \infty} \Pr\{\text{CIR}_j | \text{nonhomogeneity}\} = 1$ and $\lim_{n \to \infty} \Pr\{\bar{r}_j = j' | \text{nonhomogeneity}\} = 0$, then it follows that $\lim_{n \to \infty} \text{bias}(\bar{r}_j) = 0$. If the rate for which the power of permutation tests goes to 1 was relatively fast with respect to the increase of sample sizes then the bias drawback would not be so relevant for a finite sample. We aim at evaluating this issue via Monte Carlo simulation studies in Chapter 3, Section 3.2.

Theorem 1.4

(asymptotic estimated ranking under nonhomogeneity). *Assume that nonhomogeneity holds, that is* $\exists\, r_j = r_h$, $j,h = 1,\ldots,C$, $j \neq h$. *It follows that* $\lim_{n \to \infty} Pr\{\bar{r} \equiv r \,|$ *nonhomogeneity*$\} = \lim_{n \to \infty} Pr\{\bar{r}_j \equiv r_j \,|$ *nonhomogeneity*$\} = 1$.

Proof

The proof is trivial because by definition of $\bar{r}_j$ the events $\{\bar{r} \equiv r \,|\, \text{nonhomogeneity}\}$ and $\{\bar{r}_j \equiv r_j \,|\, \text{nonhomogeneity}\}$ occur when under the global alternative hypothesis H_1 (see Expression 1.3) all the true pairwise rejections are observed. From the consistency property of combined permutation tests (see Theorem 1.8), it follows that

$\lim_{n\to\infty}\Pr\{\bar{r} \equiv r \mid \text{nonhomogeneity}\} = \lim_{n\to\infty}\Pr\{\bar{r}_j \equiv r_j \mid \text{nonhomogeneity}\} = \lim_{n\to\infty}\Pr\{$all true rejections are observed$|H_1\} = \lim_{n\to\infty}\Pr\{s_{(j,h)} = 1\ \forall j,h = 1,\ldots,C, j \neq h$ and $\forall$ true $H_{1(jh)} | H_1\} = 1$,

where $n = \min_j n_j$ and $s_{(j,h)}$ are the 0-or-1 scores in Expression 1.9. ■

We demonstrated that as the sample sizes increase the probability that all false pairwise null hypotheses will be rejected tends to 1 so that both the downward and upward ranking estimators and the revised estimator as well will tend to match exactly the true ranking.

Thanks to Theorem 1.4, it is worth noting that under the assumption of nonhomogeneity, the ranking estimator can also be said to be a *consistent classifier*, that is, a procedure such that the probability of incorrect ranking classification gets arbitrarily close to the lowest possible risk as the sample size goes to infinity (Bousquet et al., 2004). In particular, it is a consistent classifier from two points of view: when the sample size increases or when the number of informative variable increases, keeping fixed the sample sizes (so-called *finite-sample consistency*; see Pesarin and Salmaso, 2010b).

1.2.3 Properties of Combined Permutation Tests

Let us denote a p-dimensional data set by $\mathbf{Y} = \{\mathbf{Y}_j, j = 1,\ldots,C\} = \{\mathbf{Y}_{ij}, i = 1,\ldots,n_j, j = 1,\ldots,C\} = \{\mathbf{Y}_{jk}, j = 1,\ldots,C, k = 1,\ldots,p\} = \{Y_{ijk}, i = 1,\ldots,n_j, j = 1,\ldots,C, k = 1,\ldots,p\}$. The response $\mathbf{Y}$ takes its value on the p-dimensional sample space Y, for which a σ-algebra A and a (possibly not specified) nonparametric family P of nondegenerate distributions are assumed to exist. The groups are presumed to be related to C populations and the data $\mathbf{Y}_j$ are supposed i.i.d. with distributions $P_j \in$ P, $j = 1,\ldots,C$.

Focussing on the kth univariate response and on the shth pair of populations, let us assume that under the nonhomogeneity hypothesis the distribution of Y_{sk} is shifted by a random quantity Δ_k (without loss of generality we assume that random effects are nonnegative) with respect to that of Y_{hk} and so two respective cumulative distribution functions (CDFs) are such that $F_{sk}(y) \leq F_{hk}(y)$, $y \in \mathbb{R}^1$, showing stochastic dominance. To represent data sets in the nonhomogeneity hypothesis we use the notation $\mathbf{Y}_{jk}(\Delta_k) = \{Y_{ijk} = \mu_k + \Delta_{ijk} + Z_{ijk}, i = 1,\ldots,n_j, j = s,h\}$, where μ_k is a finite nuisance quantity common to all units, and the random deviates Z_{ijk} are exchangeable and with unknown distribution $P \in$ P. Since μ_k is a nuisance quantity common to all units and thus inessential for comparing Y_{sk} to Y_{hk}, we may model data sets as $\mathbf{Y}_k(\Delta_k) = (\mathbf{Z}_{sk} + \Delta_k, \mathbf{Z}_{hk})$, where $\Delta_k = (\Delta_{11},\ldots,\Delta_{1n1})$. The latter notation emphasizes that effects Δ_k are assumed to be active only on units of the first sample. Connected with such a notation we may equivalently express the hypotheses of homogeneity versus nonhomogeneity as H_{0k}: $\Pr\{\Delta_k = 0\} = 1$ and H_{1k}: $\Pr\{\Delta_k > 0\} > 0$.

The notion of unbiasedness for a test statistic T_k is related to its comparative rejection behaviour in H_{1k} with respect to that in H_{0k}. To this end, let us consider a nondegenerate test statistic T_k: $\mathcal{Y}^n \rightarrow \mathbb{R}^1$, where typically it is $T_k(\mathbf{Y}_k) = S_1(\mathbf{Y}_{sk}) - S_2(\mathbf{Y}_{hk})$ corresponding to the comparison of two nondegenerate sampling (measurable) statistics such as $\sum_i \varphi(Y_{ijk})/n_j$ or $Md(Y_{jk})$, and so forth (a list of the most used test statistics is given in the next section). To be suitable for evaluating the related sampling divergence, we assume that functions S are invariant with respect to rearrangements of data input (same as within-sample permutation) and monotonic nondecreasing, so that large values of T_k are evidence against H_{0k}.

A test statistic T is said to be *conditionally* or *permutationally unbiased* if related p-values $\lambda_T(\mathbf{Y}) = \Pr\{T(\mathbf{Y}^*) \geq T(\mathbf{Y})|\mathcal{Y}_{/\mathbf{Y}}\}$ are such that, for each $\mathbf{Y} \in \mathcal{Y}^n$ and whatever $\Delta \in H_1$,

$$\Pr\{\lambda_T(\mathbf{Y}(\Delta)) \leq \alpha_a | \mathcal{Y}_{/\mathbf{Y}(\Delta)}\} \geq \Pr\{\lambda_T(\mathbf{Y}(0)) \leq \alpha_a | \mathcal{Y}_{/\mathbf{Y}(0)}\} = \alpha_a,$$

where $\mathbf{Y}(\Delta) = (\mathbf{Z}_s + \Delta, \mathbf{Z}_h)$, $\mathbf{Y}(0) = (\mathbf{Z}_s, \mathbf{Z}_h)$; $\mathbf{Y}^*$ is interpreted as a random permutation of $\mathbf{Y}$; and α_a is any attainable α-value. A sufficient condition for conditional unbiasedness is

$$\lambda_T(\mathbf{Y}(\Delta)) = \Pr\{T(\mathbf{Y}^*(\Delta)) \geq T^o(\Delta) = T(\mathbf{Y}(\Delta))|\mathcal{Y}_{/\mathbf{Y}(\Delta)}\} \leq \Pr\{T(\mathbf{Y}^*(0)) \geq T^o(0) = T(\mathbf{Y}(0))|\mathcal{Y}_{/\mathbf{Y}(0)}\} = \lambda_T(\mathbf{Y}(0)).$$

To clarify better the relation between the different notions of conditionally or permutationally unbiasedness and the traditional notion of unbiasedness, also called unconditional or population unbiasedness, we refer readers to Pesarin and Salmaso (2012, 2013).

In what follows we prove the uniform conditional unbiasedness of associative statistics while for general statistics, including nonassociative forms, a slightly more difficult proof is given in Pesarin and Salmaso (2010a, pp. 86–88).

Theorem 1.5

(uniform conditional unbiasedness of T_k). *Permutation tests for random shift alternatives* ($\Delta_k \overset{d}{>} 0$) *based on divergence of associative statistics of nondegenerate measurable nondecreasing transformations of the data, that is,* $T^*_{\varphi k}(\Delta_k) = \sum_i \varphi[Y^*_{isk}(\Delta_k)] - \sum_i \varphi[Y^*_{ihk}(\Delta_k)]$, *are conditionally unbiased for every attainable* $\alpha \in \Lambda^{(n)}_{\mathbf{Y}(0)}$, *every population distribution P, and uniformly for all data sets* $\mathbf{Y} \in \mathcal{Y}^n$. *In particular,*

$$\Pr\{\lambda(\mathbf{Y}(\Delta)) \leq \alpha | \mathcal{Y}_{/\mathbf{Y}(\Delta)}\} \geq \Pr\{\lambda(\mathbf{Y}(0)) \leq \alpha | \mathcal{Y}_{/\mathbf{Y}(0)}\} = \alpha.$$

Proof

To prove the theorem we need to look further into the conditional (permutation) unbiasedness of a test statistic T_k, so that let us consider its permutation structure, that is, the behaviour of values it takes in the observed data set **Y** and in any of its permutations $\mathbf{Y}^* \in \mathcal{Y}^n_{/\mathbf{Y}}$ under both H_{0k} and H_{1k}. Since $Y^*_{ijk}(\Delta_k) = Z^*_{ijk} + \Delta^*_{ijk}$, with obvious notation, we note that $\sum_i[\varphi(Z^*_{ijk} + \Delta^*_{ijk}) - \varphi(Z^*_{ijk})] = \sum_i d_{Tk}(Z^*_{ijk}, \Delta^*_{ijk}) = D_{Tk}(Z^*_{jk}, \Delta^*_{jk}) \geq 0$, $j = s,h$, where $d_{Tk}(Z^*_{ijk}, \Delta^*_{ijk}) = \varphi(Z^*_{ijk} + \Delta^*_{ijk}) - \varphi(Z^*_{ijk}) \geq 0$, because some Δ^*_{ijk} are positive quantities and φ is nondegenerate and nondecreasing by assumption. Of course, if $\Delta^*_{ijk} = 0$, $d_{Tk}(Z^*_{ijk}, 0) = 0$, whereas $D_{Tk}(Z^*_{jk}, \Delta^*_{jk}) = 0$ implies $\Delta^*_{ijk} = 0$, $i = 1,\ldots,n_j$, $j = s,h$. The two observed values are then $T^*_k(0) = \sum_i\varphi(Z_{isk}) - \sum_i\varphi(Z_{ihk})$ and $T^o_k(\Delta_k) = \sum_i\varphi(Z_{isk} + \Delta_{isk}) - \sum_i\varphi(Z_{ihk})$ because $\Delta_{ihk} = 0$, $i = 1,\ldots,n_h$. Moreover, the two permutation values are $T^*_k(0) = \sum_i\varphi(Z^*_{ijk}) - \sum_i\varphi(Z^*_{ihk})$ and $T^*_k(\Delta_k) = \sum_i\varphi(Z^*_{isk} + \Delta^*_{isk}) - \sum_i\varphi(Z^*_{ihk}) = T^*_k(0) + D_{Tk}(Z^*_{sk}, \Delta^*_{sk}) - D_{Tk}(Z^*_{hk}, \Delta^*_{hk})$. In addition, the following pointwise relations clearly occur: $D_{Tk}(Z^*_{sk}, \Delta^*_{sk}) \leq D_{Tk}(\mathbf{Z}_{sk}, \Delta_{sk})$, because for $u^*_i > n_s$ and $i \leq n_s$ the corresponding $\Delta^*_{isk} = \Delta(u^*_i) = 0$; and $D_{Tk}(Z^*_{hk}, \Delta^*_{hk}) \geq 0$, because for $u^*_i \leq n_s$ and $i > n_s$ the corresponding $\Delta^*_{ihk} = \Delta(u^*_i) \geq 0$.

Therefore, the related p-values are such that

$$\lambda_{Tk}(\mathbf{Y}(\Delta)) = \Pr\left\{T^*(\Delta_k) \geq T^o(\Delta_k) \middle| \mathcal{Y}_{/\mathbf{Y}(\Delta)}\right\}$$

$$= \Pr\Big\{T^*_k(0) + D_{Tk}(\mathbf{Z}^*_{sk}, \Delta^*_{sk}) - D_{Tk}(\mathbf{Z}^*_{hk}, \Delta^*_{hk})\mathcal{Y}_{/\mathbf{Y}(\Delta)} - D_{Tk}(\mathbf{Z}_{sk}, \Delta_{sk})$$

$$\geq T^o_k(0) \Big| \mathcal{Y}_{/\mathbf{Y}(0)}\Big\}$$

$$\leq \Pr\left\{T^*_k(0) \geq T^o_k(0) \middle| \mathcal{Y}_{/\mathbf{Y}(0)}\right\} = \lambda_{TK}(\mathbf{Y}(0)),$$

from which we see that p-values in every alternative of H_{1k} are not larger than in H_{0k}, which holds for whatever underlying distribution P, whatever associative test statistic T_k, and uniformly for all data sets $\mathbf{Y} \in \mathcal{Y}^n$, because $D_{Tk}(\mathbf{Z}^*_{sk}, \Delta^*_{sk}) - D_{Tk}(\mathbf{Z}^*_{hk}, \Delta^*_{hk})\mathcal{Y}_{/\mathbf{Y}(\Delta)} - D_{Tk}(\mathbf{Z}_{sk}, \Delta_{sk})$ is pointwise nonpositive and because $\Pr\{T^*_k - W \geq t\} \leq \Pr\{T^*_k < t\}$ for any $W \geq 0$. ■

Often in testing for complex hypotheses, when a set of $p > 1$ response variables are jointly involved it is convenient to first process data using a finite set of $q \geq p > 1$ different partial tests (note that the number q of subproblems is not necessarily equal to the dimensionality p of responses; in this connection see the so-called multiaspect testing strategy in Corain and Salmaso, 2013). Such partial tests, possibly after adjustment for multiplicity (Basso et al., 2009), may be useful for marginal or separate inferences. But if they are jointly considered, they provide information on a general overall (same as global) hypothesis, which actually represents the objective of our multivariate ranking problem where the necessity to take account of all available information through the combination of p tests in one combined test naturally arises.

Let us now consider a suitable class C of continuous nondecreasing real functions ψ: $(0,1)^p \rightarrow \mathbb{R}^1$, that applied to the p-values λ_k of partial tests T_k, $k = 1,\ldots,p$, defines the second-order global (multivariate) test T''_ψ: $T''_\psi = \psi(\lambda_1,\ldots,\lambda_p)$, provided that the following conditions hold:

- ψ is nonincreasing in each argument: $\psi(\ldots,\lambda_k,\ldots) \geq \psi(\ldots,\lambda'_k,\ldots)$, if $\lambda_k \leq \lambda'_k$, $k \in \{1,\ldots,p\}$.
- ψ attains its supremum value $\bar{\psi}$, possibly not finite, even when only one argument attains zero: $\psi(\ldots,\lambda_k,\ldots) \rightarrow \bar{\psi}$ if $\lambda_k \rightarrow 0$, $k \in \{1,\ldots,p\}$.
- ψ attains its infimum value $\underline{\phi}$, possibly not finite, when all its arguments attain one: $\psi(\ldots,\lambda_k,\ldots) \rightarrow \underline{\psi}$ if $\lambda_k \rightarrow 1$, $k = 1,\ldots,p$.
- $\forall\alpha > 0$, the acceptance region is bounded: $\underline{\psi} < T''_{\alpha/2} < T'' < T''_{1-\alpha/2} < \bar{\psi}$.

To achieve the unbiasedness of combined tests T''_ψ, $\forall\psi \in C$, first let us consider the Lemma 1.1.

Lemma 1.1

Let X and W be two random variables defined on the same univariate probability space. If X is stochastically larger than W, so that their CDFs satisfy $F_Y(t) \leq F_W(t)$, $\forall t \in \mathbb{R}^1$, and if φ is a nondecreasing measurable real function, then $\varphi(Y)$ is stochastically larger than $\varphi(W)$.

The proof of this lemma is based on the following straightforward relationships:

$$\Pr\{\varphi(Y) \leq \varphi(t)\} = F_Y(t) \leq F_W(t) = \Pr\{\varphi(W) \leq \varphi(t)\}, \ \forall t \in \mathbb{R}^1.$$

Note that the measurability of φ is relevant for the above probability statements to be well defined.

Theorem 1.6

(uniform conditional unbiasedness of T''). *If, given a data set* $\mathbf{Y}$ *and any* $\alpha > 0$, *partial permutation tests* $\mathbf{T} = \{T_k, k = 1,\ldots,p\}$ *are all marginally unbiased for respectively* H_{0k} *against* H_{1k}, $k = 1,\ldots,p$, *so that their associated p-values* λ_k, $k = 1,\ldots,p$, *are positively dependent, then* $T''_\psi = T''_\psi = \psi(\lambda_1,\ldots,\lambda_p)$, $\forall\psi \in C$, *is an unbiased combined test for* H_0:$\{\bigcap_k H_{0k}\}$ *against* H_1:$\{\bigcup_k H_{1k}\}$.

Proof

The marginal unbiasedness and positive dependence (Shaked, 1982) properties of partial tests T_k imply, with obvious notation, that $\Pr\{\lambda_k \le z | \mathrm{Y}_{/\mathbf{Y}(0)}\} \le \Pr\{\lambda_k \le z | \mathrm{Y}_{/\mathbf{Y}(\Delta)}\}, \forall z \in (0,1)$, $k = 1,\ldots,p$, because p-values $\lambda_k(\mathbf{Y}(\Delta))$ are stochastically smaller than $\lambda_k(\mathbf{Y}(0))$ (see Theorem 1.5). Thus, by the nondecreasing property of combination function ψ and Lemma 1.1, $\psi(\ldots,\lambda_k(\mathbf{Y}(\Delta)),\ldots)$ is stochastically larger than $\psi(\ldots,\lambda_k(\mathbf{Y}(0)),\ldots)$. Hence, by iterating from $k = 1$ to p, the unbiasedness of T''_ψ is achieved. ■

Let us move to power functions and consistency of both univariate and combined permutation tests. We consider a test statistic T_k applied to a data set $\mathbf{Y}_{jk}(\Delta_k) = \{Y_{ijk} = \mu_k + \Delta_{ijk} + Z_{ijk}, i = 1,\ldots,n_j, j = s,h\}$, and deviates $\mathbf{Z}$, because we are arguing conditionally, are assumed to have the role of unobservable fixed quantities. With obvious notation, the *conditional power function* is defined as

$$W_k[(\Delta_k,\alpha,T_k)|\mathrm{Y}_{/\mathbf{Y}}] = \Pr\{\lambda_T(\mathbf{Y}(\Delta_k)) \le \alpha | \mathrm{Y}_{/\mathbf{Y}(\Delta k)}\} = \mathbb{E}\{\mathbb{I}[\lambda(\mathbf{Y}^\dagger(\Delta_k)) \le \alpha] | \mathrm{Y}_{/\mathbf{Y}\dagger(\Delta k)}\},$$

where its dependence on T_k, α, Δ_k, n and $\mathbf{Z}$ is self-evident, and the mean value is taken with respect to all possible $\mathbf{Y}^\dagger(\Delta_k)$. It should be emphasized that, due to Theorem 1.5, $\Delta_k < \Delta'_k$ implies that $W_k[(\Delta_k,\alpha,T_k)|\mathrm{Y}_{/\mathbf{Y}}] \le W_k[(\Delta'_k,\alpha,T_k)|\mathrm{Y}_{/\mathbf{Y}}]$ for every $\mathbf{Y} \in \mathrm{Y}$ and any $\alpha \in \Lambda_{\mathbf{Y}}$.

If the distribution $Q(\Delta_k|\mathbf{Y})$ of random effects $\Delta_k \in \Omega$, given the observable data $\mathbf{Y} \in \mathrm{Y}$ and the population distribution $P^n[\mathbf{Y}(\Delta_k)]$, is known, the unconditional power function $W_k(\Delta_k,\alpha,T_k,P,Q,n)$ with random effects is defined as

$$W_k(\Delta_k,\alpha,T_k,P,Q,n) = \int_{\mathrm{Y}^n \times \Omega} \mathbb{I}\left[\lambda_T\left(\mathbf{Y}(\Delta_k)\right) \le \alpha \,|\, \mathrm{Y}_{/\mathbf{Y}}\right] dQ(\Delta_k \,|\, \mathbf{Y}) dP^n\left(\mathbf{Y}(\Delta_k)\right).$$

When for any fixed α, T_k and P and any given $\Delta_k > 0$, the limit as $n \to \infty$ of the unconditional power function is $\lim_{n\to\infty} W_k(\Delta_k,\alpha,T_k,P,n) = 1$, the test statistic T_k is said to be consistent in the traditional sense.

Theorem 1.7

(consistency of permutation test T_k). *Permutation tests for random shift alternatives* ($\Delta_k \overset{d}{>} 0$) *are consistent tests.*

Proof

Proof of traditional consistency may easily be obtained by considering that, in conditions for asymptotically finite critical values T_α, test statistics of the form $T_k[\mathbf{Y}^{(n)*}(\Delta_k)] = \sum_i Y^*_{ik}(\Delta_k)/\sqrt{n} = \sum_i Y^*_{ik}/\sqrt{n} + \Delta_k/\sqrt{n}$ are such that $\forall\alpha > 0$,

$$\lim_{n\to\infty} \Pr\{T_k[\mathbf{Y}^{(n)*}(\Delta_k)] \geq T_\alpha[\mathbf{Y}^{(n)}(\Delta_k)] \mid \mathbf{Y}^{(n)}(\Delta_k)\} = 1. \quad \blacksquare$$

Let us assume that when n goes to infinity, then so also do sample sizes of all groups, that is, $n \to \infty$ implies $\min_j(n_j) \to \infty$ and p is fixed. Consistency of combined tests T''_ψ, $\forall \psi \in C$, can be proved by Theorem 1.8.

Theorem 1.8

(consistency of combined tests T''). *If partial permutation tests* $\mathbf{T} = \{T_k, k = 1,\ldots,p\}$ *are marginally unbiased at least one is strongly consistent respectively for respectively* H_{0k} *against* H_{1k}, $k = 1,\ldots,p$, *then* $T''_\psi = \psi(\lambda_1,\ldots,\lambda_p)$, $\forall \psi \in C$, *is a strongly consistent combined test for* $H_0{:}\{\bigcap_k H_{0k}\}$ *against* $H_1{:}\{\bigcup_k H_{1k}\}$.

Proof

To be strongly consistent, a combined test must reach its critical region with probability 1 if at least one subalternative H_{1k}, $k = 1,\ldots,p$, is true. Suppose that H_{1k} is true, then $\lambda_k \to 1$, with probability one as $n \to \infty$. Thus, by properties of the combined functions, $T'' \to \bar{\psi} > T''_\alpha$ with probability one, $\forall \alpha > 0$. $\blacksquare$

The combined tests are also consistent from a quite different point of view, that is, when sample sizes are fixed and the number of variables to be analyzed is much larger than sample sizes. Under very mild conditions, the power function of permutation tests based on both associative and non-associative statistics monotonically increases as the related standardized noncentrality functional increases. This is true also for multivariate situations, in particular, for any added variable the power does not decrease if this variable makes larger standardized noncentrality (so-called *finite-sample consistency*; see Pesarin and Salmaso, 2010b). Numerical proofs of the finite-sample consistency can be found in the simulation studies presented in Corain and Salmaso (2013, 2015).

1.2.4 Parametric Counterpart

In this section we present a new approach for the problem of ranking several multivariate normal populations that controls the risk of false ranking classification under the hypothesis of population homogeneity while under the nonhomogeneity alternatives it is expected that the true rank can be estimated with satisfactory accuracy, especially for the 'best' populations (Carrozzo et al., 2015).

Let us consider the set $\{\Pi_1, \Pi_2, \ldots, \Pi_C\}$ related to C multivariate p-dimensional normal populations $\mathbf{Y}_j \sim N(\mu_j, \Sigma)$, $j = 1,\ldots,C$, where the variance/covariance matrix Σ is assumed to be known so that the C normal populations may differ only with respect to their location parameters $\boldsymbol{\mu}_j$. Assume that the ranking of the C populations can be established by an additive rule and assume also that the rule 'the higher the better' holds for all the p components, so that let $\theta_j = \sum_{k=1}^{p} \mu_{jk}/\sigma_k$ the 'true' ranking parameter related to the jth population. Accordingly, the 'true' ranking can be defined as

$$r(\Pi_j) = r_j = 1 + \{\#(\theta_j < \theta_h), h = 1,\ldots,C, j \neq h\}, j = 1,\ldots,C,$$

where the symbol # means 'number of times'. Note that if the rule 'the lower the better' was valid instead, then when defining the ranking we should only reverse the direction of the inequality, that is,

$$r_j = 1 + \{\#(\theta_j > \theta_h), h = 1,\ldots,C, j \neq h\}, j = 1,\ldots,C.$$

In case some components have to be interpreted with the first rule and some others with the second rule, a suitable transformation such as $1/Y$ or $-Y$ should be applied initially in order that all components can share the same underlying interpretation (obviously, in this case we should assume that the transformed components are multivariate normal differing only in the location parameter).

Consider that a random sample of size n_j is available from the jth population and let $\hat{\theta}_j = \sum_{k=1}^{p} \bar{Y}_{jk}/\sigma_k$ be the natural estimator for θ_j, where $\bar{Y}_{jk} = \sum_{i=1}^{n_j} Y_{ijk}/n_j$ is the kth univariate sample mean for the jth population. From standard calculations on transformations of normal random variables it can be proved that

$$\hat{\theta}_j \sim N\left(\theta_j; \left[p + 2\sum_{k<s} \rho_{ks}\right]/n_j\right), j = 1,\ldots,C.$$

In case the variance/covariance matrix Σ cannot be assumed as known, the previous formula is expected to be valid as approximated distribution, that is

$$\hat{\theta}_j \xrightarrow{d} N(\theta_j; \left[p + 2\sum_{k<s} \widehat{\rho_{ks}}\right]/n_j), j = 1,\ldots,C.$$

To calculate $\hat{r}_j$, that is, to provide an estimate of r_j, it is clear that we need to develop an inference on the pairwise differences $(\theta_h - \theta_j)$, $j,h = 1,\ldots,C$, $j \neq h$. For this goal let us define as

$$\mathrm{LSD}(\theta_h, \theta_j) = z_{\alpha^*/2}\sqrt{\left[p + 2\sum_{k<s} \rho_{ks}\right](1/n_h + 1/n_j)}.$$

It is clear that LSD (θ_h, θ_j) represents the least significance difference between any given pair of estimated ranking parameters, where $z_{\alpha^*/2}$ is adjusted by multiplicity standard normal percentile at the desired α – level. In this way, the natural estimator of r_j can be defined as

$$\hat{r}_j = 1 + \left\{ \# \left(\hat{\theta}_h - \hat{\theta}_j \right) > \mathrm{LSD}(\theta_h, \theta_j), h = 1, \ldots, C,\ j \neq h \right\}, j = 1, \ldots, C.$$

When performing pairwise comparisons it is well known that results may be affected by the so-called intransitivity problem (Dayton, 2003), that is, the possible inconsistency arising from pairwise results. For example, in case of three multivariate normal populations $\mathbf{Y}_1$, $\mathbf{Y}_2$ and $\mathbf{Y}_3$, assume that $\theta_3 < \theta_1 < \theta_2$ so that the true ranks are $r_1 = 2$, $r_2 = 3$ and $r_3 = 1$, but inferential results support only one (the greatest) significance difference out of three pairwise comparisons, that is, $\left|\hat{\theta}_1 - \hat{\theta}_3\right| < \mathrm{LSD}(\theta_1, \theta_3)$ and $\left|\hat{\theta}_1 - \hat{\theta}_2\right| < \mathrm{LSD}(\theta_1, \theta_2)$, but $\left(\hat{\theta}_3 - \hat{\theta}_2\right) > \mathrm{LSD}(\theta_2, \theta_3)$. In this case the estimated ranks will be $\{\hat{r}(\mathbf{Y}_1) = 1, \hat{r}(\mathbf{Y}_2) = 2, \hat{r}(\mathbf{Y}_3) = 1\}$, which is clearly an inconsistent ranking. In fact, from the logical perspective, the first two comparisons suggesting $\mathbf{Y}_1 = \mathbf{Y}_2$ and $\mathbf{Y}_2 = \mathbf{Y}_3$ should imply $\mathbf{Y}_1 = \mathbf{Y}_3$ but on the contrary empirical data support as conclusion that $\mathbf{Y}_1 \neq \mathbf{Y}_3$. To overcome the intransitivity issue let us define a new ranking estimator $\bar{r}$ defined as

$$\bar{r}_j = 1 + \left\{ \# \left(\hat{r}_j + \hat{r}'_j \right) / 2 > \left(\hat{r}_h + \hat{r}'_h \right) / 2,\ h = 1, \ldots, C,\ j \neq h \right\},\ j = 1, \ldots, C,$$

where

$$\hat{r}'_j = 1 + \left\{ \# \left(C - \left\{ \# \hat{\theta}_j - \hat{\theta}_h > \mathrm{LSD}(\theta_j, \theta_h),\ h = 1, \ldots, C,\ j\,h \right\} \right) > \left(C - \# \left\{ \left(\hat{\theta}_{j'} - \hat{\theta}_h \right) \right. \right. \right.$$
$$\left. \left. \left. > \mathrm{LSD}(\theta_{j'}, \theta_h),\ h = 1, \ldots, C, j' \neq h \right\} \right),\ j' = 1, \ldots, C,\ j' \neq j \right\},\ j = 1, \ldots, C.$$

When applied to the previous example, it follows that $\bar{r}_1 = 2$, $\bar{r}_2 = 3$ and $\bar{r}_3 = 1$, because $\hat{r}'_1 = 2$, $\hat{r}'_2 = 3$ and $\hat{r}_3 = 1$. In general, the revised ranking estimator $\bar{r}$ fully overcame the intransitivity problem and can be viewed as the average rank from the ranks derived from two types of counting: the significant observed inferiorities (i.e., $\hat{\theta}_h - \hat{\theta}_j > \mathrm{LSD}$) and the significant observed superiorities (i.e., $\hat{\theta}_j - \hat{\theta}_h > \mathrm{LSD}$).

It is worth noting that, under the hypothesis of homogeneity of all populations, that is, $\boldsymbol{\mu}_1 = \boldsymbol{\mu}_2 = \ldots = \boldsymbol{\mu}_C$, by definition all true ranking position r_j would necessarily be equal to 1, hence they would be in a full ex aequo situation, that is,

$$r_j = \{1 + \#(\theta_j < \theta_h), h = 1, \ldots, C, j \neq h\} = 1, \forall j.$$

When performing an inference on r_j via pairwise differences $(\theta_j - \theta_h)$, under the hypothesis of full ex aequo the probability of estimating the Correct Global Ranking (CGR) and the Correct Individual Ranking (CIR) are such that

$$\Pr\{\text{CGR}|\text{homogeneity}\} = \Pr\{\bar{r}_j = 1, \forall j\} = 1 - \alpha,$$
$$\Pr\{\text{CIR}|\text{homogeneity}\} = \Pr\{\bar{r}_j = 1\} \geq 1 - \alpha^{*}, j = 1,\ldots,C,$$

where α and α^* are respectively the significance level and the adjusted by multiplicity level chosen in the testing procedure.

Under the alternative hypothesis of nonhomogeneity, that is, $\exists\, \boldsymbol{\mu}_j \neq \boldsymbol{\mu}_h\ j,h = 1,\ldots,C, j \neq h$, the following expression exists:

$$\text{if } \theta_j > \theta_h \text{ then } \Pr\{\bar{r}_j > \bar{r}_h|\text{nonhomogeneity}\} > \alpha^{*}, j = 1,\ldots,C, j \neq h.$$

In particular, we can expect that the greater is the relative distance among ranking parameters the greater will be both Pr{CGR|nonhomogeneity} and Pr{CIR|nonhomogeneity}; however, because under the alternative the populations at the extreme ranking positions have a greater probability to be declared as superior/inferior, it is clear that the highest individual rates will be referred to as the true 'best' populations.

Because as the sample sizes increase Pr{CIR|nonhomogeneity} increases as well, it is worth noting that under the assumption of nonhomogeneity, the ranking estimator can be said to be also a *consistent classifier* that is a procedure such that the probability of incorrect ranking classification gets arbitrarily close to the lowest possible risk as the sample size goes to infinity (Bousquet et al., 2004).

1.3 Literature Review of the Ranking Problem

As the problem of ranking has been addressed in the literature from many different points of view, in this chapter we review the basic procedures proposed in the literature, classifying them within the main reference field where they have been developed, that is, statistics and operations research.

1.3.1 Statistical Approaches

There are many situations when we are faced with inferential problems of comparing several – more than two – populations and the goal is not just to accept or reject the so-called homogeneity hypothesis, that is, the equality of all populations, but an effort is provided to try to rank the populations according to some suitable criterion.

Multiple Comparison Procedures (MCPs) have been proposed just to determine which populations differ after obtaining a significant omnibus test result, such as the ANOVA *F*-test. However, when MCPs are applied with the goal of ranking populations they are at best indirect and less efficient, because they lack protection in terms of a guaranteed probability against picking out the 'worse' population. This drawback motivated the foundation of the so-called ranking and selection methods (Gupta and Panchapakesan, 2002), the formulations of which provide more realistic goals with respect to the need to rank or select the 'best' populations. A further class of procedures with some connection with the ranking problem are the constrained – or order restricted – inference methods (Robertson et al., 1988; Silvapulle and Sen, 2005). Finally, the ranking problem has been addressed in the literature from the point of view of investigating and modelling the variability of sampling statistics used to rank populations, that is, the empirical estimators whose rank transformation provides the estimated ranking of the populations of interest (Hall and Miller, 2009, 2010; Hall and Schimek, 2012).

1.3.1.1 Multiple Comparison Procedures

The reference to the so-called MCPs occurs when one considers a set of statistical inferences simultaneously, for example, when a set, or family, of testing procedures is considered simultaneously, in particular when we wish to compare more than two populations (treatments, groups, etc.) to find possible significant differences between them within the *C*-samples location testing problem (Westfall et al., 2011). Because incorrect rejection of the null hypothesis is more likely when the family as a whole is considered, the main issue and goal of MCPs is to prevent this from happening, allowing significance levels for single and multiple comparisons to be directly compared. These techniques generally require a stronger level of evidence to be observed for an individual comparison to be deemed 'significant', so as to compensate for the number of inferences being made.

Some contributions proposed in the field of MCPs have more or less directly to do with the ranking problem. Hsu and Peruggia (1994) critically reviewed the graphical representations of Tukey's multiple comparison method behind which we can clearly see Tukey's attempt to rank the populations from the 'best' to the 'worst'. The popular Tukey's underlining representation prescribes that after ordering the populations according to the increasing values of their estimated means, all subgroups of populations that cannot be declared different are underlined by a common line segment. After that, one can infer at least as many groups that are strictly not the best and in this way argue which population can be overall considered the best, the second best, and so forth. In fact, because the set of all pairwise orderings is equivalent to a set of rankings, from a pairwise decision-theoretic subset selection procedure on the possible significances and from the specific directions in which each significance occurs, it is possible to specify the subset of rankings selected from the set of all possible rankings (for details we refer to Bratcher and Hamilton, 2005; Hamilton et al., 2008). Bratcher and Hamilton (2005) proposed a Bayesian decision-theoretic model for producing, via all pairwise comparisons, a set of possible rankings for a given number of normal means. They performed a simulation study in which they proved the superiority of their model to popular frequentist methods used to rank normal means, including Tukey's method and the Benjamini–Hochberg procedure (1995).

Referring to the so-called global performance indexes and with the goal of ordering several multivariate populations, Arboretti Giancristofaro et al. (2010a) proposed a permutation-based method using simultaneous pairwise confidence intervals. In this connection, Arboretti Giancristofaro et al. (2010b) compared two ranking parameters in a simulation study that highlighted some differences between the parametric and nonparametric approaches.

Some additional MCP techniques focus on estimating and testing which specific population can be inferred as the best one among a set of several populations. This situation is called Multiple Comparisons with the Best (MCB; Hsu, 1992). In the same direction but in the framework of the order restricted inference, the so-called testing for umbrella alternatives (Mack and Wolfe, 1981) aims at pairwise testing and simultaneously estimating among a set of a priori ordered populations which one can be considered as the 'peak' group where the response reaches the maximum (or the minimum) value of its location parameter.

1.3.1.2 Selection and Ranking

The selection and ranking approach, also known as multiple decision procedures, arose from the need of enabling to answer natural questions regarding the selection of the 'best' populations within the framework of C-sample testing problem (Gupta and Panchapakesan, 2002). Depending on the formulation of the procedures two basic approaches have been developed: the indifference zone (IZ) formulation, originally proposed by Bechhofer (1954), and the subset selection (SS) formulation, established by Gupta (1965). The IZ formulation aims to select one of the C populations $\Pi_1,\ldots, \Pi_C$ as the best one and if the selected population is truly the best, then a correct selection (CS) is said to occur. A guaranteed minimum probability of a CS is required when the best and the second best populations $\Pi_{[1]}$ and $\Pi_{[2]}$, that is, those associated with the largest two estimated ranking parameters, $\hat{\theta}_{[1]}$ and $\hat{\theta}_{[2]}$, are sufficiently apart, that is, $\theta_{[1]} - \theta_{[2]} > \delta$, where the term ranking parameter refers to a population parameter of interest, often the location parameter $\theta = \mu$, whose rank transformation defines the true population ranking. The IZ approach can be also applied in case interest is focussed on completely ranking a set of populations (CR-IZ), that is, from 'best', 'second best',..., down to the 'worst' (Beirlant et al., 1982). In the SS approach for selecting the best population, the goal is to select a nonempty subset of the C populations so that the selected subset includes the best (which event defines a correct selection-CS) with a guaranteed minimum probability. Provided certain distributional assumptions on populations are met, these methods usually guarantee that the probability of a correct selection will be at least some pre-specified value P^* that should be specified in advance by the experimenter, that is, $\Pr\{CS\} \geq P^*$.

A few selection and ranking proposals are concerned with ranking of several multivariate populations. Under the assumption of multivariate normal distributions, several real-valued functions θ of population parameter (μ,Σ) have been adopted to rank the populations: (1) Mahalanobis distance, (2) generalized variance, (3) multiple correlation coefficient, (4) sum of bivariate product-moment correlations, and (5) coefficient of alienation (Gupta and Panchapakesan, 2002). In case the population distribution functions are not specified, several nonparametric solutions have been proposed; those procedures are based on more general ranking parameters such as the rank correlation coefficient and the probability of concordance (Govindarajulu and Gore, 1971).

1.3.1.3 Order-Restricted Inference and Stochastic Ordering

Prior information regarding a statistical model frequently constrains the shape of the parameter set and can often be quantified by placing inequality constraints on the parameters. The use of such ordering information increases the efficiency of procedures developed for statistical inference (Dykstra et al., 1986). On the one hand, such constraints make the statistical inference procedures more complicated, but on the other hand, such constraints contain statistical information as well, so that if properly incorporated they would be more efficient than their counterparts wherein such constraints are ignored (Silvapulle and Sen, 2005). Davidov and Peddada (2011) extended the order-restricted inference paradigm to the case of multivariate binary response data under two or more naturally ordered experimental conditions. In such situations one is often interested in using all binary outcomes simultaneously to detect an ordering among the experimental conditions. To make such comparisons they developed a general methodology for testing for the multivariate stochastic order between the multivariate binary distributions. Conde et al. (2012) developed a classification procedure in the case of ordered populations that exploit the underlying order among the mean values of several groups by using ideas from order-restricted inference and incorporating additional information to Fisher's linear discriminant rule (Fisher, 1936). However, it should be noted that the work of Conde et al. (2012) aims to classify individual observations into populations by exploiting restriction constraints on the parameters, while the objective of our book is to classify the populations in an ordered sequence according to sampling information and having a priori no restriction on population parameters.

1.3.1.4 Ranking Models

The ranking problem has been also addressed in the literature from the point of view of investigating and modelling the variability of sampling statistics used to rank populations, that is, the empirical estimators whose rank transformation provides the estimated ranking of the populations of interest. The distributions of ranking probabilities have been investigated by Gilbert (2003) within the one-way ANOVA layout under the assumption that parameter estimates are well approximated by a normal distribution, with possible intergroup heteroscedasticity and correlation. Gilbert (2003) proposed several methods for estimating the true (objective, frequentist) ranking probability distribution given historical data and for developing inferences about the ranking probabilities. Hall and Miller (2009) proposed using bootstrap to handle with the variability of empirical ranking and they discuss both the theoretical and the numerical properties of bootstrap estimators of the distributions of rankings. The same authors (Hall and Miller, 2010) prove that a light- or heavy-tailed underlying distribution of population variables may weakly or strongly affect the reliability of empirical rankings. Considering the problem in which C items are judged by assessors using their perceptions of a set of performance criteria, or alternatively by technical devices, Hall and Schimek (2012) considered methods and algorithms that can be used to address this problem. They studied their theoretical and numerical properties in the case of a model based on nonstationary Bernoulli trials.

Another approach to ranking models is that of formally defining a suitable model underlying the process of ranking C items, often referred to a behavioural issue where a subject based on its own individual preferences is willing to order C objects. In this connection ranking models proposed so far in the literature fall into four categories: the Thurstonian models, multistage models, models induced from paired comparison, and distance-based models (Xu, 2000). The Thurstonian models (Daniels, 1950; Mosteller, 1951) extend Thurstone's theory of paired comparison to the full ordering of several items (Thurstone, 1927). Multistage models split the ranking process into $C - 1$ stages. Starting with the full set of C items, at the first stage, one item is selected and assigned rank 1; at the second stage, another item is selected from the remaining items and assigned rank 2; and so on. The last remaining item is assigned rank C by default. One such model is based on Luce's theory of choice behaviour (Luce, 1959). Babington-Smith (1950) suggested inducing a ranking model from a set of arbitrary paired comparison probabilities. To reduce the number of parameters of Babington-Smith's model, Bradley and Terry (1952) introduced a specific condition on the paired comparison probabilities, while substituting Bradley–Terry probabilities into the Babington-Smith model leads to the well-known Mallows–Bradley–Terry MBT model (Mallows, 1957). Distance-based models were first suggested by Mallows (1957); they are based on the assumption that there is a modal ranking from which the ranking probabilities are the same. Mallows proposed two metrics used for the distance, namely to the concordance measures, Kendall's (1948) tau and Spearman's (1904) rho, respectively.

1.3.1.5 Extreme Profile Ranking Method

In this section we define an appropriate synthesis indicator of a set of p informative ordered categorical variables representing judgements on a specific aspect under evaluation (see Sections 7.1 and 7.4).

Let us denote the responses as a p-dimensional variable $\mathbf{Y} = [Y_1,\ldots,Y_p]$, where each marginal variable can assume S ordered discrete scores, $s = 1,\ldots,S$, $S \in \mathbf{N}\backslash\{0\}$, $S > 1$, and large values of s correspond to higher satisfaction rates. For application reasons these variables are given different (nonnegative) degrees of importance: $(0 < w_k \leq 1, k = 1,\ldots,p)$. Such weights are thought to reflect the different role of the variables in representing indicators of the specific aspect under evaluation (for example, in Section 7.1, indicators of a PhD researcher's success in entering the labour market or academic field), and are provided by responsible experts or by results of surveys previously carried out in the specific context.

Arboretti et al. (2007) addressed the methodological problem of finding a global satisfaction index or a global ranking of C items starting from p dependent rankings on the same C items, each representing a specific aspect under evaluation. Two main aspects should be considered when facing the problem of finding a global index or a global ranking of satisfaction:

1. The search for a suitable combining function of two or more indicators or rankings
2. Consideration of extreme units of the global ranking. Bird et al. (2005) pointed out that 'the principle that being ranked lowest does not immediately equate with genuinely inferior performance should be recognized and reflected in the method of presentation of ranking'.

The Nonparametric Combination (NPC) of dependent rankings (Lago and Pesarin, 2000) provides a solution for aspect 1. The main purpose of the NPC ranking method is to obtain a single ranking criterion for the statistical units under study, which summarizes many partial (univariate) rankings.

Let us consider a multivariate phenomenon whose variables $\mathbf{Y}$ are observed on C statistical units. Starting from component variables Y_k, $= k = 1,\ldots,p$, each one providing information about a partial aspect, we wish to construct a *global index* or *combined ranking* T:

$$T = \phi(Y_1,\ldots,Y_p; w_1,\ldots,w_p),\ \phi: \mathbb{R}^{2p} \rightarrow \mathbb{R}^1,$$

where ϕ is a real function that allows us to combine the partial dependent rankings and $(w_1,\ldots,w_p)$ is a set of weights that takes the relative degrees of importance among the p aspects of $\mathbf{Y}$ into account.

We introduce a set of minimal reasonable conditions related to variables Y_k, $k = 1,\ldots,p$:

1. For each of the p informative variables a partial ordering criterion is well established, that is to say, 'large is better'.
2. Regression relationships within the p informative variables are monotonic (increasing or decreasing).
3. The marginal distribution of each informative variable is nondegenerate.

Moreover, notice that we need not assume the continuity of Y_k, $k = 1,\ldots,p$, so that the probability of ex equo can be positive. The combining real function ϕ is chosen from class Φ of combining functions satisfying the following minimal properties:

1. ϕ must be continuous in all $2p$ arguments, in that small variations in any subset of arguments imply small variation in the ϕ-index.
2. ϕ must be monotone nondecreasing with respect to each argument:

$$\phi\left(\ldots Y_k,\ldots;w_1,\ldots,w_p\right) \geq \phi\left(\ldots,Y_k',\ldots;w_1,\ldots,w_p\right) \text{ if } 1 > Y_{ki} > Y_k' > 0, k = 1,\ldots,p;$$

3. ϕ must be symmetric with respect to permutations of the arguments, in that if, for instance, $u_1,\ldots,u_p$ is any permutation of $1,\ldots,p$ then:

$$\phi(Y_{u_1},\ldots,Y_{u_p};w_1,\ldots,w_p) = \phi(Y_1,\ldots,Y_p;w_1,\ldots,w_p).$$

Property 1 is obvious; property 2 means that if, for instance, two subjects have exactly the same values for all Ys, except for the kth, then the one with $Y_k > Y_k'$ must have assigned at least the same satisfaction ϕ-index. Property 3 states that any combining function ϕ must be invariant with respect to the order in which informative variables are processed.

For example, Fisher's combining function: $\phi = -\sum_{k=1}^{p} w_p \times \log(1-Y_k)$ (Pesarin and Salmaso, 2010a) can be useful for quality assessment. Of course, other combining functions (see Section 3.1.5) may be of interest for the problem of quality assessment. Here we simply point out that Fisher's combining function seems to be more sensitive when assessing the best quality than when assessing lower quality, in the sense that small differences in the lower quality region seem to be identified with greater difficulty than those in the best quality region.

For aspect 2 (see the two-item list on page 39, we propose an extension of the NPC ranking method to the case of ordered categorical variables based on extreme satisfaction profiles. Extreme satisfaction profiles are defined a priori on a hypothetical frequency distribution of variables Y_k, $k = 1,\ldots,p$. Let us consider data **Y**, where the rule 'large is better' holds for all variables. Observed values for the p variables are denoted as y_{jk}, $j = 1,\ldots,C$; $k = 1,\ldots,p$. Examples of extreme satisfaction profiles are given below.

The *strong* satisfaction profile is defined as follows:

- The maximum satisfaction is obtained when all subjects have the highest value of satisfaction for all variables:

$$f_{ks} = \begin{cases} 1 & \text{for } s = S \\ 0 & \text{otherwise} \end{cases}, \ \forall k,\ k = 1,\ldots,p$$

  where f_{ks} are the relative frequencies of categories s, $s = 1,\ldots,S$, for variable Y_k, $k = 1,\ldots,p$.

- The minimum satisfaction is obtained when all subjects have the smallest value of satisfaction for all variables:

$$f_{ks} = \begin{cases} 1 & \text{for } s = 1 \\ 0 & \text{otherwise} \end{cases}, \ \forall k,\ k = 1,\ldots,p.$$

The *weak* satisfaction profile is defined as follows:

- The maximum satisfaction is obtained when the same relative frequency (say 70%) of subjects has the highest value of satisfaction for all variables:

$$f_{ks} = \begin{cases} u & \text{for } s = S \\ u_s & \text{otherwise, where } \sum_{s=1}^{S-1} u_s = (1-u) \end{cases} \quad \forall k, k = 1,\ldots,p.$$

- The minimum satisfaction is obtained when the same relative frequency (say 70%) of subjects has the smallest value of satisfaction for all variables:

$$f_{ks} = \begin{cases} l & \text{for } s = 1 \\ l_s & \text{otherwise, where } \sum_{s=2}^{S} l_s = (1 - l) \end{cases} \quad \forall k,\ k = 1,\dots,p.$$

Another way to define weak satisfaction profiles is obtained when

- The maximum satisfaction is obtained when subjects have the highest value of satisfaction with relative frequencies varying across the variables:

$$f_{ks} = \begin{cases} u_k & \text{for } s = S \\ u_{ks} & \text{otherwise, where } \sum_{s=1}^{S-1} u_{ks} = (1 - u_k) \end{cases} \quad k = 1,\dots,p.$$

- The minimum satisfaction is obtained when subjects have the smallest value of satisfaction with relative frequencies varying across the variables:

$$f_{ks} = \begin{cases} l_k & \text{for } s = S \\ l_{ks} & \text{otherwise, where } \sum_{s=2}^{S} l_{ks} = (1 - l_k) \end{cases} \quad k = 1,\dots,p.$$

To include the extreme satisfaction profiles in the analysis, we transform original values s, $s = 1,\dots,S$. At first, we separate the values of s corresponding to a judgement of satisfaction, say the last t, $1 \le t, \le S$, from those values corresponding to judgements of dissatisfaction, that is, $(S - t)$. For the last t values of s corresponding to a judgement of satisfaction, the transformed values of s are defined as

$$s + f_{ks} \times 0.5 \quad s = S - t + 1,\dots,S; \quad k = 1,\dots,p.$$

For the first $(S - t)$ values of s corresponding to judgements of dissatisfaction, the transformed values of s are defined as

$$s + (1 - f_{ks}) \times 0.5 \quad s = 1,\dots,S - t; \quad k = 1,\dots,p.$$

Such a transformation is equivalent to the assignment to original values s, $s = 1,\ldots,S$ of additive degrees of importance, which depend on relative frequencies f_{ks} and which increase the original values s up to $s + 0.5$.

Let us suppose, for example, that $s = 1, 2, 3, 4$ and values 3 and 4 correspond to judgements of satisfaction. By applying the preceding transformation, the value of 3 tends to the upper value of 4, which represents higher satisfaction, when f_{k3} increases. On the contrary, the value of 1 tends to 2 (less dissatisfaction), when f_{k1} decreases. Figure 1.3 displays the example.

The transformation of values s, $s = 1,\ldots,S$ weighted by relative frequencies f_{ks}, is applied to observed values y_{jk}, $j = 1,\ldots,C$; $k = 1,\ldots,p$. For the last t values of s corresponding to a judgement of satisfaction, the transformed values of y_{jk} are defined as

$$z_{jk} = y_{jk} + \sum_{s=S-t+1}^{S} \mathbf{I}_s(y_{jk}) \times f_{ks} \times 0.5, \quad k = 1,\ldots,p;\ j = 1,\ldots,C,$$

with

$$\mathbf{I}_s(y_{jk}) = \begin{cases} 1 & \text{if} \quad y_{jk} = s \\ 0 & \text{if} \quad y_{jk} \neq s. \end{cases}$$

For the first $(S - t)$ values of s corresponding to judgements of dissatisfaction, the transformed values of y_{jk} are defined as:

$$z_{jk} = y_{jk} + \sum_{s=1}^{S-t} \mathbf{I}_s(y_{jk}) \times (1 - f_{ks}) \times 0.5, \quad k = 1,\ldots,p;\ j = 1,\ldots,C.$$

In this setting, we can consider the following transformations (partial rankings):

$$\lambda_{jk} = \frac{(z_{jk} - z_{k\min}) + 0.5}{(z_{k\max} - z_{k\min}) + 1}, \quad k = 1,\ldots,p;\ j = 1,\ldots,C,$$

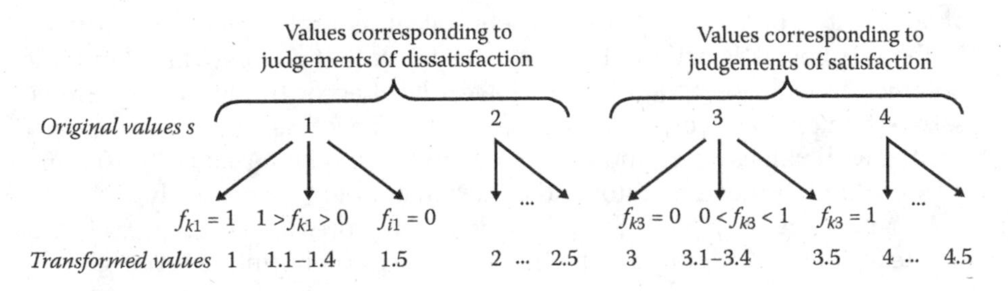

FIGURE 1.3
Transformation of original s values.

where $z_{k\,\min}$ and $z_{k\,\max}$ are obtained according to an extreme satisfaction profile. If we consider the strong satisfaction profile we have

$$z_{k\min} = y_{jk} + \sum_{s=1}^{S-t} \mathbf{I}_s(y_{jk}) \times (1 - f_{ks}) \times 0.5 = 1$$

$$\text{where } f_{ks} = 1 \text{ and } y_{ks} = s = 1, \quad k = 1,\ldots,p$$

$$z_{k\max} = y_{jk} + \sum_{s=S-t+1}^{S} \mathbf{I}_s(y_{jk}) \times f_{ks} \times 0.5 = S + 0.5$$

$$\text{where } f_{ks} = 1 \text{ and } y_{ji} = s = S, \quad k = 1,\ldots,p.$$

If we consider a weak satisfaction profile, with $u = 0.7$ and $l = 1$, we have

$$z_{k\min} = y_{jk} + \sum_{s=1}^{S-t} \mathbf{I}_s(y_{jk}) \times (1 - f_{ks}) \times 0.5 = 1$$

$$\text{where } f_{ks} = 1 \text{ and } y_{ks} = s = 1, \quad k = 1,\ldots,p$$

$$z_{k\max} = y_{jk} + \sum_{s=S-t+1}^{S} \mathbf{I}_s(y_{jk}) \times f_{ks} \times 0.5 = S + 0.35$$

$$\text{where } f_{ks} = 0.7 \text{ and } y_{ji} = s = S, \quad k = 1,\ldots,p.$$

It is worth noting that $z_{k\max}$ represents the preferred value for each variable, and it is obtained when satisfaction is at its highest level accordingly to the extreme satisfaction profile; $z_{k\min}$ represents the worst value, and it is obtained when satisfaction is at its lowest level accordingly to the extreme satisfaction profile. Scores λ_{jk}, $k = 1,\ldots,p$, $j = 1,\ldots,C$ are one-to-one increasingly related with values y_{jk}, z_{jk} and are defined in the open interval (0, 1) (+0.5 and +1 are added in the numerator and denominator of y_{jk} respectively).

To synthesize the p partial rankings based on scores λ_{jk}, $k = 1,\ldots,p$, $j = 1,\ldots,C$, by means of the NPC ranking method, we use a combining function ϕ:

$$[T_j = \phi(\lambda_{j1},\ldots,\lambda_{jp}; w_1,\ldots,w_p), j = 1,\ldots,C].$$

To calculate the global index varying in the interval [0, 1] we put:

$$S_j = \frac{T_j - T_{\min}}{T_{\max} - T_{\min}}, \; j = 1, \ldots, C,$$

where

$$T_{\min} = \phi(\lambda_{1\min}, \ldots, \lambda_{p\min}; \; w_1, \ldots, w_p),$$
$$T_{\max} = \phi\left(\lambda_{1\max}, \ldots, \lambda_{p\max}; \; w_1, \ldots, w_p\right),$$

and $\lambda_{k\min}$ and $\lambda_{k\max}$ are obtained accordingly to the extreme satisfaction profiles:

$$\lambda_{k\min} = \frac{(z_{k\min} - z_{k\min}) + 0.5}{(z_{k\max} - z_{k\min}) + 1}, \quad k = 1, \ldots, p$$
$$\lambda_{ikax} = \frac{(z_{k\max} - z_{k\min}) + 0.5}{(z_{k\max} - z_{k\min}) + 1}, \quad k = 1, \ldots, p.$$

Note that value $T_{\min}$ represents the *unpreferred value* of the satisfaction index because it is calculated from $(\lambda_{1\,\min}, \ldots, \lambda_{p\,\min})$, while $T_{\max}$ represents the *preferred value* because it is calculated from $(\lambda_{1\,\max}, \ldots, \lambda_{p\,\max})$. $T_{\min}$ and $T_{\max}$ are reference values used to evaluate the 'distance' of the observed satisfaction values from the situation of highest satisfaction defined according to the extreme satisfaction profile.

1.3.2 Operations Research Literature on the Ranking Problem

Using the information on the degree of preference of a set of alternatives to be compared and starting from a more algorithmic perspective, the ranking problem can be seen as the search for an 'optimal' order, that is, what satisfies predetermined criteria of optimality. In this perspective, operations research is the discipline that deals with the problem of finding an optimal deterministic ranking. We use the term deterministic to emphasize the fact that in this context there is no reference to any population or to samples drawn from populations, that is, in summary, there is no underlying inference. It follows that it makes no sense to speak of uncertainty of the procedure of determining the ranking so that it is essentially a deterministic process in nature.

To solve the ranking problem within the operations research literature two main approaches have emerged: multiple-criteria decision making and group ranking. The two approaches have focussed on the optimal synthesis of a multiplicity of preferences respectively referred to a set of criteria and to a group of subjects. In practice, whereas the former emphasizes the multidimensional nature of the items to be ranked, the second focusses on the multiplicity of individuals who have expressed the evaluations.

The great amount of work developed around the problem of algorithmic ranking gained a large boost from two important theoretical results: Arrow's impossibility theorem (Arrow, 1963), which inspired the group ranking approach, and the analytic hierarchy process (AHP) proposed by Saaty (1977, 1980), which became a leading approach to multicriteria decision making. With reference to the issue of voting and elections, a prominent 'impossibility' result is Arrow's (1963) fundamental theorem proving that no voting scheme can guarantee five natural fairness properties: universal domain, transitivity, unanimity, independence with respect to irrelevant alternatives here referred to as rank reversal, and nondictatorship. Kemeny and Snell (1962) proposed an axiomatic approach for dealing with preference ranking that models the problem as minimizing the deviation from individual rankings defined by the distance between two complete rankings. In the AHP proposed by Saaty (1977, 1980), the decision problem is modelled as a hierarchy of criteria, subcriteria, and alternatives. The method features a decomposition of the problem to a hierarchy of simpler components, extracting experts' judgements and then synthesizing those judgements. After the hierarchy is constructed, the decision maker assesses the intensities in a pairwise comparison matrix.

Hochbaum and Levin (2006) proved that there is a modelling overlap between the problems of multicriteria decision making and aggregate ranking, although these two issues have often been pursued separately and traditionally are considered distinct. Authors proposed a framework that unifies several streams of research and offers an integrated approach for the group-ranking problem and multicriteria decision making.

An important role in all approaches of operations research to the ranking problem has been played by the Perron–Frobenius theorem (Hofuku and Oshima, 2006; Keener, 1993), which asserts that a real square matrix with positive entries has a unique largest real eigenvalue and that the corresponding eigenvector has strictly positive components, and also asserts a similar statement for certain classes of nonnegative matrices. In fact, the idea of using a square matrix A, often called preference matrix, to find a ranking vector has been around for some time and the idea of powering the matrix A to find a ranking vector was initiated by Wei (1952) and Kendall (1955) and revisited often (e.g., Saaty, 1987).

1.3.2.1 Multiple-Criteria Decision-Making Approach

Numerous contributions to the ranking problem have been proposed within the management science and operations research literature focussing on the optimization point of view and referring to behavioural issues and decision theory. Methods based on the so-called Multiple-Criteria Decision-Making (MCDM) approach aim at solving decision-making problems in which more actions of a set of individuals are compared to determine which alternative (among a given set) is the best or to establish a ranking (Köksalan et al., 2011). Among such kinds of techniques proposed in the literature, essentially three methods are considered: aggregation methods using utility functions, interactive methods and outranking methods. The dominance relation associated with a multicriteria problem is based on the unanimity of the point of view; however, this is usually so poor that it cannot be used for solving real problems and therefore many authors have proposed outranking methods to enrich the dominance relation. The most popular methods in this area are ELECTRE I, II, III and IV. However, ELECTRE methods are rather intricate because they require a large number of parameters, the values of which are to be fixed to the decision-maker and the analyst. To avoid these difficulties a modified approach called PROMETHEE was proposed (Brans and Vincke, 1985).

MCDM or Multiple-Criteria Decision Analysis (MCDA) is a sub-discipline of operations research that explicitly considers multiple criteria in decision-making environments. The main concern of MCDM is to structure and solve decisions, and to plan problems that involve multiple criteria. The purpose of MCDM is to support decision makers facing these types of problems. Typically, there is no unique optimal solution for such problems, and therefore it is necessary to use the decision maker's preferences to differentiate between solutions.

'Solving' can be interpreted in different ways. It could correspond to choosing the 'best' alternative from a set of available alternatives (where 'best' can be interpreted as 'the most preferred alternative' for a decision maker). Another interpretation of 'solving' could be choosing a small set of good alternatives, or grouping alternatives into different preference sets. An extreme interpretation could be to find all 'efficient' or 'nondominated' alternatives (which we define shortly).

The difficulty of the problem originates from the presence of more than one criterion. There is no longer a unique optimal solution to an MCDM problem that can be obtained without incorporating preference information. The concept of an optimal solution is often replaced by the set of nondominated solutions. A nondominated solution has the property that it is not possible to move away from it to any other solution without sacrificing at least one criterion. Therefore, it makes sense for the decision maker to choose a solution from the nondominated set. Otherwise, he could do better in terms of some or all of the criteria, and not do worse in any of them. Generally, however, the set of nondominated solutions is too large to be presented to the decision maker for his final choice. Hence we need tools that help the decision maker focus on his preferred solutions (or alternatives). Normally one has to 'trade off' certain criteria for others.

1.3.2.2 Group-Ranking Approach

Still with reference to operations research, another class of solutions for the ranking problem is based on the so-called group-ranking methods which refer to the group decision-making theory (also known as collaborative decision making): a situation faced when individuals collectively make a choice from the alternatives that have been submitted to them. The problem of 'group-ranking', also known as 'rank-aggregation', has been studied in contexts varying from sports, to decision making, to machine learning, to ranking Web pages, and to behavioural issues (Hochbaum and Levin, 2006). The essence of this problem is how to consolidate and aggregate decision makers' rankings to obtain a group ranking that is representative of 'better coherent' ordering for the decision makers' rankings (Chen and Cheng, 2009). According to the completeness of preference information provided by decision makers, the group ranking problem can be roughly classified into two major approaches, the total ranking approach and the partial ranking approach. The former needs individuals to appraise all alternatives, while the latter requires only a subset of alternatives. Roughly speaking, the goal of most total ranking methods is to determine a full ordering list of items that expresses the consensus achieved among a group of decision makers. Therefore, the advantage of these researches is that no matter how much users' preferences conflict, an ordering list of all items to represent the consensus is always produced. Unfortunately, this advantage is also a disadvantage, because when there is no consensus or only slight consensus on items' rankings, the previous approach still generates a total ordering list using their ranking algorithms. In such a situation, what we obtain is really not a consensus list, but merely the output of algorithms. Traditionally, there are three formats to express users' preferences about items in the total ranking approach. These formats include weights/scores of items, set of pairwise comparisons on the items and ranking lists of items.

Moreover, the group ranking problem can be classified according to the format to express users' preferences. Depending on the input format used to express preferences, they can be classified into weights/scores of items, set of pairwise comparisons and ranking lists of items. The first kind of format requires each individual to provide weights/scores for all items. Thus, the accuracy of this approach would be affected by personal differences in scoring behaviour. The second format needs individuals to provide set of pairwise comparisons on all items. This kind of format is a general way in expressing users' preference about the items. However, providing these comparisons becomes an awful work, in case of large number of items. The last format is to ask users to provide lists of ranking items. When items are many, it is not easy for users to determine a full ordering list.

1.4 Specificities and Advantages of the Permutation Approach for the Multivariate Ranking Problem

A critical revision of the literature on the ranking problem highlights that despite the fact the problem has been extensively treated from many perspectives, none of these seems to be close enough to the procedure we propose from either the point of view of the purpose or the method used. Even if MCPs indirectly address the problem of ranking treatment groups, there is generally no clear indication on how to deal with the information from pairwise comparisons, especially in the case of a multivariate response variable. At first sight the indifference zone formulation of ranking and selection approach in multiple decision theory, as can be seen in Gupta and Panchapakesan (2002), which contains an extensive discussion on the whole theory, looks more or less close to the goal of our methodology. However, there are two issues that suggest that ranking and selection methods cannot be regarded as a counterparts of our approach to the ranking problem. First, ranking and selection focusses mainly on the parametric-based decision theory while we are much more oriented to a nonparametric less demanding approach toward explorative analysis of multivariate data. Second, to guarantee the probability of right decisions, the ranking and selection procedures need to specify the probability of correct classification and how far two ordered populations should be from each other. These two elements have nothing to do with our framework so a direct comparison or application to the same problems is actually not possible.

Although constrained statistical inference and stochastic ordering apparently seem to have a connection with the ranking problem as we intend, it should be noted that the way the C-sample testing problem is usually faced in this framework does not allow one to rank populations at all, because the alternative (or sometimes the null) hypothesis is always specified on the basis of a given constraint/prior knowledge on the parameters of interest. Conversely, our proposed procedure goes in a different direction: the alternative hypothesis is expressed in general form to determine which of all possible alternatives is the most supported by the results of the pairwise comparisons.

Finally, the vast literature in operations research cannot be of any help at all for our purposes because even if it comes to establish decision-making ranking algorithms the reference context is never suitable to make inferences, so that within this research area there is no connection with the uncertainty of estimating a true underlying unobservable ranking among populations.

As mentioned in Section 1.2, where we presented the formalization of the multivariate ranking problem, it is clear that its general solution requires a key element: an hypothesis testing procedure for directional multivariate alternatives, which must be applied for any kind of responses irrespective of the sample sizes. To the best of our knowledge, the only method proposed in the literature that achieves this goal is the nonparametric combination of dependent permutation tests, the so-called NPC methodology (Pesarin and Salmaso, 2010a). The main advantages of using the permutation and combination approach to classify and rank several multivariate populations is that it allows us to derive multivariate directional p-value statistics that can be calculated also when the number of response variables is much larger than the number of replications (so-called *finite-sample consistency* of combined permutation tests; see Chapter 2). It is worth noting that in this situation, which can be common in many real applications, traditional parametric and nonparametric testing procedures are not appropriate at all. The NPC approach has several nice features: it always provides an exact solution for whatever sample size, is very low demanding in terms of assumptions and finally is quite flexible because it allows handling jointly with all types of response variables, that is, numeric, binary and ordered categorical even in the presence of any noninformative or informative missing data (missing completely at random or not at random).

1.5 Multivariate Ranking and Quality Improvement

From the point of view of applied research, the multivariate ranking problem can be viewed as a tool of quality improvement. In fact, as we mentioned in the first section of this chapter, we remarked that the necessity and usefulness of defining an appropriate ranking of items quite often occurs as a natural conclusion of many researches in the real world of managerial and engineering activities. It is worth noting that in this context the ranking problem takes on a connotation of an effort aimed at obtaining some form of progress or improvement of an organizational aspect or a business practice or a process/product technology. In other words, it can be viewed as a quality improvement task. To support this argument, let us reconsider the two guideline examples: the second one was related to a real case study in the field of indoor environmental quality evaluations where a sample of pupils from a primary school have been enrolled to fulfil a questionnaire and at the same time several instrumental measures have been recorded. The goal of the study is to compare and rank three classrooms where pupils attend lessons in terms of subjective and objective well-being of indoor environmental quality, related to microclimatic conditions and building related factors. Actually the three classrooms differ from each other because of their specific layout, exposure to the sun, position of the windows, and so forth. The classroom ranking in terms of environmental quality evaluations will be useful in linking the wellness to their specific characteristics to derive in this way useful information for improving the quality of indoor environments. Let us reconsider also the first more technical/technological guideline example in the field of new product development for industry. When developing new products the ranking problem can be viewed as a comparative study where the interest is not only to establish if the products are equal or different but also to find out a suitable ranking (possibly by including some ties) able to evaluate the relative degree of preference of a given product (treatment/condition) with respect to all others. Note that the case study arises from a well-designed experiment so that the underlying inference is truly effective. On the contrary, in the indoor quality survey we are facing an observational case study so that the related inferences are necessarily weaker.

It is worth noting that quite often the items of interest to be ranked are multivariate in nature, meaning that many aspects of the items can be simultaneously observed on the same unit/subject. In the indoor environmental quality study several different subjective evaluations (thermal sensation, acoustic comfort, light intensity, etc.) and objective measures (temperature, humidity, etc.) are jointly recorded. Similarly, when performing an industrial experiment, because the benefits are simultaneously evaluated on several different outputs, the response variable is multivariate in nature.

This consideration leads us to try to define and characterize a new class of multivariate ranking problems. Assuming that a true underlying ranking of the items under investigation does exist, using some suitable multivariate sampling information we would like to take a decision about the possible equality of all items (that is, all items are tied) versus a procedure of estimation of a suitable ranking. This goal represents a nonstandard statistical inferential problem where hypothesis testing and classification are both involved. At the same time, because we are referring to a multivariate setting this issue also represents a complex problem so that a flexible nonparametric solution is advisable.

In the next chapter we suggest a general nonparametric permutation and combination-based theoretical framework (Pesarin and Salmaso, 2010a) where new solutions for the ranking problem can be developed.

References

Arboretti Giancristofaro R., Pesarin F., Salmaso L. (2007). Nonparametric approaches for multivariate testing with mixed variables and for ranking on ordered categorical variables with an application to the evaluation of PhD programs. In

S. Sawilowsky (ed.), *Real Data Analysis*, pp. 355–385. Quantitative Methods in Education and the Behavioural Sciences: Issues, Research and Teaching (Ronald C. Serlin, series ed.). Charlotte, NC: Information Age Publishing.

Arboretti Giancristofaro R., Corain L., Gomiero D., Mattiello F. (2010a). Nonparametric multivariate ranking methods for global performance indexes. *Quaderni di Statistica*, 12, 79–106.

Arboretti Giancristofaro R., Corain L., Gomiero D., Mattiello F. (2010b). Parametric vs. nonparametric approach for interval estimators of multivariate ranking Parameters. In *Proceedings of the 2010 JSM – Joint Statistical Meetings*, 31 July–5 August, 2010, Vancouver, Canada, pp. 894–908.

Arboretti Giancristofaro R., Bonnini S., Corain L., Salmaso L. (2014). A permutation approach for ranking of multivariate populations. *Journal of Multivariate Analysis*, 132, 39–57.

Arrow K.J. (1963). *Social Choice and Individual Values*. New York: John Wiley & Sons.

Babington-Smith B. (1950). Discussion of Professor Ross' paper. *Journal of the Royal Statistical Society, Series B*, 12, 153–162.

Bacelli F., Makowski A. M. (1989). Multidimensional stochastic ordering and associated random variables. *Operations Research*, 37, 478–487.

Basso D., Pesarin F., Salmaso L., Solari A. (2009). *Permutation Tests for Stochastic Ordering and ANOVA: Theory and Applications with R*. Lecture Notes in Statistics. New York: Springer Science+Business Media.

Bechhofer R. E. (1954). A single-sample multiple decision procedure for ranking means of normal populations with known variances. *The Annals of Mathematical Statistics*, 25(1), 16–39.

Beirlant J., Dudewicz E. J., van der Meulen E. C. (1982). Complete statistical ranking of populations, with tables and applications. *Journal of Computational and Applied Mathematics*, 8(3), 187–201.

Benjamini Y., Hochberg Y. (1995). Controlling the false discovery rate. *Journal of the Royal Statistical Society*, 57, 289–300.

Bertoluzzo F., Pesarin F., Salmaso L. (2013). On multi-sided permutation tests. *Communications in Statistics: Simulation and Computation*, 42(6), 1380–1390.

Bird S. M., Cox D., Farewell V. T., Goldstein H., Holt T., Smith P. C. (2005). Performance indicators: Good, bad and ugly. *Journal of the Royal Statistical Society, Series A*, 168(1), 1–27.

Bonnini S., Corain L., Cordellina A., Crestana A., Musci R., Salmaso L. (2009). A novel global performance score with application to the evaluation of new detergents. In M. Bini, P. Monari, D. Piccolo, L. Salmaso (eds.), *Statistical Methods for the Evaluation of Educational Services and Quality of Products*, pp. 161–179. Contributions to Statistics. Heidelberg: Springer-Verlag.

Bousquet O., Boucheron S., Lugosi G. (2004). Introduction to statistical learning theory. In O. Bousquet, U.v. Luxburg, G. Rätsch (eds.), *Advanced Lectures in Machine Learning*, Springer, 169–207.

Bradley R. A., Terry M. E. (1952). Rank analysis of incomplete block designs. *Biometrika*, 39, 324–335.

Brans J. P., Vincke P. (1985). A preference ranking organisation method: The PROMETHEE method for MCDM. *Management Science*, 31(6), 647–656.

Bratcher T. L., Hamilton C. (2005). A Bayesian multiple comparison procedure for ranking the means of normally distributed data. *Journal of Statistical Planning and Inference*, 133, 23–32.

Carrozzo E., Corain L., Musci R., Salmaso L., Spadoni L. (2015). A new approach to rank several multivariate normal populations with application to life cycle assessment. *Communications in Statistics – Simulation and Computation*, doi: 10.1080/03610918.2014.925926.

Chen Y.-L., Cheng L.-C. (2009). Mining maximum consensus sequences from group ranking data. *European Journal of Operational Research*, 198, 241–251.

Conde D., Fernández M. A., Rueda C., Salvador B. (2012). Classification of samples into two or more ordered populations with application to a cancer trial. *Statistics in Medicine*, 31,

Corain L., Salmaso L. (2007). A nonparametric method for defining a global preference ranking of industrial products. *Journal of Applied Statistics*, 34(2), 203–216.

Corain L., Salmaso L. (2013). Nonparametric permutation and combination-based multivariate control charts with applications in microelectronics. *Applied Stochastic Models in Business and Industry*, 24(4), 334–349.

Corain L., Salmaso L. (2015). Improving power of multivariate combination-based permutation tests. *Statistics and Computing*, 25(2), 203–214.

Daniels H. E. (1950). Rank correlation and population models. *Journal of the Royal Statistical Society, Series B*, 12(2), 171–191.

Davidov O., Peddada S. (2011). Order restricted inference for multivariate binary data with application to toxicology. *Journal of American Statistical Association*, 106(496), 1394–1404.

Dayton M. C. (2003). Model comparisons using information measures. *Journal of Modern Applied Statistical Methods*, 2(2), 281–292.

Dudewicz E. J., Taneja V. S. (1978). Multivariate ranking and selection without reduction to a univariate problem. In *Proceeding WSC '78: Proceedings of the 10th Conference on Winter Simulation*, Vol 1, pp. 207–210.

Dykstra R., Robertson T., Wright F. T. (1986). *Advances in Order Restricted Statistical Inference*. Lecture Notes in Statistics, Vol. 37. Heidelberg: Springer-Verlag.

Finos L., Salmaso L., Solari A. (2007). Conditional inference under simultaneous stochastic ordering constraints. *Journal of Statistical Planning and Inference*, 137, 2633–2641.

Finos L., Pesarin F., Salmaso L., Solari A. (2008). Exact inference for multivariate ordered alternatives. *Statistical Methods and Applications*, 17, 195–208.

Fisher R. A. (1936). The use of multiple measurements in taxonomic problems. *Annals of Eugenics*, 7(2), 179–188.

Gilbert S. (2003). Distribution of rankings for groups exhibiting heteroscedasticity and correlation. *Journal of the American Statistical Association*, 98(461), 147–157.

Govindarajulu Z., Gore A. P. (1971). Selection procedures with respect to measures of association. In *Proceedings of Symposium on Statistical Decision Theory Related Topics*, Purdue University 1970, pp. 313–345.

Gupta S. S. (1965). On some multiple decision (selection and ranking) rules. *Technometrics*, 7(2), 225–245.

Gupta S. S., Panchapakesan S. (1987). Statistical selection procedures in multivariate models. In A. K. Gupta (ed.), *Advances in Multivariate Statistical Analysis*, pp. 141–160. Theory and Decision Library, Vol. 5.

Gupta S. S., Panchapakesan S. (2002). *Multiple Decision Procedures: Theory and Methodology of Selecting and Ranking Populations*. Philadelphia: Society for Industrial and Applied Mathematics. Netherlands: Springer.

Hall P., Miller H. (2009). Using the bootstrap to quantify the authority of an empirical ranking. *The Annals of Statistics*, 37(6B), 3929–3959.

Hall P., Miller H. (2010). Modeling the variability of rankings. *The Annals of Statistics*, 38(5), 2652–2677.

Hall P., Schimek M. (2012). Moderate-deviation-based inference for random degeneration in paired rank lists. *Journal of the American Statistical Association*, 107(498), 661–672.

Hamilton C., Bratcher T. L., Stamey J. D. (2008). Bayesian subset selection approach to ranking normal means. *Journal of Applied Statistics*, 35(8), 847–851.

Hochbaum D. S., Levin A. (2006). Methodologies and algorithms for group-rankings decision. *Management Science*, 52(9), 1394–1408.

Hofuku I., Oshima K. (2006). Rankings methods for various aspects based on Perron-Frobenius theorem. *Information*, 9, 37–52.

Hsu, J. C. (1992). Stepwise multiple comparisons with the best. *Journal of Statistical Planning and Inference*, 3(2), 197–204.

Hsu J. C., Peruggia M. (1994). Graphical representations of Tukey's multiple comparison method. *Journal of Computational and Graphical Statistics*, 3, 2, 143–161.

Keener J. P. (1993). The Perron–Frobenius theorem and the ranking of football teams. *SIAM Review*, 35(1), 80–93.

Kemeny J. G., Snell J. L. (1962). Preference ranking: An axiomatic approach. In *Mathematical Models in the Social Sciences*, pp. 9–23. Boston: Ginn.

Kendall M. G. (1948). *Rank Correlation Methods*. London: Charles Griffin and Co.

Kendall M. G. (1955). Further contributions to the theory of paired comparisons. *Biometrics*, 11, 43.

Klingenberg B., Solari A., Salmaso L., Pesarin F. (2009). Testing marginal homogeneity against stochastic order in multivariate ordinal data. *Biometrics*, 65, 452–462.

Köksalan M., Wallenius J., Zionts S. (2011). *Multiple Criteria Decision Making: From Early History to the 21st Century*. Singapore: World Scientific.

Lago A., Pesarin F. (2000). Nonparametric combination of dependent rankings with application to the quality assessment of industrial products. *Metron*, LVIII(1–2), 39–52.

Lehmann E. L. (1966). Some concepts of dependence. *Annals of Mathematical Statistics*, 37, 1137–1153.

Luce R. D. (1959). *Individual Choice Behavior*. New York: John Wiley & Sons.

Mack G. A., Wolfe D. A. (1981). K-sample rank tests for umbrella alternatives. *Journal of the American Statistical Association*, 76, 175–181.

Mallows C. L. (1957). Non-null ranking models. *Biometrika*, 44, 114–130.

Minhajuddin A. T. M., Frawley W. H., Schucany W. R., Woodward W. A. (2007). Bootstrap tests for multivariate directional alternatives. *Journal of Statistical Planning and Inference*, 137(7), 2302–2315.

Mosteller E. (1951). Remarks on the method of paired comparisons. I: The least squares solution assuming equal standard deviations and equal correlations. *Psychometrika*, 16, 3–9.

Pesarin F. (2001). *Multivariate Permutation Tests with Applications in Biostatistics*. Chichester, UK: John Wiley & Sons.

Pesarin F., Salmaso L. (2010a). *Permutation Tests for Complex Data: Theory, Applications and Software*. Wiley Series in Probability and Statistics. Chichester, UK: John Wiley & Sons.

Pesarin F., Salmaso L. (2010b). Finite-sample consistency of combination-based permutation tests with application to repeated measures designs. *Journal of Nonparametric Statistics*, 22(5), 669–684.

Pesarin F., Salmaso L. (2012). A review and some new results on permutation testing for multivariate problems. *Statistics and Computing*, 22(2), 639–646.
Pesarin F., Salmaso L. (2013). On the weak consistency of permutation tests. *Communications in Statistics – Simulation and Computation*, 42(6), 1368–1379.
Robertson T., Wright F. T., Dykstra R. L. (1988). *Order Restricted Statistical Inference*. Wiley Series in Probability and Statistics, Chichester, UK: John Wiley & Sons.
Saaty T. S. (1977). A scaling method for priorities in hierarchical structures. *Journal of Mathematical Psychology*, 15, 234–281.
Saaty T. S. (1980). *The Analytic Hierarchy Process*. New York: McGraw-Hill.
Saaty T. S. (1987). Rank according to Perron: A new insight, *Mathematics Magazine*, 60, 211.
Santos E. N. F, Ferreira D. F. (2012). Multivariate multiple comparisons by bootstrap and permutation tests. *Biometric Brazilian Journal*, 30(3), 381–400.
Scarsini M., Shaked M. (1990). Stochastic ordering for permutation symmetric distributions. *Statistics & Probability Letters*, 9, 217–222.
Shaffer J. P. (1986). Modified sequentially rejective multiple test procedure. *Journal of the American Statistical Association*, 81, 826–831.
Shaffer J. P. (2002). Multiplicity, directional (type III) errors, and the null hypothesis. *Psychological Methods*, 7(3), 356–369.
Shaked M. (1982). A general theory of some positive dependence notions. *Journal of Multivariate Analysis*, 12(2), 199–218.
Shaked M., Shanthikumar J. G. (2007). *Stochastic Orders*. Springer Series in Statistics. New York: Springer Science+Business Media.
Silvapulle M. J., Sen P. K. (2005). *Constrained Statistical Inference: Order, Inequality, and Shape Constraints*. Wiley Series in Probability and Statistics. Chichester, UK: John Wiley & Sons.
Spearman C. (1904). The proof and measurement of association between two things. *American Journal of Psychology*, 15, 88.
Thurstone L. L. (1927). A law of comparative judgment. *Psychological Review*, 34, 273–286.
Wei T. H. (1952). *The Algebraic Foundations of Ranking Theory*. Cambridge, UK: Cambridge University Press.
Westfall P. H., Tobias R. D., Rom D., Wolfinger R. D. (2011). *Multiple Comparisons and Multiple Tests Using SAS*, 2nd Edition. Cary, NC: SAS Institute.
Xu L. (2000). A multistage ranking model. *Psychometrika*, 65(2), 217–231.

2

Permutation Tests and Nonparametric Combination Methodology

In Chapter 1 we presented the theoretical background on ranking of multivariate populations where the main issue of the proposed method was determining a suitable set of multivariate pairwise one-sided hypothesis testing procedures. In this chapter we show that, thanks to its componentwise nature and its flexibility in handling numeric and ordered categorical data, the combined permutation tests represent a valid solution for the problem at hand. This chapter also aims to introduce readers to the theory of univariate and multivariate permutation tests, showing the advantage of using such nonparametric procedures instead of traditional parametric solutions.

2.1 Introduction to Permutation Tests

The importance of the permutation approach in resolving a large number of inferential problems is well documented in the literature, where the relevant theoretical aspects, as well as the extreme effectiveness and flexibility, emerge from an applicatory point of view (Basso et al., 2009; Edgington and Onghena, 2007; Good, 2010; Pesarin and Salmaso, 2010a).

When compared with the more traditional parametric or nonparametric rank-based solutions, the main advantages of using the permutation approach in hypothesis testing problems are that in general permutation tests require fewer and easy to justify assumptions, are exact in nature and offer flexible solutions in dealing with complex problems. In this respect, the permutation-based solution for a complex problem such as the comparison of interventions in Group Randomized Trials (GRTs) is able to maintain a nominal test size thanks to its intrinsic exactness also in case of small sample sizes that usually occur in real applications but are not sufficient to make possible using asymptotic approximations (Braun and Feng, 2001). A simulation study described in the same paper, proves that in the case of the usual realistic sample sizes some traditional asymptotic-based procedures for the problem at hand (Generalized Estimating Equations [GEEs]) have liberal sizes; that is, they do not maintain the nominal level. Moreover, when considering a suitable model-based testing procedure (Penalized Quasi-Likelihood [PQL]) even if it is slightly more powerful than the permutation tests when the model of the simulated data exactly corresponds to that assumed, it is outperformed by the permutation tests when there are too few clusters to support asymptotic methods. In summary, permutation tests for GRTs are appropriate and solutions are more general and powerful than asymptotic counterparts in that they require fewer distributional assumptions. The use of permutation tests is becoming increasingly popular in biomedical research thanks in part to the effective debate within the community of biostatisticians. In a popular paper, Ludbrook and Dudley (1998) argued that, because randomization rather than random sampling is the norm in biomedical research and because group sizes are usually small, exact permutation or randomization tests for differences in location should be preferred over t or F-tests. In this connection, when selecting the more appropriate test statistic and in the planning of the size of a study, Weinberg and Lagakos (2000) derived the asymptotic distribution of permutation tests under a general contiguous alternative, and then investigated the implications for test selection and study design for several diverse areas of biomedical applications.

del Castillo and Colosimo (2011) proposed a permutation test for detecting differences in shape for the analysis of experiments where the response is the geometric shape of a manufactured part. They showed that the permutation test provides higher power for 2D circular profiles than the traditional F-based methods used in manufacturing practice, which are based on the circularity form of errors. The authors highlight that the proposed permutation test does not require the error assumptions that are needed in the F-test, which may be too restrictive in practice. Still, in the context of the shape analysis, but more from a biological and morphometric point of view, Iaci et al. (2008) proposed a permutation-based significance test for a new general index based on Kullback–Leibler information that measures relationships between multiple sets of random vectors.

In the context of statistical analysis for spatial point patterns Ute (2012) proposed a studentized permutation test for the null hypothesis that two (or more) observed point patterns are realizations of the same spatial point process model. The proposed test performs convincingly well in terms of empirical level and appears more powerful than a bootstrap-type competitor proposed in the literature. The superiority of the permutation tests toward bootstrap solutions is proved also by Troendle et al. (2004) when testing the equality of two multivariate distributions for small sample sizes and in the case of high dimension such as when analysing microarray data. When the interest is in detecting genes that are possibly expressed in only a part of the cases or expressed at different levels among the cases (so-called overexpression), van Wieringen et al. (2008) proposed a new permutation-type test based on the mixing proportion in a nonparametric mixture that minimizes a weighted distance function. They proved by a simulation study that this permutation test is indeed more powerful than the two-sample t-test and the Cramér–von Mises test.

Sometimes problems of interest are so complex that an asymptotic procedure is too complicated to be developed and the related rate of convergence too difficult to determine and thus it is preferred to propose a permutation test, which thanks to its simplicity and flexibility can often offer a possible effective solution. For example Cook and Yin (2001) suggest a permutation test as a means of making inference for dimension reduction in discriminant analysis. Hothorn et al. (2006) proposed a new theoretical framework for permutation tests that opens up the way to a unified and generalized view, emphasizing the flexibility of permutation tests as conditioned testing procedures, where conditioning is with respect to the observed data which are always sufficient statistics in the null hypothesis for any underlying distribution (Pesarin and Salmaso, 2010a). Even when some normal theory-type solution for a given multivariate testing problem such as MANOVA F-test does exist, the assumption of multivariate normality is often violated in practice, and the impact of such a violation on the validity of tests may be greater when the sample size is smaller (Zeng et al., 2011). Thus, for most sample sizes of practical interest, the relative lack of efficiency of permutation solutions may sometimes be compensated by the lack of approximation of parametric asymptotic counterparts. In addition, assumptions regarding the validity of parametric methods (such as normality and random sampling) are rarely satisfied in practice, so that consequent inferences, when not improper, are necessarily approximated, and their approximations are often difficult to assess.

For any general testing problem, in the null hypothesis (H_0), which usually assumes that data come from only one (with respect to groups) unknown population distribution P, the whole set of observed data $\mathbf{y}$ is considered to be a random sample, taking values on sample space $\mathcal{Y}^n$, where $\mathbf{y}$ is one observation of the n-dimensional sampling variable $\mathbf{Y}^{(n)}$ and where this random sample does not necessarily have independent and identically distributed (i.i.d.) components; in fact it suffices that data are exchangeable. We note that the observed data set $\mathbf{y}$ is always a set of sufficient statistics in H_0 for any underlying distribution.

Given a sample point $\mathbf{y}$, if $\mathbf{y}^* \in \mathcal{Y}^n$ is such that the likelihood ratio $f_P^{(n)}(\mathbf{y})/f_P^{(n)}(\mathbf{y}^*) = \rho(\mathbf{y}, \mathbf{y}^*)$ is not dependent on f_P for whatever $P \in \mathcal{P}$, then $\mathbf{y}$ and $\mathbf{y}^*$ are said to *contain essentially the same amount of information with respect to P*, so that they are equivalent for inferential purposes. The set of points that are equivalent to $\mathbf{y}$, with respect to the information contained, is called the orbit associated with $\mathbf{y}$, and is denoted by $\mathcal{Y}^n_{/\mathbf{y}}$, so that $\mathcal{Y}^n_{/\mathbf{y}} = \bigcup_{u^* \in \Pi(u)} \left(\mathbf{Y}u_i^*, i = 1, \ldots, n; n_1, n_2 \right)$ where $\Pi(\mathbf{u})$ is the set of all permutations of $\mathbf{u} = (1, \ldots, n)$.

The same conclusion is obtained if $f_P^{(n)}(\mathbf{y})$ is assumed to be invariant with respect to permutations of the arguments of $\mathbf{y}$; that is, the elements $(y_1, \ldots, y_n)$. This happens when the assumption of independence for observable data is replaced by that of *exchangeability*, $f_P^{(n)}(y_1, \ldots, y_n) = f_P^{(n)}\left(y_{u_1^*}, \ldots, y_{u_n^*}\right)$, where

$\left(u_1^*,\ldots,u_n^*\right)$ is any permutation of (1,…,n). Note that, in the context of permutation tests, this concept of exchangeability is often referred to as the *exchangeability of the observed data with respect to groups*. Orbits $Y^{n}_{/\mathbf{y}}$ are also called *permutation sample spaces*. It is important to note that orbits $Y^{n}_{/\mathbf{y}}$ associated with data sets $\mathbf{y}\in Y^n$ always contain a finite number of points, as n is finite.

Since, in the null hypothesis and assuming exchangeability, the conditional probability distribution of a generic point $\mathbf{y}'\in Y^n$, for any underlying population distribution $P\in$P, is P-independent, permutation inferences are invariant with respect to P in H_0. Some authors, emphasizing this invariance property, prefer to give them the name of invariant tests. However, owing to this invariance property, permutation tests are distribution free and nonparametric.

Formally, let $Y^{n}_{/\mathbf{y}}$ be the orbit associated with the observed vector of data $\mathbf{y}$. The points of $Y^{n}_{/\mathbf{y}}$ can also be defined as $\mathbf{y}^*$: $\mathbf{y}^* = \pi\mathbf{y}$ where π is a random permutation of indexes $1,2,\ldots,n$. Define a suitable test statistic T on $Y^{n}_{/\mathbf{y}}$ for which large values are significant for a right-handed one-sided alternative: The support of $Y^{n}_{/\mathbf{y}}$ through T is the set T that consists of S elements (if there are no ties in the given data). Let

$$T^*_{(1)} \leq T^*_{(2)} \leq,\ldots,\leq T^*_{(S)}$$

be the ordered values of T. Let T^{obs} be the observed value of the test statistic, $T^{obs} = T(\mathbf{y})$. For a chosen attainable significance level $\alpha \in \{1/S, 2/S,\ldots,(S-1)/S\}$, let $k = S(1-\alpha)$. Define a permutation test, the function $\phi^* = \phi(T^*)$, for a one-sided alternative

$$\phi^*(T) = \begin{cases} 1 \text{ if } T^{obs} \geq T^*_{(k)} \\ 0 \text{ if } T^{obs} < T^*_{(k)} \end{cases}.$$

Permutation tests have general good properties such as exactness, unbiasedness and consistency (see Hoeffding, 1952; Pesarin and Salmaso, 2010a).

2.2 Multivariate Permutation Tests and Nonparametric Combination Methodology

In this section, we provide details on the construction of multivariate permutation tests via the nonparametric combination approach. Consider, for instance, two multivariate populations within the usual one-way MANOVA layout and the related two-sample multivariate hypothesis testing problem where p (possibly dependent) variables are considered.

Let us denote a p-dimensional data set by $\mathbf{Y} = \{\mathbf{Y}_j, j = 1,2\} = \{\mathbf{Y}_{ij}, i = 1,\ldots,n_j, j = 1,2\} = \{\mathbf{Y}_{jk}, j = 1,2, k = 1,\ldots,p\} = \{Y_{ijk}, i = 1,\ldots,n_j, j = 1,2, k = 1,\ldots,p\}$. The response $\mathbf{Y}$ takes its value on the p-dimensional sample space Y, for which a σ-algebra A and a (possibly not specified) nonparametric family P of nondegenerate distributions are assumed to exist. The groups are presumed to be related to two populations and the data $\mathbf{Y}_j$ are supposed i.i.d. with distributions $P_j \in$ P, $j = 1, 2$. Each univariate component response Y can be of the continuous or binary or ordered categorical; moreover, the multivariate response can also be mixed (some univariate components are continuous/binary and some others are ordered categorical).

The main difficulties when developing a multivariate hypothesis testing procedure arise because of the underlying dependence structure among variables, which is generally unknown and more complex than linear. Moreover, a global answer involving several dependent variables (aspects) is often required, so the question is how to combine the information related to the p variables (aspects) into one global test.

To better explain the proposed approach let us denote an $n \times p$, $n = n_1 + n_2$, data set with $\mathbf{Y}$:

$$\mathbf{Y} = \begin{bmatrix} \mathbf{Y}_1 \\ \mathbf{Y}_2 \end{bmatrix} = [Y_1, Y_2, \ldots, Y_p] = \begin{bmatrix} y_{11} & y_{12} & \cdots & y_{1p} \\ y_{21} & y_{22} & \cdots & y_{2p} \\ \cdots & \cdots & \cdots & \cdots \\ y_{n1} & y_{n2} & \cdots & y_{np} \end{bmatrix},$$

where $\mathbf{Y}_1$ and $\mathbf{Y}_2$ are the $n_1 \times p$ and the $n_2 \times p$ samples drawn from the first and second population respectively. In the framework of Nonparametric Combination (NPC) of Dependent Permutation Tests we suppose that, if the global null hypothesis $H_0 : \mathbf{Y}_1 \stackrel{d}{=} \mathbf{Y}_2$ of equality of the two populations is true, the hypothesis of exchangeability of random errors holds. Hence, the following set of mild conditions should be jointly satisfied:

a. We suppose that for $\mathbf{Y} = [\mathbf{Y}_1, \mathbf{Y}_2]$ an appropriate probabilistic p-dimensional distribution P exists, $P_j \in$ P, $j = 1, 2$, belonging to a (possibly nonspecified) family P of nondegenerate probability distributions.
b. The null hypothesis H_0 states the equality in distribution of the multivariate distribution of the p variables in two groups:

$$H_0 : [P_1 = P_2] = \left[Y_1 \stackrel{d}{=} Y_2\right].$$

The null hypothesis H_0 implies the exchangeability of the individual data vector with respect to two groups. Moreover, according to Roy's Union Intersection Criterion (1953), H_0 is supposed to be properly decomposed into p sub-hypotheses H_{0k}, $k = 1,\ldots,p$, each appropriate for partial (univariate) aspects, thus H_0 (multivariate) is true if all the H_{0k} (univariate) are jointly true:

$$H_0 : \left[\bigcap_{k=1}^{p} Y_{1k} \overset{d}{=} Y_{2k} \right] = \left[\bigcap_{k=1}^{p} H_{0k} \right].$$

H_0 is called the *global* or *overall null hypothesis*, and H_{0k}, $k = 1,\ldots,p$, are called the *partial null hypotheses*.

c. The alternative hypothesis H_1 is represented by the union of partial H_{1k} subalternatives:

$$H_1 : \left[\bigcup_{k=1}^{p} H_{1k} \right],$$

so that H_1 is true if at least one of subalternatives is true.

In this context, H_1 is called the *global* or *overall alternative*, and H_{1k}, $k = 1,\ldots,p$, are called the *partial alternatives*.

d. Let $\mathbf{T} = \mathbf{T}(\mathbf{Y})$ represent a p-dimensional vector of test statistics, $p \geq 1$, whose components T_k, $k = 1,\ldots,p$, represent the partial univariate and nondegenerate *partial test* appropriate for testing the sub-hypothesis H_{0k} against H_{1k}. Without loss of generality, all partial tests are assumed to be marginally unbiased, significant for large values, and at least one of them is consistent (for more details and proofs see Pesarin and Salmaso, 2010a).

At this point, to test the global null hypothesis H_0 and the p univariate hypotheses H_{0k}, the key idea comes from the partial (univariate) tests that are focussed on kth partial aspects, and then, combining them with an appropriate combining function, first to test H_{0k}, $k = 1,\ldots,p$, and finally to test the global (multivariate) test which is referred to as the global null hypothesis H_0.

We should observe that in most real problems when the sample sizes are large enough, there is a clash over the problem of computational difficulties in calculating the conditional permutation space. This means that in practice it is not possible to calculate the exact p-value of the observed statistic T_{k0}. This limitation is brilliantly overcome by using the Conditional Monte Carlo Procedure (CMCP). The CMCP on the pooled data set $\mathbf{Y}$ is a random simulation of all possible permutations of the same data under H_0 (for more details refer to Pesarin and Salmaso, 2010a). Hence, to obtain an estimate of the permutation distribution under H_0 of all test statistics, a CMCP can be used. Every resampling without replacement $\mathbf{Y}^*$ from the data set $\mathbf{Y}$ actually consists of a random attribution of the individual block data vectors to the C groups. In every Y_r^* resampling, $r = 1,\ldots,B$, the k partial tests are calculated to obtain the set of values $[\mathbf{T}_{ir}^* = \mathbf{T}(\mathbf{Y}_{ir}^*),\ i = 1,\ldots,k;\ r = 1,\ldots,B]$, from the B independent random resamplings. It should be emphasized that CMCP considers permutations only of individual data vectors, so that all underlying dependence relations that are present in the component variables are preserved. From this point of view, the CMCP is essentially a multivariate procedure.

Without loss of generality, let us suppose that all partial tests are significant for large values. More formally, the steps of the CMC procedure are described as follows:

1. Calculate the p-dimensional vectors of statistics, each one related to the corresponding partial tests from the observed data:

$$\underset{p\times 1}{\mathbf{T}^{\text{obs}}} = \mathbf{T}(\mathbf{Y}) = \left[T_k^{\text{obs}} = T_k(\mathbf{Y}), k = 1,\ldots,p\right],$$

2. Calculate the same vectors of statistics for the permuted data:

$$\mathbf{T}_b^* = \mathbf{T}\left(\mathbf{Y}_b^*\right) = \left[T_{bk}^* = T_k\left(\mathbf{Y}_b^*\right), k = 1,\ldots,p\right],$$

3. Independently repeat the previous step B times. We denote with $\left\{\mathbf{T}_b^*, b = 1,\ldots,B\right\}$ the resulting sets from the B conditional resamplings. Each element represents a random sample from the p-variate permutation cumulative distribution function (CDF) $F_T(z|\mathbf{Y})$ of the test vector $\mathbf{T}(\mathbf{Y})$.

The resulting estimates are:

$$\hat{F}_T\left(\mathbf{z} \mid \mathbf{Y}\right) = \left[\frac{1}{2} + \sum_{b=1}^{B} \mathbb{I}\left(\mathbf{T}_b^* \le \mathbf{z}\right)\right] \Big/ (B+1), \forall \mathbf{z} \in \mathbb{R}^p,$$

$$\hat{L}_{T_k}\left(z \mid \mathbf{Y}\right) = \left[\frac{1}{2} + \sum_{b=1}^{B} \mathbb{I}\left(\mathbf{T}_{bk}^* \ge z\right)\right] \Big/ (B+1), \forall z \in \mathbb{R}^1,$$

$$\hat{\lambda}_k = \hat{L}_{T_k}\left(T_k^{\text{obs}} \mid \mathbf{Y}\right) = \left[\frac{1}{2} + \sum_{b=1}^{B} \mathbb{I}\left(\mathbf{T}_{bk}^{*} \geq T_k^{\text{obs}}\right)\right] \Big/ (B+1), k = 1,\ldots,p,$$

where $\mathbb{I}(.)$ is the indicating function and where with respect to the traditional empirical distribution function (EDF) estimators, 1/2 and 1 have been added respectively to the numerators and denominators to obtain estimated values in the open interval (0, 1), so that transformations by inverse CDF of continuous distributions are continuous and so are always well defined.

If $\hat{\lambda}_k < \alpha$ the null hypothesis corresponding to the kth variable (H_{0k}) is rejected at a significance level equal to α (adjusted for multiplicity). Moreover, choice of partial tests has to provide that

1. All partial tests T_k are marginally unbiased, formally:

$$\Pr\{T_k \geq z \mid \mathbf{Y}, H_{0k}\} \leq \mathbb{P}\{T_k \geq z \mid \mathbf{Y}, H_{1k}\}, \forall z \in \mathbb{R}^1.$$

2. At least one partial tests is consistent, that is,

$$\Pr\{T_k \geq T_{k\alpha} \mid H_{1k}\} \to 1, \forall \alpha > 0 \text{ as } n \to \infty,$$

where $T_{k\alpha}$, is a finite α-level for T_k.

Let us now consider a suitable continuous nondecreasing real function, ϕ: $(0, 1)^p \to \mathbb{R}$, that applied to the p-values of partial tests T_k defines the second-order global (multivariate) test T'':

$$T'' = \phi(\lambda_1,\ldots,\lambda_p),$$

provided that the following conditions hold:

- ϕ is nonincreasing in each argument: $\phi(\ldots,\lambda_k,\ldots) \geq \phi(\ldots,\lambda_k',\ldots)$, if $\lambda_k \leq \lambda_k'$, $k \in \{1,\ldots,p\}$.
- ϕ attains its supremum value $\bar{\phi}$, possibly not finite, even when only one argument attains zero: $\phi(\ldots,\lambda_k,\ldots) \to \bar{\phi}$ if $\lambda_k \to 0$, $k \in \{1,\ldots,p\}$.
- ϕ attains its infimum value $\underline{\phi}$, possibly not finite, when all arguments ϕ attain one: $\phi(\ldots,\lambda_k,\ldots) \to \underline{\phi}$ if $\lambda_k \to 1$, $k = 1,\ldots,p$.
- $\forall \alpha > 0$, the acceptance region is bounded: $\underline{\phi} < T''_{\alpha/2} < T'' < T''_{1-\alpha/2} < \bar{\phi}$.

Frequently used combining functions are

- Fisher combination: $\phi_F = -2\sum_k \log(\lambda_k)$,
- Tippet combination: $\phi_T = \max_{1 \leq k \leq p}(1 - \lambda_k)$,

- Direct combination: $\phi_D = \sum_k T_k$,
- Liptak combination: $\phi_L = \sum_k \Phi^{-1}(1 - \lambda_k)$,

where $k = 1,\ldots,p$ and Φ is the standard normal CDF.

It can be seen that under the global null hypothesis the CMC procedure allows for a consistent estimate of the permutation distributions, marginal, multivariate and combined, of the k partial tests. Usually, Fisher's combination function is considered (Pesarin and Salmaso, 2010a), mainly for its finite and asymptotic good properties. Of course, it would also be possible to take into consideration any other combining function (Lancaster, Mahalanobis, etc.; see Folks, 1984; Pesarin and Salmaso, 2010a). In the stated conditions the combined test is also unbiased and consistent. For a detailed description we refer to Pesarin and Salmaso (2010a).

2.3 Hypothesis Testing for Curve Comparison

Important inferential problems usually occur when the data of interest are a collection of scalars or vectors that can be viewed as samples drawn from a population of curves or trajectories. Thanks to modern technologies such kinds of data are more and more frequently observed in many different areas and contexts. Let us think of examples such as spectrometric curves, radar waveforms, gene expressions, and so forth. From the statistical point of view this type of data can be viewed as either longitudinal data from repeated measures on the same units/subjects (Diggle et al., 2002) or observations, called functional data, lying to infinite dimensional spaces commonly called functional spaces (Ferraty and Vieu, 2006). Functional data are characterized in general by data collected regularly at a high frequency, while longitudinal data are usually more sparse in time and collected at irregular intervals (Rice, 2004). Consequently, functional data analysis focuses more on dimension reduction. In addition, longitudinal data analysis is more model based and inferential, while functional data analysis has a more exploratory and nonparametric point of view with a focus on describing the data (with principal components, smoothing, etc.).

2.3.1 Literature Review on Testing for Curve Comparison

Until recently, functional data analysis (FDA) and longitudinal data analysis (LDA) have been viewed as distinct enterprises. As of a 2004 special issue of *Statistica Sinica*, it is seen that endeavours have been made to reconcile the two lines of research. It is worth noting that although functional data analysis and longitudinal data analysis are both devoted to analysing curves/

trajectories on the same subjects, FDA and LDA are also intrinsically different (Davidian et al., 2004). Longitudinal data are involved in follow-up studies (common in biomedical sciences) that usually require several (few) measurements of the variables of interest for each individual during the period of study. They are often treated by multivariate parametric techniques that study the variation among the means during the time controlled by a number of covariates. In contrast, functional data are frequently recorded by mechanical instruments (more common in engineering and physical sciences, although also in an increasing number of biomedical problems) that collect many repeated measurements per subject. Its basic units of study are complex objects such as curves (commonly), images or shapes (information along the time of the same individual is jointly considered). Conceptually, functional data can be considered sample paths of a continuous stochastic process (Valderrama, 2007) where the usual focus is studying the covariance structure. In addition, the infinite dimensional structure of the functional data makes the links with standard nonparametric statistics (in particular with smoothing techniques) particularly strong (González-Manteiga and Vieu, 2007). Despite these differences, which involve mainly the viewpoints and ways of thinking about the data of both fields, Zhao et al. (2004) connected them, illustrating the ideas in the context of a gene expression study example, introducing LDA to the FDA viewpoint.

As pointed out by Sirski (2012), despite the fact that the comparison between the curves is not only of methodological interest but also has important practical implications especially for technological and biomedical applications, much of the literature in the field of FDA is concerned with describing the data (with principal components, smoothing, etc.) as opposed to formal hypothesis testing, including assessing statistical significance between groups of curves. Even if the emphasis is on describing curves, the FDA literature proposes some solutions for the testing problem of comparing population of curves. Beyond the FDA proposals several other testing solutions can be also identified in the field of LDA. From a critical review on hypothesis testing solutions for curve comparisons it appears that in the literature there are basically two main approaches that may be briefly described as basis function approximation solutions and overall tests. The first class of procedures is concerned with testing the equality of the coefficients from a basis function approximation while the second one is aimed at developing a global test that compares the population of curves using a suitable test statistic defined in the whole domain of the functional response. Similar to other testing problems, both the FDA and LDA literature contain parametric and nonparameteric solutions for the two approaches.

Let us begin our review with the basis function approximations. The rationale behind this approach for hypothesis testing on curves is based on the principle of preliminarily transforming and reducing data to reduce the dimensionality of the problem by using a suitable transformation such as Fourier, wavelet, principal component analysis, and so forth and then testing for the equality among groups of the related coefficients. This principle applies both for the parametric and for the nonparametric framework. From a parametric point of view, a class of basis function solutions may be referred to for the thresholding methods for testing problems. Fan and Lin (1998) developed some adaptive Neyman tests for curve data from stationary Gaussian processes using orthogonal transformations such as Fourier and wavelet to preprocess the data and compress signals (Fan, 1996). If the comparison involves more than two sets of curves, authors call the proposed solutions HANOVA, that is, the analysis of variance when the dimensionality is high.

Within a nonparametric framework for curve comparisons using basis function approximations, Eubank (2000) proved that among the different ways of combining the coefficients into a test statistic, the L2 norm, a simple sum of the squared coefficients, is asymptotically equivalent to the uniformly most powerful test when the grid size goes to infinity. Zhang and Chen (2007) proposed an L2-norm-based global test statistic based on the local polynomial kernel smoothing technique demonstrating how to reconstruct individual functions from a discrete functional data set using local polynomial smoothing. They show that, under some mild conditions, the effects of substitutions of the functions with their local polynomial reconstructions can be ignored asymptotically.

In a generalized nonparametric regression framework Behseta and Kass (2005) proposed two methods referred to as the so-called Bayesian adaptive regression splines (BARS). The first method uses Bayes factors, and the second method uses a modified Hotelling-type test. Behseta et al. (2007) extend the application of BARS to the likelihood ratio tests based on the asymptotic distribution of BARS fits, where the reference distributions of the test statistics are derived asymptotically or via bootstrap.

Let us move the focus to the overall test solutions for curves comparison. In this connection and with respect to the parametric proposals, a class of solutions has been inspired by the attempt to extend the ANOVA techniques in the context of functional data. In their classical textbook reference on FDA, Ramsay and Silverman (2005) suggested calculating the F-test statistic at each time point, but do not address how to deal with the resulting statistics. Shen and Faraway (2004) suggested a functional F-test for comparing nested functional linear models, the null distribution of which is derived by approximation. The reference model for the Shen and Faraway solution is the so-called varying-coefficient models with time fixed covariates which is a special case for the general varying-coefficient model with both time varying or time fixed covariates (Wu and Yu, 2002). Shen and Faraway applies their functional F-test to some data from ergonomics and investigate its nominal size and power by a simulation study including some competing tests such as bootstrap methods as described in Faraway (1997), the traditional multivariate log likelihood ratio test on raw data and a test based on a B-spline basis function representation. Their simulations show that the power of the tests is data-dependent and argue that the F-test has the benefit of avoiding the risk that the other two comparison tests have with being influenced by 'unimportant directions of variation'. A simulation study on the comparison between the functional F-test with the multivariate likelihood ratio and B-spline–based tests is also presented in Shen and Xu (2007) where results confirms that the better performance of the functional F-test is setting-dependent. In the same direction Yang et al. (2007) illustrates how to apply the Shen and Faraway functional F-test to the setting of longitudinal data. They conduct a simulation study to investigate the statistical power for the F-test compared with the Wilks' likelihood ratio test and the linear mixed-effects model using AIC. Results confirms the general good behaviour of the functional F-test and its power is proved one more time to be setting-dependent, specifically the covariance structure of the error process affects the power of the test. Xu et al. (2011) propose a quasi F-test for functional linear models with functional covariates and outcomes which is an extension of the Shen and Faraway's F-statistic. From simulations to study the size and the power of the quasi F-test when comparing it with a linear mixed effects model approach, Xu et al. observe that their proposed test is more powerful than the linear mixed effects method only in case of large effects. Zhang (2011) proposed a new functional F-type test to reduce the bias in the approximate null distribution of the Shen and Faraway's F-test. In a simulation study Zhang demonstrated that the bias-reduced method and the naive method perform similarly when the data are highly or moderately correlated, but the former outperforms the latter significantly when the data are nearly uncorrelated. Always in the spirit of ANOVA techniques in the context of functional data, Cuevas et al. (2004) suggested an asymptotic version of the well-known ANOVA F-test in the case of functional data. To overcome some difficulties in practice with handling the asymptotic distribution of the test statistic they proposed a numerical Monte Carlo approach. Along the same line as the proposal of Cuevas et al., Martinez-Camblor and Corral (2011) suggested a generalization of the classical ANOVA F-ratio for repeated measures to the functional setting. Both the parametric and the nonparametric approaches are considered to derive the asymptotic and the resampling distribution of the test statistic. Within a nonparametric framework, asymptotic distributions of ANOVA-type test statistics are presented in Wang and Akritas (2010, 2011). The test statistics are in the form of a difference of two quadratic forms and have a limiting chi-square and normal distribution.

For nonparametric methods whose goal is to develop a global test suitable for the problem of comparing curves, there are two main approaches in the literature: permutation-based solutions and nonparametric regression solutions. Let us first consider the permutation approach, which seems very popular for the testing problem at hand, probably owing to the flexibility of permutation solutions in handling complex testing problems, especially when the asymptotic distributions are difficult to derive and/or the parametric assumptions are hard to justify (Pesarin and Salmaso, 2010a). From the notion of similarity between two curves and to test the null hypothesis of no difference, Munoz-Maldonado et al. (2002) suggested three permutation statistics, that is, pooled-mean, pooled-variance and the ratio between them. A similar solution is proposed in the work of Sturino et al. (2010); after quantifying the distance between mean or median curves from two treatments, apply a permutation test using as the test statistic either the difference of the means (or medians) or the difference of the areas between mean curves. They validate via simulations the proposed solutions, including also an additional permutation statistic employing functional principal component analysis. Using a different strategy based on pairwise differences between individual curves, Sirski (2012) proposed a set of permutation statistics and compared them by simulation with a collection of other permutation tests proposed in the literature, a test based on the functional principal components scores (Sturino et al., 2010), the adaptive Neyman test (Fan and Lin, 1998) and the functional *F*-test (Shen and Faraway, 2004). Sirski's simulation results suggest that the solution with the best power performance is the test based on the mean of the pairwise L1 norms, while the worst solution is the functional *F*-test which has a poor performance in case of nonnormal errors and of small sample sizes. The joint use of permutation tests and the nonparametric combination methodology allowed Pesarin and Salmaso (2010a) to propose a global test they called time-to-time permutation test that is able to combine properly in a suitable global test the set of the dependent univariate permutation tests performed in all observation points. A permutation solution inspired from a multiple comparison procedure was proposed by Cox and Lee (2008), applying a set of pointwise *t*-tests on smooth functional data. Similarly, Ramsay et al. (2009) addressed the issue of hypothesis testing, proposing a permutation approach based on the absolute value of the test statistic similar in form to the *t*-test statistic at each point in time. They also proposed a functional *F* statistic and used a permutation test based on the maximum *F* value.

Alternative nonparametric solutions are proposed in the literature within the nonparametric regression framework. In this connection Cardot et al. (2007) suggested a permutation approach to check if a real covariate has a significant effect on a functional response in a regression setting using two test statistics, that is, an adapted *F*-statistic and a statistic based on the kernel smoother applied to the residuals. Zhang and Lin (2003) proposed a solution for testing the equivalence of two nonparametric functions in semiparametric additive mixed models for correlated non-Gaussian data. This test extends the previous work of Zhang et al. (2000). Neumeyer and Dette (2003) proposed a new test for the comparison of two regression curves that is based on a difference of two marked empirical processes based on residuals that is applicable in the case of different design points and heteroscedasticity. Finally, from a functional data analysis and nonparametric regression perspective Hall and Van Keilegom (2007) suggested a Cramer–von Mises type test and took up the issue of studying how the data preprocessing interferes with the performance of two-sample statistical tests.

2.3.2 Time-to-Time Permutation Solution

To formalize the problem of comparing two or more populations of curves within the permutation and combination methodology let us note that functional data can also be interpreted as discrete or discretized stochastic processes for which at most a countable set of data points is observed. In this way an observed curve is nothing more than a set of *repeated measures* in which each subject/unit (a given curve) is observed on a finite or at most a countable number of occasions so that successive responses are dependent (Pesarin and Salmaso, 2010a). In practice, responses of one unit may be viewed as obtained by a discrete or discretized stochastic process.

With reference to each specific subject, repeated observations are also called the *response profiles*, and may be viewed as a multivariate variable. In the context of the repeated measurements designs, an existing permutation solution has already been proposed by Pesarin and Salmaso (2010a) that essentially employs the method of nonparametric combination of dependent permutation tests, each obtained by a partial analysis of data observed on the same ordered occasion (so-called *time-to-time analysis*).

Without loss of generality, we discuss general problems that can be referred to in terms of a one-way MANOVA layout for response profiles. Hence, we refer to testing problems for treatment effects when

a. Measurements are typically repeated a number of times on the same units.
b. Units are partitioned into C groups or samples, there being C levels of a treatment.
c. The hypotheses being tested aim to verify whether the observed profiles do or do not depend on treatment levels.
d. It is presumed that responses may depend on time, space, and so forth and that related effects are not of primary interest.

For simplicity, from here onwards we refer to time occasions of observation, where time means any sequentially ordered entity including space, lexicographic ordering, and so forth.

Let us assume that the permutation testing principle holds; in particular, in the null hypothesis, in which treatment does not induce differences with respect to levels, we assume that the individual response profiles are exchangeable with respect to groups. To be more specific, let us refer to a problem in which n units are partitioned into C groups and a univariate variable X is observed. Groups are of size $n_j \geq 2$, $j = 1,\ldots,C$, with $n = \sum_j n_j$. Units belonging to the jth group are presumed to receive a treatment at the jth level. All units are observed at N fixed ordered occasions $\tau_1, \ldots, \tau_N$, where N is an integer. For simplicity, we refer to time occasions by using t to mean τ_t, $t = 1, \ldots, N$. Hence, for each unit, we observe the discrete or discretized profile of a stochastic process, and profiles related to different units are assumed to be stochastically independent. Thus, within the hypothesis that treatment levels have no effect on response distributions, profiles are exchangeable with respect to groups.

Let us refer to a univariate stochastic time model with additive effects, covering a number of practical situations. Extensions of the proposed solution to multivariate response profiles are generally straightforward.

The symbol $\mathbf{X} = \{X_{ji}(t), i = 1, \ldots, n_j, j = 1, \ldots, C, t = 1, \ldots, N\}$ indicates that the whole set of observed data is organized as a two-way layout of univariate observations. Alternatively, especially when effects due to time are not of primary interest, $\mathbf{X}$ may be organized as a one-way layout of profiles, $\mathbf{X} = \{\mathbf{X}_{ji}(t), i = 1, \ldots, n_j, j = 1, \ldots, C\}$, where $\mathbf{X} = \{X_{ji}(t), t = 1, \ldots, N\}$ indicates the jith observed profile.

The general additive response model referred to this context is

$$X_{ji}(t) = \mu + \eta_j(t) + \varepsilon_{ji}(t),$$

where $\varepsilon_{ji}(t) = \Delta_{ji}(t) + \sigma(\eta_j(t)) \cdot Z_{ji}(t)$, $i = 1, \ldots, n_j$, $j = 1, \ldots, C$, $t = 1, \ldots, N$. In this model, $Z_{ji}(t)$ are generally non-Gaussian error terms distributed as a stationary stochastic process with null mean and unknown distribution P_Z (i.e., a generic white noise process); these error terms are assumed to be exchangeable with respect to units and treatment levels but, of course, not independent of time. Moreover, μ is a population constant; the coefficients $\eta_j(t)$ represent the *main treatment effects* and may depend on time through any kind of function, but are independent of units; the quantities $\Delta_{ji}(t)$ represent the so-called *individual effects*; and $\sigma(\eta_j(t))$ are time-varying scale coefficients which may depend, through monotonic functions, on main treatment effects η_j provided that the resulting CDFs are pairwise ordered so that they do not cross each other, as in $X_j(t) \overset{d}{<} (\text{or} \overset{d}{>}) X_r(t)$, $t = 1, \ldots, N$ and $j \neq r = 1, \ldots, C$. When $\Delta_{ji}(t)$ are stochastic, we assume that they have null mean values and distributions which may depend on main effects, units and treatment levels. Hence, random $\Delta_{ji}(t)$ are determinations of an unobservable stochastic process or, equivalently, of a k-dimensional variable $\Delta = \{\Delta(t), t = 1, \ldots, N\}$. In this context, we assume that $\Delta_j \sim \mathcal{D}_k\{0, \beta(\eta_j)\}$, where $\mathcal{D}_k$ is any unspecified distribution with null mean vector and unknown dispersion matrix β, indicating how unit effects vary with respect to main effects $\eta_j = \{\eta_j(t), t = 1, \ldots, N\}$. Regarding the dispersion matrix β, we assume that the resulting treatment effects are pairwise stochastically ordered, as in $\Delta_j(t) \overset{d}{<} (\text{or} \overset{d}{>}) \Delta_r(t)$, $t = 1, \ldots, N$ and $j \neq r = 1, \ldots, C$. Moreover, we assume that the underlying bivariate stochastic processes $\{\Delta_{ji}(t), \sigma(\eta_j(t)) \cdot Z_{ji}(t)),$ $t = 1, \ldots, N\}$ of individual stochastic effects and error terms, in the null hypothesis, are exchangeable with respect to groups. This property is easily justified when subjects are randomized to treatments.

This setting is consistent with a general form of dependent random effects fitting a very large number of processes that are useful in most practical situations. In particular, it may interpret a number of the so-called *growth processes*. Of course, when $\beta = \mathbf{0}$ with probability 1 for all t, the resulting model has fixed effects.

To appreciate the inherent difficulties in statistical analysis of real problems when repeated observations are involved, see, for example, Diggle et al. (2002). In particular, when dispersion matrices Σ and β have no known simple structure, the underlying model may not be identifiable and thus no parametric inference is possible. Also, when $N \geq n$, the problem cannot admit any parametric solution (see Blair et al., 1994, in which heuristic solutions are suggested under normality of errors $\mathbf{Z}$ and for fixed effects).

Among the many possible specifications of models for individual effects, one of these assumes that terms $\Delta_{ji}(t)$ behave according to an AR(1) process:

$$\Delta_{ji}(0) = 0;\ \Delta_{ji}(t) = \gamma(t)\cdot\Delta_{ji}(t-1) + \beta(\eta_j(t))\cdot W_{ji}(t),$$

$i = 1,\ldots,n_j$, $j = 1,\ldots,C$, $t = 1,\ldots,N$, where $W_{ji}(t)$ represent random contributions interpreting deviates of individual behaviour; $\gamma(t)$ are autoregressive parameters that are assumed to be independent of treatment levels and units, but not time; $\beta(\eta_j(t))$, $t = 1,\ldots,N$, are time-varying scale coefficients of autoregressive parameters, which may depend on the main effects. By assumption, the terms $W_{ji}(t)$ have null mean value and unspecified distributions and are possibly time dependent, so that they may behave as a stationary stochastic process.

A simplification of the previous model considers a regression-type form such as

$$\Delta_{ji}(t) = \gamma_j(t) + \beta(t)\cdot W_{ji}(t),\ i = 1,\ldots,n_j,\ j = 1,\ldots,C,\ t = 1,\ldots,N.$$

Note that many other models of dependence errors might be taken into consideration, including situations where matrices Σ and β are both full.

Within the layout presented in the preceding text, the hypotheses of interest we wish to test are

$$H_0 : \left\{\mathbf{X}_1 \overset{d}{=} \ldots \overset{d}{=} \mathbf{X}_C\right\} = \left\{X_1(t) \overset{d}{=} \ldots \overset{d}{=} X_C(t), t = 1,\ldots,N\right\}$$
$$= \left\{\bigcap_{t=1}^{N}\left[X_1(t) \overset{d}{=} \ldots \overset{d}{=} X_C(t)\right]\right\} = \left\{\bigcap_{t=1}^{N} H_{0t}\right\},$$

against H_1:$\{\bigcup_t[H_{0t}$ is not true$]\} = \{\bigcup_t H_{1t}\}$, in which a decomposition of the global hypotheses into N sub-hypotheses according to time is highlighted. This decomposition corresponds to the so-called *time-to-time* analysis for which an existing permutation solution has already been proposed by Pesarin and Salmaso (2010a).

Note that by decomposition into N partial sub-hypotheses, each sub-problem is reduced to a one-way ANOVA. Also note that, from this point of view, the associated two-way ANOVA, in which effects due to time are not of interest, becomes equivalent to a one-way MANOVA.

In the given conditions, N partial permutation tests $T_t^* = \sum_j n_j \cdot \left(\bar{X}_j^*\right)^2$, where $\bar{X}_j^* = \sum_i X_{ji}^*(t)/n_j$, $t = 1,\ldots,N$ are appropriate for time-to-time sub-hypotheses H_{0t} against H_{1t}. Thus, to achieve a global complete solution for H_0 against H_1, we must combine all these partial tests. Of course, owing to the complexity of the problem and to the unknown N-dimensional distribution of $(T_1,\ldots, T_N)$ (see Diggle et al., 2002), we are generally unable to evaluate all dependence relations among partial tests directly from $\mathbf{X}$. Therefore, this combination should be nonparametric and may be obtained through any combining function $\phi \in C$. Of course, when the underlying model is not identifiable, and so some or all of the coefficients cannot be estimated, this NPC becomes unavoidable. Moreover, when all observations come from only one type of variable (continuous, discrete, nominal, ordered categorical) and thus partial tests are homogeneous, a direct combination of standardized partial tests, such as $T_t^* = \sum_j n_j \cdot \left[\bar{X}_j^*(t) - \bar{X}_{\bullet}(t)\right]^2 / \sum_{ji}\left[X_j^*(t) - \bar{X}_j^*(t)\right]^2$, may be appropriate especially when N is large.

2.4 Some Properties of Univariate and Multivariate Permutation Tests

In this section we present two important properties of univariate and multivariate permutation tests, specifically the equivalence of permutation statistics and the so-called finite sample consistency.

2.4.1 Equivalence of Permutation Statistics

The concept of permutationally equivalent statistics is useful in simplifying computations and sometimes in facilitating the establishment of the asymptotic equivalence of permutation solutions with respect to some of their parametric counterparts.

Let $\mathbf{Y}$ be the given data set from a response variable Y, so that Y belongs to the sample space Ω. We have the following definition of the permutation equivalence of two statistics:

Definition

Two statistics T_1 and T_2, both mapping Ω into $\mathbb{R}^1$, are said to be permutationally equivalent when, for all points $\mathbf{Y} \in \Omega$ and $\mathbf{Y}^* \in \Omega_{/\mathbf{Y}}$, the relationship

$\{T_1(\mathbf{Y}^*) \leq T_1(\mathbf{Y})\}$ is true if and only if $\{T_2(\mathbf{Y}^*) \leq T_2(\mathbf{Y})\}$ is true, where $\mathbf{Y}^*$ indicates any permutation of $\mathbf{Y}$ and $\Omega_{/\mathbf{Y}}$ indicates the associated permutation sample space. This permutation equivalence relation is indicated by $T_1 \approx T_2$.

With reference to this definition we have the following theorem and corollaries (Pesarin, 2010). ■

Theorem 2.1

If between the two statistics T_1 and T_2 there is a one-to-one increasing relationship, then they are permutationally equivalent and $\Pr\{T_1(\mathbf{Y}^*) \leq T_1(\mathbf{Y})|\mathbf{Y}\} = \Pr\{T_2(\mathbf{Y}^*) \leq T_2(\mathbf{Y})\}|\mathbf{Y}\}$, where these probabilities are evaluated with respect to conditional distribution $P_{|\mathbf{Y}}$ induced by the sampling experiment and defined on the permutation measurable space $(\Omega_{/\mathbf{Y}}, \mathrm{B}_{/\mathbf{Y}})$. ■

Corollary 2.1

If T_1 and T_2 are related by an increasing one-to-one relationship with probability 1, then they are permutationally equivalent with probability 1, where this probability is measured in terms of population distribution P. ■

Corollary 2.2

If T_1 and T_2 are related by a decreasing one-to-one relationship, then they are permutationally equivalent in the sense that $\{T_1(\mathbf{Y}^*) \leq T_1(\mathbf{Y})\} \leftrightarrow \{T_2(\mathbf{Y}^*) \geq T_2(\mathbf{Y})\}|\mathbf{Y}\}$ for all $\mathbf{Y} \in \Omega$ and $\mathbf{Y}^* \in \Omega_{/\mathbf{Y}}$. ■

Corollary 2.3

The permutation equivalence relation is reflexive: $T_1 \approx T_1$. ■

Corollary 2.4

The permutation equivalence relation is transitive: If $T_1 \approx T_2$ and $T_2 \approx T_3$, then $T_1 \approx T_3$. ■

2.4.2 Finite Sample Consistency

A quite important problem usually occurs in some multidimensional applications when sample sizes are fixed and the number of variables that are to be analysed is much larger than sample sizes (Blair et al., 1994; Goggin, 1986). Pesarin (2001) showed that, under very mild conditions, the power function of permutation tests based on both associative and nonassociative statistics monotonically increases as the related standardized noncentrality functional increases. This is true also for multivariate situations. In particular, for any added variable the power does not decrease if this variable makes larger standardized noncentrality (*finite-sample consistency*). These results confirm and extend those presented by Blair et al. (1994) and Pesarin and Salmaso (2010b). In particular, Blair et al. (1994) presents an exhaustive power simulation study comparing permutation tests and Hotelling's T^2 test when the number of variables increases with respect to fixed sample sizes and shows a better behaviour of the permutation tests.

To present the finite-sample consistency we refer to one-sided two-sample designs for nonnegative alternatives. Extensions to nonpositive and/or two-sided alternatives, and multisample designs are straightforward. Let $\mathbf{Y}_j = \{Y_{ij}; i = 1,\ldots,n_j\} \in \mathcal{Y}^{n_j}$ be the independent and identically distributed (i.i.d.) sample data of size n_j from $(Y, \mathcal{Y}, P_j \in \mathcal{P})$, $j = 1, 2$, where Y is the variable of interest taking values in the sample space $\mathcal{Y}$ according to the distribution P_j. A notation for data sets with independent samples is $\mathbf{Y} = \{Y_{11},\ldots,Y_{1n_1},Y_{21},\ldots,Y_{2n_2}\} \in \mathcal{Y}^n$, where $n = n_1 + n_2$. To denote data sets in the permutation context it is sometimes convenient to use the unit-by-unit representation: $\mathbf{Y} = \{Y_i; i = 1,\ldots,n; n_1, n_2\}$, where it is intended that the first n_1 data in the list belong to the first sample and the rest to the second. In practice, denoting by $(u_1^*,\ldots,u_n^*)$ a permutation of $(1,\ldots,n)$, $\mathbf{Y}^* = \{Y_i^* = Y(u_i^*, i = 1,\ldots,n; n_1, n_2\}$ is the related permutation of $\mathbf{Y}$, so that $\mathbf{Y}_1^* = \{Y_{1i}^* = Y(u_i^*), i = 1,\ldots,n_1\}$ and $\mathbf{Y}_2^* = \{Y_{2i}^* = Y(u_i^*), i = n_1 + 1,\ldots,n_2\}$ are the two permuted samples, respectively.

Here we discuss testing problems for stochastic dominance alternatives as are generated by treatments with nonnegative shift effects δ. In particular, the alternative assumes that treatments produce an effect δ so that $\delta > \mathbf{0}$. Thus, the hypotheses are

$$H_0 : \left[\mathbf{Y}_1 \overset{d}{=} \mathbf{Y}_2 \overset{d}{=} \mathbf{Y}\right] \equiv [P_1 = P_2] \text{ vs. } H_1 : \left[\mathbf{Y}_1 + \boldsymbol{\delta} \overset{d}{>} \mathbf{Y}_2\right].$$

Note that under H_0 data of two samples are exchangeable, in accordance with the notion that subjects are randomized to treatments. Since effects δ may depend on null responses Y_1, stochastic dominance $(Y_1 + \delta) > Y_2 \overset{d}{=} Y$ is compatible with nonhomoscedasticities in the alternative. Thus, the null hypothesis may also be written as $H_0 : [\delta = 0]$. Sometimes, to emphasize the role of effects we use

$$\mathbf{Y}(\delta) = \{Y_{11} + \delta,\ldots,Y_{1n_1} + \delta, Y_{21},\ldots,Y_{2n_2}\},$$

to denote data sets, and so $\mathbf{Y}(0)$ denotes data in H_0. In this context, it is also worth noting that observed variable Y, sample space $\mathcal{Y}$, and effect δ are p-dimensional, with $p \geq 1$.

In this chapter we consider test statistics based on comparison of sampling indicators such as $T^*(\boldsymbol{\delta}) = S(\mathbf{Y}_1^*(\boldsymbol{\delta})) - S(\mathbf{Y}_1^*)$, where $S(\mathbf{Y}_j^*(\boldsymbol{\delta})) : \mathcal{Y}^{nj} \rightarrow \mathbb{R}^1$, is any symmetric function, that is, invariant with respect to rearrangements of entry arguments. These kinds of statistics include associative forms such as

$$T^*(\boldsymbol{\delta}) = 1/n_1 \,\Sigma_i\, \varphi\left[\mathbf{Y}_{1j}^*(\boldsymbol{\delta})\right] - 1/n_2 \,\Sigma_i\, \varphi\left[\mathbf{Y}_{2j}^*\right],$$

where φ is any nondegenerate measurable nondecreasing function of the data and so T^* corresponds to the comparison of sampling φ-means: $T^* = \bar{\varphi}_1^* - \bar{\varphi}_2^*$ say. Moreover, they also include nonassociative statistics such as $T^*(\boldsymbol{\delta}) = \varphi\left[\tilde{\mathbf{Y}}_1^*(\boldsymbol{\delta})\right] - \varphi\left[\tilde{\mathbf{Y}}_2^*\right]$ as, for instance, the comparison of sampling medians: $T^* = \tilde{\varphi}_1^* - \tilde{\varphi}_2^*$, and so forth.

Suppose also that effects diverge to the infinity according to whatever monotonic sequence $\{\delta_p, p \geq 1\}$, the elements of which are such that $\delta_p \leq \delta_{p'}$ for any pair $p < p'$. If those conditions are satisfied, then the permutation (conditional) rejection rate of T converges to 1 for all α-values not smaller than the minimum attainable α_a; thus, T is conditional and unconditional finite-sample consistent. Furthermore, suppose that effects δ are such that there exists a function $\rho(\delta) > 0$ of effects δ the limit of which is 0 as δ goes to the infinity, T is any test statistic as previously, and the data set is obtained by considering the transformation $\mathbf{X}(\delta) = \rho(\delta)\mathbf{Y}(\delta)$. If $\lim_{\delta\uparrow\infty} \boldsymbol{\delta}\rho(\boldsymbol{\delta}) = \tilde{\boldsymbol{\delta}} > 0$, then the unconditional rejection rate converges to 1 for all α-values not smaller than the minimum attainable α_a; and thus T is weak unconditional finite-sample consistent (Pesarin and Salmaso, 2010b). The extension of these results to random effects Δ, $0 \leq \Pr\{\Delta = \mathbf{0}\} < 1$, is also shown in Pesarin and Salmaso (2010b).

For instance, suppose a problem in which the p-dimensional data set is

$$\mathbf{Y}(\delta) = (\delta_k + \sigma_k Z_{k1i},\ i = 1,\ldots,n_1;\ \sigma_k Z_{k2i},\ i = 1,\ldots,n_2;\ k = 1,\ldots,p),$$

where δ_k and σ_k are the fixed effect and the scale coefficient of the kth component variable, respectively, the hypotheses are

$$H_0 : \left[\mathbf{Y}_1 \overset{d}{=} \mathbf{Y}_2\right] = [\boldsymbol{\delta} = \mathbf{0}] \text{ against } H_1 : \left[\mathbf{Y}_1 \overset{d}{>} \mathbf{Y}_2\right] = [\boldsymbol{\delta} > \mathbf{0}].$$

Suppose also that the test statistic is $T_k^*(\boldsymbol{\delta}) = 1/p \sum_{k \leq p}\left[\bar{Y}_{1k}^*(\delta_k) - \bar{Y}_{2k}^*(\delta_k)\right]/S_k$, where $\bar{Y}_{jk}^*(\delta_h) = \Sigma_i[Y_{ijk}^*(\delta_h)/n_j$ is the permutation mean of jth sample and S_k a permutation invariant statistic indicator for the kth scale coefficient σ_k, that is, a function $S[Y_{ijk}(\delta_k),\ i = 1,\ldots,n_j,\ j = 1, 2]$ of pooled data such as for instance $S_k = Md\left[\left|Y_{ijh} - \tilde{Y}_k\right|, i = 1,\ldots,n_j, j = 1,2\right]$ the median of absolute deviations from the median specific to the kth variable. It can be proved that a sufficient condition for finite-sample consistency of $T_k^*(\delta)$ is that all population means μ_k exist finite. Thus, when some of the multivariate components do not possess finite mean value, a test based on comparisons of sampling means is not finite-sample consistent. It is worth noting that $T_k^*(\delta)$ represents the direct combination (Pesarin and Salmaso, 2010a) of p partial tests $\left[\bar{Y}_{1k}^*(\delta_k) - \bar{Y}_{2k}^*\right]$.

To extend finite-sample consistency to nonassociative statistics, let us briefly introduce the notion of conditional (permutation) unbiasedness for any kind of statistics $T^*(\boldsymbol{\delta}) = S\left(\mathbf{Y}_1^*\right)(\boldsymbol{\delta})) - S\left(\mathbf{Y}_2^*\right)$. To this end and with clear meaning of the symbols, let us observe that

$T^o(0) = S(\mathbf{Z}_1) - S(\mathbf{Z}_2)$, that is, the null observed value of statistic T.

$T^o(\delta) = S(\mathbf{Z}_1 + \delta) - S(\mathbf{Z}_2) = S(\mathbf{Z}_1) + D_S(\mathbf{Z}_1, \delta) - S(\mathbf{Z}_2) = T^o(0) + D_S(\mathbf{Z}_1, \delta)$, where $D_S(\mathbf{Z}_1, \delta) \geq 0$.

$T^*(0) = S\left(\mathbf{Z}_1^*\right) - S\left(\mathbf{Z}_2^*\right)$, that is, the value of T in the permutation $\mathbf{u}^* = u_1^*, \ldots, u_n^*$.

$T^*(\boldsymbol{\delta}) = S\left(\mathbf{Z}_1^* + \boldsymbol{\delta}^*\right) - S\left(\mathbf{Z}_2^*\right) = T^*(0) + D_S\left(\mathbf{Z}_1^*, \boldsymbol{\delta}^*\right) - D_S\left(\mathbf{Z}_2^*\right)$.

$D_S\left(\mathbf{Z}_1^* \boldsymbol{\delta}^*\right) \geq D_S\left(\mathbf{Z}_2^*, 0\right) = 0 = D_S(\mathbf{Z}_2, 0)$, because effects $\boldsymbol{\delta}_{2i}^*$ coming from first group are nonnegative.

$D_S\left(\mathbf{Z}_1^*, \boldsymbol{\delta}^*\right) \leq D_S\left(\mathbf{Z}_1^*, \boldsymbol{\delta}\right)$ pointwise, because in $D_S\left(\mathbf{Z}_1^*, \boldsymbol{\delta}\right)$ there are nonnegative effects assigned to units coming from group 2; for example, suppose $n_1 = 3$, $n_2 = 3$, and $\mathbf{u}^* = (3, 5, 4, 1, 2, 6)$, then $\left(\mathbf{Z}_1^*, \boldsymbol{\delta}^*\right) = [(Z_{13}, \boldsymbol{\delta}_{13}), (Z_{22}, 0), (Z_{21}, 0)]$, and so

$$\left(\mathbf{Z}_1^*, \boldsymbol{\delta}\right) = [(Z_{13}, \boldsymbol{\delta}_{13}), (Z_{22}, \boldsymbol{\delta}_{11}), (Z_{21}, \boldsymbol{\delta}_{12})],$$

or

$$\left(\mathbf{Z}_1^*, \boldsymbol{\delta}_1\right) = [(Z_{13}, \boldsymbol{\delta}_{13}), (Z_{22}, \boldsymbol{\delta}_{12}), (Z_{21}, \boldsymbol{\delta}_{11})];$$

it is to be emphasized that $Y\left(u_i^*\right) = Z\left(u_i^*\right) + \boldsymbol{\delta}\left(u_i^*\right)$ if $u_i^* \leq n_1$, that is, units coming from the first group maintain their effects, whereas the rest of effects are randomly assigned to units coming from the second group.

$D_S\left(\mathbf{Z}_1^*, \boldsymbol{\delta}\right) \overset{d}{=} D_S(\mathbf{Z}_1, \boldsymbol{\delta})$, because $\Pr\left\{\mathbf{Z}_1^* \middle| X_{/\mathbf{Y}(0)}\right\} = \Pr\left\{\mathbf{Z}_1^* \middle| X_{/\mathbf{Y}(0)}\right\}$ (see Pesarin and Salmaso, 2010a).

Thus $D_S\left(\mathbf{Z}_1^*, \boldsymbol{\delta}^*\right) - D_S\left(\mathbf{Z}_2^*\right) \leq D_S(\mathbf{Z}_1, \boldsymbol{\delta})$ in permutation distribution and so

$$\begin{aligned}
\lambda_T(\mathbf{X}(\boldsymbol{\delta})) &= \Pr\left\{T(\mathbf{X}^*(\boldsymbol{\delta})) \geq T(\mathbf{X}(\boldsymbol{\delta})) \middle| X_{/\mathbf{Y}(\boldsymbol{\delta})}\right\} \\
&= \Pr\left\{T^*(0) + D_S\left(\mathbf{Z}_1^*, \boldsymbol{\delta}^*\right) - D_S\left(\mathbf{Z}_2^*\right) - D_S(\mathbf{Z}_1, \boldsymbol{\delta}) \geq T^o(0) \middle| X_{/\mathbf{Y}(0)}\right\} \\
&\leq \Pr\left\{T^*(0) \geq T^o(0) \middle| X_{/\mathbf{Y}(0)}\right\} = \lambda_T(\mathbf{Y}(0)),
\end{aligned}$$

which establishes the dominance in permutation distribution of $\lambda_T(\mathbf{Y}(\delta))$ with respect to $\lambda_T(\mathbf{Y}(0))$, uniformly for all data sets $\mathbf{Y}\in\mathcal{Y}^n$, for all underlying distributions P and for all associative and nonassociative statistics $T = S(\mathbf{Y}_1) - S(\mathbf{Y}_2)$.

These results allow us to prove the following:

Theorem 2.2

Suppose that in a two-sample problem there are $p \geq 1$ nonhomoscedastic variables $Y = (Y_1,\ldots,Y_p)$, the observed data set is $\mathbf{Y}(\delta) = (\delta_k + \sigma_k Z_{i1k}, i = 1,\ldots,n_1, \sigma_k Z_{i2k}, i = 1,\ldots,n_2, k = 1,\ldots,p)$ and the hypotheses are

$$H_0 : \left[\mathbf{Y}_1 \overset{d}{=} \mathbf{Y}_2\right] = [\boldsymbol{\delta} = \mathbf{0}] \text{ against } H_1 : \left[\mathbf{Y}_1 \overset{d}{>} \mathbf{Y}_2\right] = [\boldsymbol{\delta} > \mathbf{0}],$$

where $\delta = (\delta_1,\ldots,\delta_p)'$. For the testing purpose consider the statistic

$$T_k^*(\boldsymbol{\delta}) = 1/p\sum_{k\leq p}\left[\tilde{Y}_{1k}^*(\delta_k) - \tilde{Y}_{2k}^*\right]\Big/S_k,$$

where $\tilde{Y}_{ji}^*(\boldsymbol{\delta}) = \mathrm{Md}\left[Y_{ijk}^*(\boldsymbol{\delta})\right]\Big/S_{Mk}$, $k = 1,\ldots,p]$, $i = 1,\ldots,n_j$, $j = 1, 2$, is the median vector of p scale-free variables specific to ith subject, and $S_{Mk} = MAD_k = \mathrm{Md}\left[\left|Y_{ijk} - \tilde{Y}_k\right|, i = 1,\ldots,n_j, j = 1,2\right]$ is the median of absolute deviations from the median specific to the variable Y_k.

In this setting, the test based on $T_{\mathrm{Md}}^*(\boldsymbol{\delta})$ is conditional and unconditional finite-sample consistent as far as p diverges and $\mathrm{Md}(Y_1(\delta)) > 0$ without requiring the existence of any positive moment for p variables. ■

Proof

For the nonassociative statistics the uniformly stochastic ordering of the significance level functions with respect to δ and $\mathbf{Y}$ applies, that is, for $\delta' > \delta$

$$\Pr\{\lambda_T[\mathbf{Y}(\boldsymbol{\delta}')] \leq \alpha\} \overset{d}{\leq} \Pr\{\lambda_T[\mathbf{Y}(\boldsymbol{\delta})] \leq \alpha\};$$

hence, with reference to the finite-sample consistency of the second-order combined test using the medians

$$T''^{obs} = 1/p\sum_{k\leq p}\left[\tilde{Y}_{1k}(\delta_k) - \tilde{Y}_{2k}(0)\right]\Big/S_k = 1/p\sum_{k\leq p}\left[\tilde{Y}_{1k}(0) + \delta_k - \tilde{Y}_{2k}^*(0)\right]\Big/S_k$$
$$= 1/p\sum_{k\leq p}\left[\tilde{Y}_{1k}(0) - \tilde{Y}_{2k}(0)\right]\Big/S_k + 1/p\sum_{k\leq p}\delta_k/S_k.$$

It should be noted that the quantity $1/p\sum_{k\leq p}\left[\tilde{Y}_{1k}(0)-\tilde{Y}_{2k}(0)\right]/S_k$ is nothing else than the arithmetic mean of p sample differences which are all measurable, given that all p involved variables are nondegenerate by assumption (i.e., $S_k > 0$; $k = 1,\ldots,p$) and, provided that $\min(n_1, n_2)$ is not too small, are all finite (for instance, with the *Pareto* distribution if its parameter is $\gamma \geq [\min(n_1, n_2)/2]$, where $[\cdot]$ is the integer part of $(\cdot)$ and the first moment $E_Y(Y, \gamma)$ is finite; it is noticeable that $E_Y(Y, \gamma)$ does not exist $\gamma \leq 1$). Thus, by the law of large numbers for sequences of dependent variables, as p diverges it converges weakly to a constant, not necessarily null. If the induced standardized global noncentrality $1/p\sum_{k\leq p}\delta_k/S_k$, which is itself a mean of nonnegative and measurable quantities, converges, it does so to a positive quantity but it could be let free to diverge as well. ■

References

Basso D., Pesarin F., Salmaso L., Solari A. (2009). *Permutation Tests for Stochastic Ordering and ANOVA: Theory and Applications with R*. Lecture Notes in Statistics. New York: Springer Science+Business Media.

Behseta S., Kass R. E. (2005). Testing equality of two functions using BARS. *Statistics in Medicine*, 24, 3523–3534.

Behseta S., Kass R. E., Moorman D. E., Olson C. R. (2007). Testing equality of several functions: Analysis of single-unit firing-rate curves across multiple experimental conditions. *Statistics in Medicine*, 26, 3958–3975.

Blair R. C., Higgins J. J., Karniski W., Kromrey J. D. (1994). A study of multivariate permutation tests which may replace Hotelling's t^2 test in prescribed circumstances. *Multivariate Behavioral Research*, 29, 141–163.

Braun T., Feng Z. (2001). Optimal permutation tests for the analysis of group randomized trials. *Journal of the American Statistical Association*, 96, 1124–1132.

Cardot H., Prchal L., Sarda P. (2007). No effect and lack-of-fit permutation tests for functional regression. *Computational Statistics*, 22, 371–390.

Cook R. D., Yin X. (2001). Dimension-reduction and visualization in discriminant analysis. *Australian and New Zealand Journal of Statistics*, 43, 147–200.

Cox D. D., Lee J. S. (2008). Pointwise testing with functional data using the WestfallYoung randomization method. *Biometrika*, 95, 621–634.

Cuevas A., Febrero M., Fraiman R. (2004). An ANOVA test for functional data. *Computational Statistics and Data Analysis*, 47, 111–122.

Davidian M., Lin X., Wang J. (2004). Introduction: Emerging issues in longitudinal and functional data analysis. *Statistica Sinica*, 14, 613–614.

del Castillo E., Colosimo B. M. (2011). Statistical shape analysis of experiments for manufacturing processes. *Technometrics*, 53(1), 1–15.

Diggle P., Heagerty P., Liang K-Y., Zeger S. (2002). *Analysis of Longitudinal Data*, 2nd ed. Oxford: Oxford University Press.

Edgington E. S., Onghena P. (2007). *Randomization Tests*, 4th ed. London: Chapman and Hall.

Eubank R. L. (2000). Testing for no effect by cosine series methods. *Scandinavian Journal of Statistics*, 27, 747–763.

Fan J. (1996). Test of significance based on wavelet thresholding and Neyman's truncation. *Journal of the American Statistical Association*, 91(434), 674–688.

Fan J., Lin S.-K. (1998). Test of significance when data are curves. *Journal of the American Statistical Association*, 93(443), 1007–1021.

Faraway J. J. (1997). Regression analysis for a functional response. *Technometrics*, 39, 254–261.

Ferraty F., Vieu P. (2006). *Nonparametric Functional Data Analysis*. New York: Springer Science+Business Media.

Folks J. L. (1984). Combinations of independent tests. In P. R. Krishnaiah, P. K. Sen (eds.), *Handbook of Statistics*, Vol. 4, pp. 113–121. Amsterdam: North-Holland.

Goggin M. L. (1986). The "too few cases/too many variables" problem in implementation research. *The Western Political Quarterly*, 39, 328–347.

González-Manteiga W., Vieu P. (2007). Statistics for functional data. *Computational Statistics & Data Analysis*, 51, 4788–4792.

Good P. (2010). *Permutation, Parametric, and Bootstrap Tests of Hypotheses*, 3rd ed. Springer Series in Statistics. New York: Springer Science+Business Media.

Hall P., Van Keilegom I. (2007). Two-sample tests in functional data analysis starting from discrete data. *Statistica Sinica*, 17, 1511–1531.

Hoeffding W. (1952). The large-sample power of tests based on permutations of observations. *Annals of Mathematical Statistics*, 23, 169–192.

Hothorn T., Hornik K., van de Wiel M. A., Zeileis A. (2006). A Lego system for conditional inference. *The American Statistician*, 60(3), 257–263.

Iaci R., Yin X., Sriram T. N., Klingenberg C. P. (2008). An informational measure of association and dimension reduction for multiple sets and groups with applications in morphometric analysis. *Journal of the American Statistical Association*, 103(483), 1166–1176.

Ludbrook J., Dudley H. (1998). Why permutation tests are superior to t and F tests in biomedical research. *The American Statistician*, 52(2), 127–132.

Martinez-Camblor P., Corral N. (2011). Repeated measures analysis for functional data. *Computational Statistics and Data Analysis*, 55, 3244–3256.

Munoz-Maldonado Y., Staniswalis J. G., Irwin L. N., Byers D. (2002). A similarity analysis of curves. *Canadian Journal of Statistics*, 30, 373–381.

Neumeyer N., Dette H. (2003). Nonparametric comparison of regression curves: An empirical process approach. *Annals of Statistics*, 31(3), 880–920.

Pesarin F. (2001). *Multivariate Permutation Tests with Application in Biostatistics*. Wiley Series in Probability and Statistics. Chichester, UK: John Wiley & Sons.

Pesarin F., Salmaso L. (2010a). *Permutation Tests for Complex Data: Theory, Applications and Software*. Wiley Series in Probability and Statistics. Chichester, UK: John Wiley & Sons.

Pesarin F., Salmaso L. (2010b). Finite-sample consistency of combination-based permutation tests with application to repeated measures designs. *Journal of Nonparametric Statistics*, 22(5), 669–684.

Ramsay J. O., Silverman, B. W. (2005). *Functional Data Analysis*, 2nd ed. New York: Springer Science+Business Media.

Ramsay J. O., Hooker G., Graves S. (2009). *Functional Data Analysis with R and Matlab*. New York: Springer Science+Business Media.

Rice J. A. (2004). Functional and longitudinal data analysis: Perspectives on smoothing. *Statistica Sinica*, 14, 631–647.

Roy S. N. (1953). On a heuristic method of test construction and its use in multivariate analysis. *Annals of Mathematical Statistics*, 24(2), 220–238.

Shen Q., Faraway J. (2004). An *F* test for linear models with functional responses. *Statistica Sinica*, 14, 1239–1257.

Shen Q., Xu H. (2007). Diagnostics for linear models with functional responses. *Technometrics*, 49, 26–33.

Sirski M. (2012). *On the Statistical Analysis of Functional Data Arising from Designed Experiments*. PhD thesis, Department of Statistics, University of Manitoba.

Sturino J., Zorych I., Mallick B., Pokusaeva K., Chang Y.-Y., Carroll R., Bliznuyk N. (2010). Statistical methods for comparative phenomics using high-throughput phenotype microarrays. *The International Journal of Biostatistics*, 6(1), Article 29.

Troendle J. F., Korn E. L., McShane L. M. (2004). An example of slow convergence of the bootstrap in high dimensions. *The American Statistician*, 58(1), 25–29.

Ute H. (2012). A studentized permutation test for the comparison of spatial point patterns. *Journal of the American Statistical Association*, 107(498), 754–764.

Valderrama M. J. (2007). An overview to modelling functional data. *Computational Statistics*, 22(3), 331–334.

van Wieringen W. N., van de Wiel M. A., van der Vaart A. W. (2008). A test for partial differential expression. *Journal of the American Statistical Association*, 103(483), 1039–1049.

Wang H., Akritas M. G. (2010). Inference from heteroscedastic functional data. *Journal of Nonparametric Statistics*, 22(2), 149–168.

Wang H., Akritas M. G. (2011). Asymptotically distribution free tests in heteroscedastic unbalanced high dimensional ANOVA. *Statistica Sinica*, 21, 1341–1377.

Weinberg J. M., Lagakos S. W. (2000). Asymptotic behavior of linear permutation tests under general alternatives, with application to test selection and study design. *Journal of the American Statistical Association*, 95(450), 596–607.

Wu C. O., Yu K. F. (2002). Nonparametric varying-coefficient models for the analysis of longitudinal data. *International Statististica Review*, 70, 373–393.

Xu H., Shen Q., Yang X., Shoptawe S. (2011). A quasi *F*-test for functional linear models with functional covariates and its application to longitudinal data. *Statistics in Medicine*, 30, 2842–2853.

Yang X., Shen Q., Xu H., Shoptaw S. (2007). Functional regression analysis using an *F* test for longitudinal data with large numbers of repeated measures. *Statistics in Medicine*, 26, 1552–1566.

Zeng C., Pan Z., MaWhinney S., Barón A. E., Zerbe G. O. (2011). Permutation and *F* distribution of tests in the multivariate general linear model. *The American Statistician*, 65(1), 31–36.

Zhang J.-T. (2011). Statistical inferences for linear models with functional responses. *Statistica Sinica*, 21, 1431–1451.

Zhang D., Lin X. (2003). Hypothesis testing in semiparametric additive mixed models, *Biostatistics*, 4(1), 57–74.

Zhang J.-T., Chen J. (2007). Statistical inferences for functional data. *The Annals of Statistics*, 35(3), 1052–1079.

Zhang D., Lin X., Sowers M. F. (2000). Semiparametric regression for periodic longitudinal hormone data from multiple menstrual cycles. *Biometrics*, 56, 31–39.

Zhao X., Marron J. S., Wells M. T. (2004). The functional data analysis view of longitudinal data. *Statistica Sinica*, 14, 789–808.

3

A Permutation Approach for Ranking of Multivariate Populations

In Chapter 1, Section 1.2 we formalized our approach to solve the problem we called *the multivariate ranking problem* – that of ranking several multivariate populations from the 'best' to the 'worst' according to a given prespecified criterion when a sample from each population is available and for each marginal univariate response there is a natural preferable direction. As the key element of our solution is a testing procedure suitable for multivariate one-sided alternatives, the Nonparametric Combination (NPC) methodology represents our main methodological reference framework. In fact, to the best of our knowledge, the nonparametric NPC permutation tests are the only method proposed in the literature that is suitable to achieve this goal. Moreover, when deriving the multivariate one-sided p-values we can also benefit from the flexibility of the method for obtaining a series of advantages: NPC methodology allows handling with all types of response variables – numeric, binary and ordered categorical – even in the presence of missing data (at random or not at random, i.e. noninformative or informative) and this can be done also when the number of response variables is much larger than that of units without the need of having to worry about the curse of dimensionality or the problem of the reduction of degrees of freedom. On the contrary, thanks to the so-called finite-sample consistency of combined permutation tests, the power function does not decrease for any added variable that makes larger standardized noncentrality (Pesarin and Salmaso, 2010b; see Chapter 2, Section 2.4.2). It is worth noting that in this situation, which can be common in many real applications, all traditional parametric and nonparametric testing procedures are not at all appropriate (also for the case in which all multivariate alternatives were of two-sided type). Finally, the NPC approach has many nice features: it is of very low demand in terms of assumptions and always provides an exact solution for whatever finite sample size whenever the permutation principle applies, that is, when the null hypothesis implies data exchangeability.

We recall that our goal is to classify and rank C multivariate populations with respect to several marginal variables in which samples from each population are available. Note that the multivariate ranking problem is essentially related to a post hoc comparative multivariate C-sample problem in which the populations of interest are treatments, groups or items to be investigated by an experimental or observation study. As we will see later, although our main reference design obviously will be the one-way MANOVA layout, thanks to the flexibility of the NPC methodology more complex design and analysis are also allowed.

This chapter is devoted mainly to describing the permutation approach for solving the multivariate ranking problem, from the initial set-up phases to the computation of the final global ranking. In the last part of this chapter, several simulation studies are presented to validate the proposed approach numerically.

3.1 Set-Up of the Multivariate Ranking Problem

In this section we illustrate in detail the steps that need to be followed to set up and solve a multivariate ranking problem. First we present the different type of designs that may be considered in this context. Then we consider situations in which the ranking process needs to take into account either confounding factors and/or of intermediate levels of aggregation of the response variables; for these, the so-called stratified and domain analyses are very useful procedures. Finally, we address some more practical questions: the choices of the test statistics and of the combining function and the specific pairwise permutation strategy to be used. We close with some issues on multiplicity control and simultaneous testing.

3.1.1 Types of Designs

Up to now the main reference design for the multivariate ranking problem has been the so-called one-way MANOVA layout whose statistical model in case of fixed effects can be represented as

$$\mathbf{Y}_{ij} = \boldsymbol{\mu}_j + \boldsymbol{\varepsilon}_{ij} = \boldsymbol{\mu} + \boldsymbol{\tau}_j + \boldsymbol{\varepsilon}_{ij},\ i = 1,\ldots,n_j,\ j = 1,2,\ldots,C, \tag{3.1}$$

where $\boldsymbol{\mu}_j$ (or $\boldsymbol{\tau}_j$) is the p-dimensional mean effect, $\boldsymbol{\varepsilon}_{ij}$~IID(0, Σ) is a p-variate random term of experimental errors with zero mean and variance/covariance matrix Σ. Extensions to more complex designs and situations are also possible. In fact, either the Multivariate Randomized Complete Block (MRCB) design, the Repeated Measures (RM) design and finally the multivariate C-sample comparison of curves or trajectories (functional data), can be taken into consideration.

- MRCB design: $\mathbf{Y}_{ij} = \boldsymbol{\mu} + \boldsymbol{\tau}_j + \boldsymbol{\beta}_i + \boldsymbol{\varepsilon}_{ij}$, where $\boldsymbol{\beta}_i$, $i = 1,\ldots,n_j$, is the block effect (see Aboretti et al., 2012a)
- RM design/functional data (with random effects): $\mathbf{Y}_{ji}(t) = \boldsymbol{\mu} + \boldsymbol{\eta}_j(t) + \boldsymbol{\varepsilon}_{ji}(t)$, where $\boldsymbol{\varepsilon}_{ji}(t) = \boldsymbol{\Delta}_{ji}(t) + \boldsymbol{\sigma}(\boldsymbol{\eta}_j(t))\cdot\mathbf{Z}_{ji}(t)$, $i = 1,\ldots,n_j$, $j = 1,\ldots,C$, $t = 1,\ldots,N$, where $\mathbf{Z}_{ji}(t)$ are non-Gaussian error terms distributed as a stationary

stochastic process with null mean and unknown distribution P_Z; $\boldsymbol{\mu}$ is a population constant; the coefficients $\boldsymbol{\eta}_j(t)$ represent the *main treatment effects* and may depend on time through any kind of function, but are independent of units; the quantities $\boldsymbol{\Delta}_{ji}(t)$ represent the so-called *individual effects*; and $\boldsymbol{\sigma}(\boldsymbol{\eta}_j(t))$ are time-varying scale coefficients that may depend, through monotonic functions, on main treatment effects $\boldsymbol{\eta}_j$ (for details see Chapter 2, Section 2.3).

RM design is similar but a bit different from MRCB design. An RM design refers to studies in which the same measures are collected multiple times for each unit/subject but under different conditions or time occasions. For instance, RMs are collected in a longitudinal study in which change over time is assessed.

From the point of view of obtaining a valid testing solution and within the NPC framework, all situations listed in the preceding text are no more than designs with nuisance parameters ($\boldsymbol{\tau}$, $\boldsymbol{\eta}$ and $\boldsymbol{\Delta}$) that can be fully removed exploiting the concept of constrained permutations (Basso et al., 2009; Pesarin and Salmaso, 2010a): because under the null hypothesis the exchangeability of observations holds only within given conditions, the null permutation distributions of the test statistics are computed allowing permutations to occur only under a given restriction. In practice, once a suitable blocking factor is defined and permutations are allowed only among samples within the same level of that blocking factor when calculating the test statistics, all nuisance parameter are implicitly removed (they vanish by applying suitable linear transformations; see Basso et al., 2009). Note that in the case of even more complex designs such as the Latin and Graeco-Latin squares (Montgomery, 2012), the concept just exposed remains valid as long as two or more blocking factors are used so as to define a single blocking factor whose levels are obtained from the combinations of the levels of the individual blocking factors.

3.1.2 Stratified Analysis

Sometimes in the multivariate ranking problem we need to take the presence of possible confounding factors into consideration; that is, there are units/subjects features that potentially have a noise effect on the problem at hand. Typical confounding factors are sex or age of the subjects. In this situation a stratified analysis can be very useful to provide, before getting a final global ranking, separate results for each level of the stratification factor. In practice, to deal with a stratification factor we use an additional classification criterion of units/subjects and then allow permutations among samples only within the same stratification factor level.

Note that even if blocking and stratification are both handled by a restriction on the exchangeability under the null hypothesis, they only apparently appear as the same situation. In fact, blocking refers to a technique aimed at removing nuisance parameters, whereas stratification provides intermediate separate results for each stratum and a final global analysis in which the possible confounding effect has been removed thanks to a constrained permutation strategy.

3.1.3 Domain Analysis

We refer to a domain as a result of a classification or grouping of marginal response variables that share some basic features with respect to the problem at hand. For example, in shape analysis, domains are subgroups of landmarks sharing anatomical, biological or locational features (Brombin and Salmaso, 2009). Very often in a multivariate problem we are facing the presence of such kinds of domains, for example, sections of a questionnaire on the consumer relevance of a product (Example 5.4) or on the type of response of a new product development experiment in industry (Example 5.1). In a similar way as in the stratified analysis, the presence of domains in the multivariate ranking problem suggests providing intermediate results for each domain before obtaining a final global analysis.

3.1.4 Choice of the Test Statistics

It is worth noting that within the NPC framework an optimal statistic cannot exist because it is a function of the population distributions, which by definition is unknown (Pesarin and Salmaso, 2010a). For this reason it is important to consider for each type of response variable a number of different test statistics. We recall that each univariate partial test statistic we are presenting must be suitable for one-sided alternatives with respect to the hypotheses $H_{0k(jh)}$ versus $H_{1k(jh)}$:

$$\begin{cases} H_{0k(jh)}: Y_{jk} \overset{d}{=} Y_{hk} \\ H_{1k(jh)}: \left(Y_{jk} \overset{d}{<} Y_{hk}\right) \bigcup \left(Y_{jk} \overset{d}{>} Y_{hk}\right) \end{cases}, j,h = 1,\ldots,C, j \neq h, k = 1,\ldots,p.$$

As far as the multivariate test statistics suitable for testing the hypotheses $H_{0(jh)}$ versus $H_{1(jh)}$ are concerned,

$$\begin{cases} H_{0(jh)}: \mathbf{Y}_j \overset{d}{=} \mathbf{Y}_h \\ H_{1(jh)}: \left(\mathbf{Y}_j \overset{d}{<} \mathbf{Y}_h\right) \bigcup \left(\mathbf{Y}_j \overset{d}{>} \mathbf{Y}_h\right), j,h = 1,\ldots,C, j \neq h. \end{cases}$$

We will apply the nonparametric combination methodology using a suitable combining function (see Chapter 2, Section 2.2).

3.1.4.1 Continuous or Binary Response Variables

When the univariate marginal response variable is continuous or binary, within the permutation framework we can use a number of test statistics suitable for one-sided alternatives. In this context, we underline that the test statistics obviously should not be permutationally equivalent (see Chapter 2, Section 2.4.1); in particular we can refer to

- Difference of sample means: ${}_{DM}T_{k(jh)} = \Sigma_i Y_{ijk}/n_j - \Sigma_i Y_{ihk}/n_h$;
- Difference of sample means in case of missing values: ${}_{DM-MV}T_{k(jh)} = n_j\Sigma_i Y_{ijk}\cdot(v_h/v_j)^{1/2} - n_h\Sigma_i Y_{ihk}\cdot(v_j/v_h)^{1/2}$, where v_h and v_j represent the actual sample sizes of valid data in the two samples (for details we refer to Pesarin and Salmaso, 2010a);
- Difference of sample medians or quartiles: ${}_{DMd}T_{k(jh)} = Md(Y_{jk}) - Md(Y_{hk})$, ${}_{DQ1}T_{k(jh)} = Q_1(Y_{jk}) - Q_1(Y_{hk})$, ${}_{DQ3}T_{k(jh)} = Q_3(Y_{jk}) - Q_3(Y_{hk})$, where $j, h = 1,...,C$, $j \neq h$, $k = 1,...,p$, and $Md = \tilde{\mu}$, $Q_1 = q_1$ and $Q_3 = q_3$ are the median, the first and the third quartile operators such that $\Pr\{Y_{jk} < \tilde{\mu}\} = \Pr\{Y_{jk} > \tilde{\mu}\}$, $3\cdot\Pr\{Y_{jk} < q_1\} = \Pr\{Y_{jk} > q_1\}$ and $\Pr\{Y_{jk} < q_3\} = 3\cdot\Pr\{Y_{jk} > q_3\}$.

  When providing the preceding list, we implicitly assume that the ranking problem was a location-type problem in which a natural preference direction for the response variable does exist. Actually, if the ranking criteria were based on the classification of several populations with respect to the variability (scale ranking problem), we may consider a number of scale-type test statistics:

- Difference of sample squares: ${}_{DS}T_{k(jh)} = \Sigma_i Y_{ijkj}^2/n_j - \Sigma_i Y_{ihk}^2/n_h$;
- Difference of sample interquartile ranges: ${}_{DIQR}T_{k(jh)} = Q_3 - Q_1$.

3.1.4.2 Ordered Categorical Response Variables

When the univariate marginal response variable is ordered categorical with S ordinal categories, within the permutation framework we can use a number of test statistics suitable for directional alternatives. The following are some examples of suitable test statistics.

- Anderson–Darling: ${}_{AD}T_{k(jh)} = \Sigma_{s=1}^{S-1} N_{hsk} \cdot [N_{\cdot sk} \cdot (n_{ik} + n_{jk} - N_{\cdot sk})]^{-\frac{1}{2}}$; where $N_{\cdot sk} = N_{jsk} + N_{hsk}$ are the cumulative frequencies;
- Multifocus: ${}_{MF}T_{ks(jh)} = \left(f_{jks} - \hat{f}_{jks}\right)^2$, $s = 1,...,S$, where f_{jks} and $\hat{f}_{jks}$ are respectively the observed and the estimated frequencies of the sth two-by-two sub-table; note that there are a number of S multifocus statistics for each univariate response variable so that an additional

combination phase is needed to obtain the kth test statistic (for details see Pesarin and Salmaso, 2010a).

- Kolmogorov–Smirnov: ${}_{KS}T_{k(jh)} = \max(F_{jsk} - F_{hsk})$, $s = 1,\ldots,S$, where F_{jsk} and F_{hsk} are the empirical distribution functions (EDFs).

As in case of ordered categorical response, when providing the preceding list we implicitly assumed that the ranking problem was similar to a shift-in-distribution problem where a natural preference direction for the ordered categorical response variable does exist. Actually, if the ranking criterion were based on the classification of several populations with respect to the variability (heterogeneity ranking problem), we may consider a number of ordered categorical scale-type test statistics:

- Difference of sample Gini indexes (Gini, 1912):

$$_{DG}T_{k(jh)} = G_{jk} - G_{hk} = \left[\sum_{s=1}^{S-1} f_{jsk} \cdot (1 - f_{jsk})\right] - \left[\sum_{s=1}^{S-1} f_{hsk} \cdot (1 - f_{hsk})\right];$$

- Difference of sample Shannon indexes (Shannon, 1948):

$$_{DS}T_{k(jh)} = H_{jk} - H_{hk} = \left[-\sum_{s=1}^{S-1} f_{jsk} \cdot \log(f_{jsk})\right] - \left[-\sum_{s=1}^{S-1} f_{hsk} \cdot \log(f_{hsk})\right],$$

where f_{jsk} and f_{hsk} are the observed frequencies from the jth and hth population, $s = 1,\ldots,S$, $k = 1,\ldots,p$.

3.1.4.3 Multiaspect

It is worth noting that within the NPC framework an optimal statistic cannot exist because it is a function of the population distributions which are unknown by definition (Pesarin and Salmaso, 2010a). More formally, when the whole data set $\mathbf{Y}$ is minimally sufficient in H_0, univariate statistics suitable for summarizing the entire information on an aspect of interest do not exist. To overcome this limitation and to reduce the loss of information associated with using only one single overall statistic, it is possible to take account of a set of statistics suitable for complementary or concurrent viewpoints, each fitted for summarizing information on a specific aspect of interest for the problem, and so to find solutions within the so-called *multiaspect strategy* (Salmaso and Solari, 2005), that is, combining different test statistics, suitable for testing different aspects related to the same univariate null hypothesis by working on the same dataset. The multiaspect strategy was originally proposed by R. A. Fisher: 'In hypotheses testing problems the experimenter might have to face not only one, but a class of hypotheses, and it may happen that he is interested in all the hypotheses of such class... It follows that different significance tests may be thought as a set of tests for testing different aspects of the same hypothesis' (Pesarin and Salmaso, 2010a, pp. 159–163). Hence, since different test statistics may be suitable and effective for testing different aspects of the same null hypothesis (i.e., univariate location problem), instead of using just one statistic per variable we may use a list of statistics and then combine all of them to get a final multivariate and multiaspect p-values. In this way we obtain two advantages: first, by using several test statistics, this allows us to possibly include the more sensitive procedures with respect to the unknown alterative hypothesis and population distributions and then, thanks to the NPC methodology, we have the chance to gain more additional power (Pesarin and Salmaso, 2010a) by using a suitable multiaspect procedure.

3.1.5 Choice of the Combining Function

To define one-sided multivariate test statistics within the combination of dependent permutation testing methodology, a suitable combining function must be chosen (Pesarin and Salmaso, 2010a). Frequently used combining functions are

- Fisher combination: $\phi_F = -2\sum_k \log(\lambda_k)$;
- Tippet combination: $\phi_T = \max_{1 \le k \le p}(1-\lambda_k)$;
- Direct combination: $\phi_D = \sum_k T_k$;
- Liptak combination: $\phi_L = \sum_k \Phi^{-1}(1-\lambda_k)$;

where $k = 1,\ldots,p$ and Φ is the standard normal cumulative distribution function (CDF).

It can be seen that under the global null hypothesis the Conditional Monte Carlo Procedure (CMCP) allows for a consistent estimation of the permutation distributions, marginal, multivariate and combined, of the k partial tests. Usually, Fisher's combination function is considered (Pesarin and Salmaso, 2010a), mainly for its finite and asymptotic good properties. Of course, it would also be possible to take into consideration any other combining function (Lancaster, Mahalanobis, etc.; see Folks, 1984). The combined test is also unbiased and consistent. For a detailed description we refer to Pesarin and Salmaso (2010a).

It is worth noting that because within the NPC framework an optimal combination function in general does not exist because each combining function has different sensitivity to different configurations of the alternative

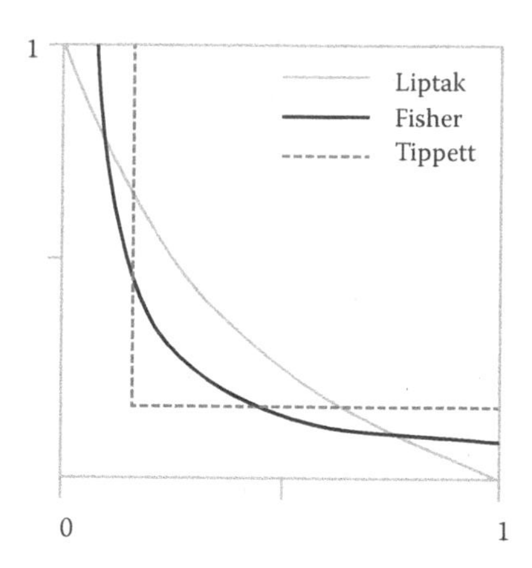

FIGURE 3.1
Critical regions of three combining functions in the case of two independent tests.

hypothesis, that is, every combining function has its own characteristics that make it preferable instead of another in a specific situation. To better understand this concept, let us consider the critical regions of the three combining functions in the case of two independent tests (Figure 3.1).

Figure 3.1 suggests that inferential results may slightly differ using different combining functions. To reduce the effect of this limitation from the computational point of view, we can try to iteratively apply different combination functions to reach a more stable result rather than applying one specific combining function. This is the so-called *iterated combination strategy*, formally described by the following algorithm.

1. Apply at least three different combining functions (e.g. Fisher, Liptak and Tippett) to the same partial p-values; in general, the obtained p-values will be slightly different.
2. Apply to the results of step 1 the same combining functions; in general, the obtained p-values will be different but slightly closer one another.
3. Iteratively repeat step 2 until all combining functions provide slightly the same resulting p-value.

Figure 3.2 reports an example of the behaviour of the iterated combination, showing that in this specific case after six iterations the resulting p-value is practically independent of the choice of the specific combining function.

3.1.6 Multiplicity Issue and Simultaneous Testing

The multiplicity issue occurs in case of simultaneous testing when a set, or family, of statistical tests is considered simultaneously. Because the

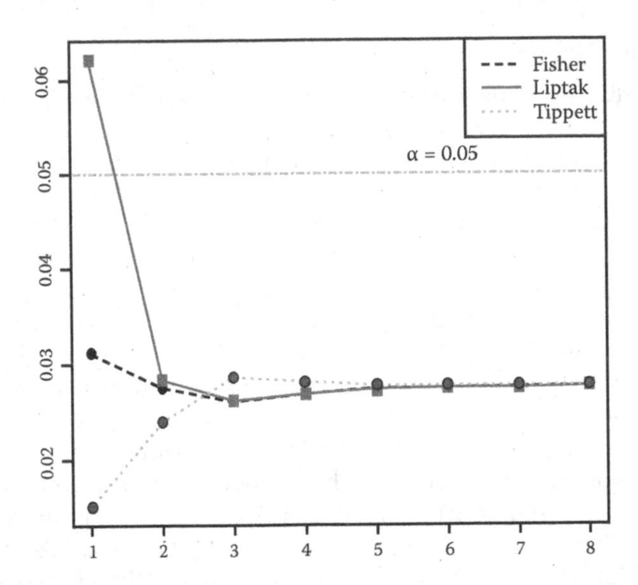

FIGURE 3.2
Behaviour of the iterated combination.

multivariate ranking problem makes use of inferential tools, obviously if falls within this context either because we are facing a multivariate response and we are considering a set of Multiple Comparison Procedures (MCPs). However, within the multivariate ranking problem the need to properly control the global type I error actually occurs only at the final stage, that is when a set of $C \times (C - 1)$ one-sided pairwise comparisons are performed via p-values.

In general, incorrect rejection of the null hypothesis is more likely when the family as a whole is considered, and failure to compensate for multiple comparisons can lead to committing serious mistakes in the process of classification and ranking. Among the many definitions of global type I error for MCPs, we take into consideration the most used, that is, Familywise Error Rate (FWER), which is the probability of rejecting at least one true null sub-hypothesis (for a review of alternative definitions and their main properties see Westfall et al., 2011).

Several statistical techniques have been developed to prevent this problem, mainly adjusting the significance levels of the considered tests. The so-called single-step procedures work singularly on each subtest and do not take the dependence structure of the tests into account. Among the most famous and most used methods are the Bonferroni and Tukey solutions. The stepwise procedures at first test only some sub-hypotheses and according to the results, they then take other sub-hypotheses into consideration, until a given condition is satisfied. Among the most used stepwise procedures are Bonferroni–Holm and the Shaffer methods (1986).

3.1.6.1 Shaffer Method

Often, the MCP hypotheses being tested are logically interrelated so that not all combinations of true and false hypotheses are possible (Shaffer, 1986). It is easily seen from the relations among the hypotheses that if any one of them is false, at least one other must be false. Thus there cannot be one false and two true hypotheses among these three. If we are testing all hypotheses of pairwise equality with more than three distributions, there are many such constraints. When there are logical implications among the hypotheses and alternatives Holm's procedure (1979) can be improved to obtain a further increase in power. Shaffer (1986) considers a class of stepwise MCP methods for achieving such improvement, based on the following idea: given that $j - 1$ hypotheses have been rejected, the denominator of α', used to adjust for the multiplicity issue, instead of being set at $n - j + 1$ for the next test as in the Holm's procedure, can be set at t_j, where t_j equals the maximum number of hypotheses that could be true, given that at least $j - 1$ hypotheses are false. Obviously, t_j is never greater than $n - j + 1$, and for some values of j it may be strictly smaller. Then the Shaffer's modified sequentially rejective Bonferroni procedure will never be less powerful (and typically will be more powerful) than the Holm's procedure while (as is proved in Shaffer, 1986) maintaining an FWER significance level $\leq\alpha$. The Shaffer's method is readily applicable to a wide variety of standard and nonstandard problems, including the problem of the control of the multiplicity for the multivariate ranking problem.

Arboretti Giancristofaro et al. (2010) proved via simulation studies that Shaffer's method can be considered slightly better than the Bonferroni–Holm method.

3.1.6.2 Closed Testing Procedure Using the Permutation minP Method (Tippett's Test)

Given a set of partial hypotheses, such as the pairwise comparisons of an MCP, the Closed Testing approach (Marcus et al., 1976) concentrates on testing the family of all the nonempty intersections of these partial hypotheses. In recent years the specialized literature has seemed to agree that the *Closed Testing* procedures are preferable for multiplicity control. Indeed these procedures prove to be easily adaptable to a wide range of experimental situations, at the same time enjoying properties consistent with the logical formulation of the analysis (control of the FWER, consistency, consonance). In particular they are easily adaptable to permutation tests.

Suppose we wish to test a set of hypotheses of primary interest, say H_1, H_2, H_3. These hypotheses might be comparisons of three treatment groups with a common control group on a single response variable (example of a multiple comparison problem), or comparisons of two groups on three distinct response variables (example of a multiple testing problem). Closed testing methods work by first testing each minimal hypothesis H_1, H_2, H_3

using an appropriate α-level partial test. We then have to compose hypotheses as necessary to form the 'closure' of the family of inferences, including all intersection hypotheses, in the specific example, hypotheses H_{12}, H_{13}, H_{23}, H_{123}. In this way the closure set consists of a hierarchy of hypotheses. After that, we test each member of the closed family using a suitable α-level combined test and choosing an appropriate test for each composite hypotheses H_{12}, H_{13}, H_{23}, H_{123}. After testing the composite hypotheses, we may reject any hypothesis H_i, with control of the FWER, provided that its corresponding test is statistically significant at level α and that every hypothesis in the family includes it in an intersection, thus implying it is rejected at level α. We can express results of a closed testing procedure in terms of adjusted p-values: for a given hypothesis H_i, the adjusted p-value is the maximum of all p-values for tests that include H_i as a special case, including the p-value for the H_i itself.

Note that different testing methods for minimal and composite hypotheses result in different closed testing procedures, so the use of powerful tests can lead to the best closed testing methods.

We shall now briefly review some tests for composite hypotheses based on the minimum p-value (minp) of the individual component tests corresponding to the minimal hypotheses included in the composite hypothesis. The Bonferroni minP method tests a composite hypothesis by comparing the minp of the individual component tests to α/g, where g is the number of components in the composite hypothesis and α is the desired FWE level. The composite hypothesis will be rejected when minp $\leq \alpha/g$, or equivalently when $g \times$ minp $\leq \alpha$, so that $g \times$ minp is the p-value for the composite test. For example, the p-value of the composite hypothesis H_{12} is $p = 2 \times \min(\lambda_1, \lambda_2)$, where λ_1, λ_2 are the p-values of the component hypotheses H_1, H_2. After testing all the composite hypotheses, we may reject any hypothesis, say H_3, with control of the FWE, if its corresponding test and all null hypotheses that include it, H_{13}, H_{23}, H_{123}, are rejected at level α. We may also obtain the adjusted p-value of H_3 as the maximum of all p-values for tests on hypotheses H_3, H_{13}, H_{23}, H_{123}. Holm demonstrated a short procedure to apply the closed testing procedure and to derive adjusted p-values based on Bonferroni's minP test that avoid calculating p-values for all the composite hypotheses. Closed testing using Bonferroni's minP test is known as the Bonferroni–Holm minP method (see Westfall et al., 2011).

The Bonferroni minP test tends to be conservative, particularly when the correlation structure among variables is high. Westfall and Young (1993) suggest comparing the observed minp for a given composite hypothesis to the actual α-quantile of the MinP null distribution, instead of α/g, where MinP represents the random variable of minp for the given composite hypothesis. Formally, this is equivalent to calculating the p-value of the composite test as $p = \Pr[\mathrm{MinP} \leq \min_{1\leq i\leq g}(\lambda_i)]$, where $\min_{1\leq i\leq g}(\lambda_i)$ is the observed value of minp for the given composite hypothesis. We then reject the composite hypothesis comparing the obtained p-value to the FWE level α. The distribution of MinP, usually unknown, can be estimated via bootstrap resampling (Westfall and Young, 1993). Alternatively, we can estimate the distribution of MinP via resampling without replacement, by a CMCP. In this case, the comparison of the observed minp for a given composite hypothesis to the estimated α-quantile of the MinP distribution, is equivalent to calculating the p-value of a composite hypothesis as $p = \Pr[\min_{1\le i\le g}(\lambda_i)^* \le \min_{1\le i\le g}(\lambda_i)]$, where $\min_{1\le i\le g}(\lambda_i)^*$ refers to the permutation distribution of MinP, and $\min_{1\le i\le g}(\lambda_i)$ is the observed minimum p-value from the given composite hypothesis.

Note that

$$p = \Pr[\min_{1\le i\le g}(\lambda_i)^* \le \min_{1\le i\le g}(\lambda_i)] = \Pr[1 - \min_{1\le i\le g}(\lambda_i)^* \ge 1 - \min_{1\le i\le g}(\lambda_i)] = \Pr[\max_{1\le i\le g}(1 - \lambda_i)^* \ge \max_{1\le i\le g}(1 - \lambda_i)].$$

So testing the composite hypotheses using the permutation minP test is equivalent to testing the composite hypotheses using the nonparametric combination of permutation partial component tests with Tippett's combining function T_T''.

As in the case of the Bonferroni–Holm minP method, not all composite hypotheses need to be tested using the permutation minP test. For example, if the minp observed for the minimal hypotheses on three variables (H_1,H_2,H_3) is the p-value of H_3 and its value is 0.00015, while the p-value of the global hypothesis H_{123} is 0.00035, the hypotheses H_{13}, H_{23} need not be tested because it is guaranteed that their p-values will be smaller than that of H_{123}. For example:

$$\Pr[\min(\lambda_1, \lambda_3)^* \le 0.00015] < \Pr[\min(\lambda_1, \lambda_2, \lambda_3)^* \le 0.00015] = 0.00035.$$

After testing the composite hypotheses, we may reject any hypothesis, say H_3, with control of the FWER, if its corresponding test and all hypotheses that include it, that is, H_{13}, H_{23}, H_{123}, are statistically significant at level α, or we may refer to the adjusted p-value of H_3 obtained as the maximum of all p-values for tests on hypotheses H_3, H_{13}, H_{23}, H_{123}.

The adjusted p-values for the permutation minP test (Tippett's test) are smaller than those of the Bonferroni–Holm method because they incorporate the underlying correlation structure among variables. Furthermore, simulation studies (Finos et al., 2003) have shown that permutation Tippett's test is particularly suitable for Closed Testing procedures it provides:

- Greater robustness of the permutation partial (related to minimal hypotheses) and combined (related to composite hypotheses) tests compared to parametric tests, especially with nonnormal distributions.
- The ability to obtain combined tests that take the dependence structure among variables into consideration without it being made formally explicit.

- The possibility to evaluate systems made up of directional and non-directional hypotheses and/or characterized by a large cardinality of the component hypotheses.

Results of a simulation study in Arboretti Giancristofaro et al. (2012a) highlight the advantages of closed testing methods and prove that they are more powerful than other classic MCPs controlling the FWER.

3.2 Some Monte Carlo Simulation Studies to Validate the Ranking Methods for Multivariate Populations

To numerically validate the permutation approach for ranking of multivariate populations we report in this section a suitable Monte Carlo simulation study. As the permutation and combination-based multivariate ranking method is not feasible in the case of unreplicated MANOVA design, but the parametric method proposed in Chapter 1, Section 1.2.4. A parametric counterpart is a valid solution in case of multivariate normal populations, we present also a Monte Carlo simulation study under several unreplicated settings.

3.2.1 Simulation Study in Case of Ordered Categorical Responses

To validate the permutation and combination-based multivariate ranking method Arboretti Giancristofaro et al. (2014) carried out a Monte Carlo simulation study focussing on multivariate ordered categorical responses. The rationale of the simulation study was to investigat the behaviour under the null hypothesis of equality of all populations and how the estimated global ranking is affected by the type of distribution and by the different strength of dependence for random errors in the case of small sample sizes commonly used in real applications. More specifically, the simulation study considered 1000 independent data generation of samples and was designed to take into account 24 different settings, defined as combinations of the following configurations:

- Two values for the number of populations: $C = 3, 5$, where the number of response variables was always kept fixed at $p = 5$; under the alternative hypothesis we shifted the true means (on all response variables) by a $\delta = 1$ in case of $C = 3$ and $\delta = 0.5$ in the case of $C = 5$.
- Two values for the sample size: $n = 6, 12$.
- Three types of multivariate distributions for random errors: (1) Normal; (2) a heavy tailed distribution, that is, Student's t with 3 degrees of freedom; and (3) a right-skewed distribution. To generate

ordered categorical variables we rounded the continuous values to the nearest integer for the first two multivariate distributions whereas for multivariate possibly correlated right-skewed errors we use the algorithm proposed by Ferrari et al. (2011) and implemented in the R package GenOrd (Barbiero and Ferrari, 2012). Figure 3.3a represents the univariate shape of these three considered distributions in case of three populations.

- Two types of variance/covariance matrices: (1) I_p (identity matrix, i.e. the case of independence where $\sigma_{jh} = 0$, $\forall\, j,h = 1,\ldots,p$) and (2) Σ_p is such that each univariate random component has $\sigma_j^2 = 1$, $\forall\, j = 1,\ldots,p$ and $\sigma_{jh} = 0.4$, $\forall\, j,h = 1,\ldots,p$. Let us note that we considered moderate strength of dependence for random errors and we assumed homoschedasticity and set the dependence among univariate components, from the situation of independence to moderate correlation, by turning the single parameter σ_{jh}. Figure 3.3b represents the effect of bivariate correlation for the three types of random errors.

Setting the α-level as equal to 0.05, the performance of the proposed method has been evaluated in terms of empirical properties of ranking estimator; more specifically we considered

- The exact overall rate, that is, the proportion of times the method simultaneously classifies *all population* in the their correct ranking position; actually, this is an estimate of Correct Global Ranking (CGR) rate.
- The individual rate, that is, the proportion of times the method classifies *a given population* in its own correct ranking position; actually, this is an estimate of Correct Individual Ranking (CIR_j) rate, $j = 1,\ldots,C$.
- The estimated bias: $\widehat{\text{bias}}(\overline{r})_j$, $j = 1,\ldots,C$
- The estimated type II and III errors (under the nonhomogeneity settings)

We considered as permutation test statistics the Anderson–Darling (Pesarin and Salmaso, 2010a) and as combination functions the Fisher's combining function (Pesarin and Salmaso, 2010a). The null permutation distribution was estimated by $B = 1000$ CMC iterations.

Simulations results are reported in the following tables where in **bold** are highlighted the cells estimating the correct individual ranking (Tables 3.1 through 3.4).

First, it is worth noting that the simulation results basically do confirm all theoretical properties for the proposed ranking estimator. In fact, under the assumption of homogeneity all settings show that both theoretical values of CGR (0.95) and CIR_j (more than 0.95) are fully matched, regardless of the number of populations, sample sizes, type of error distributions and

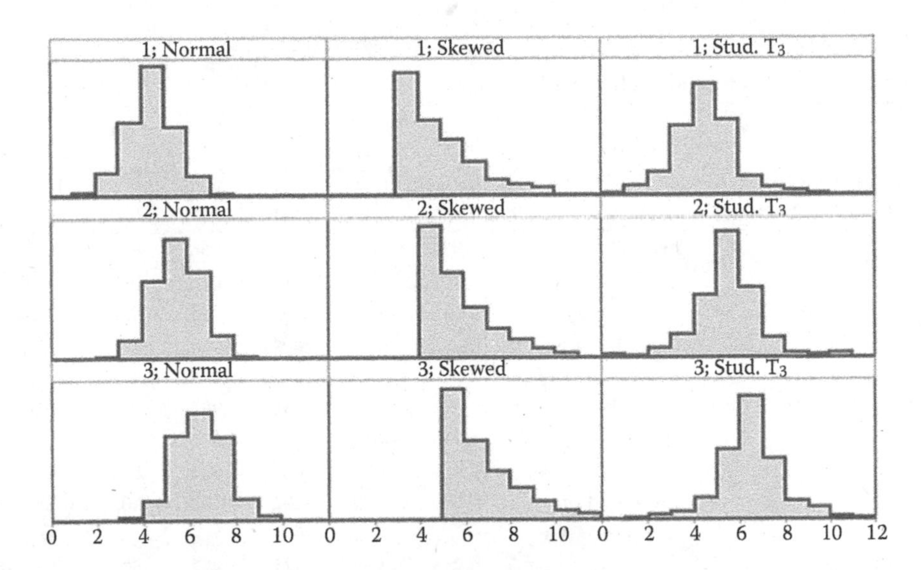

(a)

Independent errors

Y_2_Normal \ Y_1_Normal	1	2	3	4	5	6	7	8	9	10
1	.00	.00	.00	.00	.00	.00	.00	.00	.00	.00
2	.00	.00	.00	.01	.01	.00	.00	.00	.00	.00
3	.00	.00	.02	.05	.04	.02	.00	.00	.00	.00
4	.00	.00	.06	.09	.13	.05	.01	.00	.00	.00
5	.00	.00	.04	.12	.11	.05	.01	.00	.00	.00
6	.00	.01	.02	.04	.04	.02	.00	.00	.00	.00
7	.00	.00	.00	.01	.01	.00	.00	.00	.00	.00
8	.00	.00	.00	.00	.00	.00	.00	.00	.00	.00
9	.00	.00	.00	.00	.00	.00	.00	.00	.00	.00
10	.00	.00	.00	.00	.00	.00	.00	.00	.00	.00

Y_2_Skewed \ Y_1_Skewed	1	2	3	4	5	6	7	8	9	10
1	.00	.00	.00	.00	.00	.00	.00	.00	.00	.00
2	.00	.00	.00	.00	.00	.00	.00	.00	.00	.00
3	.00	.00	.00	.00	.00	.00	.00	.00	.00	.00
4	.00	.00	.00	.16	.10	.06	.03	.02	.01	.01
5	.00	.00	.00	.09	.06	.04	.03	.01	.01	.00
6	.00	.00	.00	.06	.05	.03	.02	.01	.01	.00
7	.00	.00	.00	.04	.02	.02	.01	.01	.00	.01
8	.00	.00	.00	.03	.01	.00	.01	.00	.00	.00
9	.00	.00	.00	.02	.01	.00	.00	.00	.00	.00
10	.00	.00	.00	.01	.01	.01	.00	.00	.00	.00

Y_2_Stud. T_3 \ Y_1_Stud. T_3	1	2	3	4	5	6	7	8	9	10
1	.00	.00	.00	.00	.00	.00	.00	.00	.00	.00
2	.00	.00	.01	.01	.01	.01	.00	.00	.00	.00
3	.00	.01	.02	.04	.03	.01	.01	.00	.00	.00
4	.00	.01	.03	.12	.10	.03	.01	.00	.00	.00
5	.01	.01	.04	.10	.12	.03	.01	.00	.00	.00
6	.00	.00	.03	.03	.04	.03	.01	.00	.00	.00
7	.00	.00	.01	.01	.01	.00	.00	.00	.00	.00
8	.00	.00	.00	.00	.00	.01	.00	.00	.00	.00
9	.00	.00	.00	.00	.00	.00	.00	.00	.00	.00
10	.00	.00	.00	.00	.00	.00	.00	.00	.00	.00

Correlated errors

Y_2_Normal \ Y_1_Normal	1	2	3	4	5	6	7	8	9	10
1	.00	.00	.00	.00	.00	.00	.00	.00	.00	.00
2	.00	.00	.01	.01	.01	.00	.00	.00	.00	.00
3	.00	.00	.03	.05	.03	.01	.00	.00	.00	.00
4	.00	.01	.07	.12	.11	.03	.00	.00	.00	.00
5	.00	.01	.04	.10	.14	.06	.01	.00	.00	.00
6	.00	.00	.01	.03	.05	.04	.01	.00	.00	.00
7	.00	.00	.00	.00	.00	.01	.00	.00	.00	.00
8	.00	.00	.00	.00	.00	.00	.00	.00	.00	.00
9	.00	.00	.00	.00	.00	.00	.00	.00	.00	.00
10	.00	.00	.00	.00	.00	.00	.00	.00	.00	.00

Y_2_Skewed \ Y_1_Skewed	1	2	3	4	5	6	7	8	9	10
1	.00	.00	.00	.00	.00	.00	.00	.00	.00	.00
2	.00	.00	.00	.00	.00	.00	.00	.00	.00	.00
3	.00	.00	.00	.00	.00	.02	.00	.00	.00	.00
4	.00	.00	.00	.25	.10	.04	.03	.02	.00	.00
5	.00	.00	.00	.08	.07	.04	.02	.03	.00	.00
6	.00	.00	.00	.04	.04	.04	.02	.01	.01	.00
7	.00	.00	.00	.02	.02	.02	.02	.02	.00	.00
8	.00	.00	.00	.01	.01	.01	.01	.01	.00	.00
9	.00	.00	.00	.00	.00	.00	.00	.01	.00	.00
10	.00	.00	.00	.00	.00	.00	.00	.01	.00	.00

Y_2_Stud. T_3 \ Y_1_Stud. T_3	1	2	3	4	5	6	7	8	9	10
1	.01	.00	.00	.00	.00	.00	.00	.00	.00	.00
2	.01	.00	.01	.01	.01	.00	.00	.00	.00	.00
3	.00	.01	.03	.05	.02	.01	.00	.00	.00	.00
4	.01	.01	.04	.13	.10	.02	.01	.00	.00	.00
5	.00	.00	.02	.12	.12	.04	.01	.00	.00	.00
6	.00	.00	.01	.02	.03	.03	.01	.00	.00	.00
7	.00	.00	.00	.01	.01	.01	.00	.00	.00	.00
8	.00	.00	.00	.00	.00	.00	.00	.01	.00	.00
9	.00	.00	.00	.00	.00	.00	.00	.00	.00	.00
10	.00	.00	.00	.00	.00	.00	.00	.00	.00	.00

(b)

FIGURE 3.3
(a) Shape of the univariate distributions ($C = 3$, under nonhomogeneity) and (b) effect of bivariate correlation for the three types of random errors considered in the simulation study.

TABLE 3.1

Simulation Results under Homogeneity, $C = 3$

				$n = 6$			$n = 12$		
				True Rank			True Rank		
Random Errors	**Variance**			**1st**	**2nd**	**3rd**	**1st**	**2nd**	**3rd**
Normal	$\sigma_{ij} = 0$	Estimated Rank	1st	**.986**	**.974**	**.973**	**.960**	**.968**	**.981**
			2nd	.003	.012	.019	.025	.015	.011
			3rd	.011	.014	.009	.015	.017	.008
		Estimated Bias		–.03	–.04	–.04	–.05	–.05	–.03
		Estimated CGR			.968			.955	
	$\sigma_{ij} = 0.4$	Estimated Rank	1st	**.981**	**.985**	**.988**	**.982**	**.975**	**.953**
			2nd	.010	.012	.006	.006	.021	.014
			3rd	.010	.003	.006	.012	.004	.032
		Estimated Bias		–.03	–.02	–.02	–.03	–.03	–.08
		Estimated CGR			.982			.951	
Student $t(3)$	$\sigma_{ij} = 0$	Estimated Rank	1st	**.983**	**.997**	**.984**	**.971**	**.966**	**.978**
			2nd	.012	.001	.006	.016	.016	.011
			3rd	.005	.002	.010	.013	.018	.011
		Estimated Bias		–.02	–.00	–.03	–.04	–.05	–.03
		Estimated CGR			.976			.960	
	$\sigma_{ij} = 0.4$	Estimated Rank	1st	**.970**	**.982**	**.974**	**.965**	**.973**	**.983**
			2nd	.022	.010	.015	.023	.009	.006
			3rd	.008	.008	.011	.012	.018	.011
		Estimated Bias		–.04	–.03	–.04	–.05	–.04	–.03
		Estimated CGR			.966			.959	
Right-skewed	$\sigma_{ij} = 0$	Estimated Rank	1st	**.964**	**.972**	**.976**	**.959**	**.969**	**.973**
			2nd	.034	.015	.004	.023	.015	.005
			3rd	.003	.013	.020	.019	.016	.022
		Estimated Bias		–.04	–.04	–.04	–.06	–.05	–.05
		Estimated CGR			.955			.953	
	$\sigma_{ij} = 0.4$	Estimated Rank	1st	**.963**	**.970**	**.976**	**.974**	**.977**	**.957**
			2nd	.015	.016	.011	.013	.007	.024
			3rd	.023	.014	.014	.013	.017	.019
		Estimated Bias		–.06	–.04	–.04	–.04	–.04	–.06
		Estimated CGR			.952			.955	

presence of correlation. This is an important outcome because it shows the nonparametric and exact behaviour of the proposed ranking estimator. Moreover, the estimated bias under homogeneity is relatively small and close to the chosen α.

Under the nonhomogeneity settings, irrespective of type of errors, we obtain overall satisfactory results in terms of both correct individual and global classification rates, especially recalling that we are considering very

TABLE 3.2

Simulation Results under Nonhomogeneity, $C = 3$

Random Errors	Variance			$n = 6$			$n = 12$		
				True Rank			True Rank		
				1st	**2nd**	**3rd**	**1st**	**2nd**	**3rd**
Normal	$\sigma_{ij} = 0.4$	Estimated Rank	1st	**1.00**	.120	.000	**1.00**	.009	.000
			2nd	.000	**.880**	.142	.000	**.991**	.008
			3rd	.000	.000	**.858**	.000	.000	**.992**
		Estimated Bias		.00	.12	.14	.00	.01	.01
		Est. CGR-Type II err.-Type III err.		.740	.260	.000	.991	.009	.000
	$\sigma_{ij} = 0.4$	Estimated Rank	1st	**.992**	.349	.051	**1.00**	.105	.000
			2nd	.008	**.641**	.266	.000	**.895**	.120
			3rd	.000	.010	**.683**	.000	.000	**.880**
		Estimated Bias		–.01	.34	.37	.00	.11	.12
		Est.CGR-Type II err.-Type III err.		.391	.590	.019	.779	.221	.000
Student $t(3)$	$\sigma_{ij} = 0$	Estimated Rank	1st	**.987**	.302	.046	**1.00**	.100	.000
			2nd	.014	**.689**	.255	.000	**.900**	.093
			3rd	.000	.009	**.700**	.000	.000	**907**
		Estimated Bias		–.01	.29	.35	.00	.10	.09
		Est.CGR-Type II err.-Type III err.		**.448**	.530	.022	.813	.187	.000
	$\sigma_{ij} = 0.4$	Estimated Rank	1st	**.966**	.516	.252	**.993**	.317	.044
			2nd	.034	**.460**	.250	.007	**.671**	.301
			3rd	.000	.024	**.498**	.000	.012	**.655**
		Estimated Bias		–.03	.49	.75	–.01	.31	.39
		Est.CGR-Type II err.-Type III err.		.237	.708	.056	.387	.598	.015
Right-skewed	$\sigma_{ij} = 0$	Estimated Rank	1st	**1.00**	.260	.032	**1.00**	.098	.000
			2nd	.000	**.740**	.273	.000	**.902**	.099
			3rd	.000	.000	**.696**	.000	.000	**.901**
		Estimated Bias		.00	.26	.34	.00	.10	.10
		Est.CGR-Type II err.- Type III err.		.477	.524	.000	.812	.189	.000
	$\sigma_{ij} = 0.4$	Estimated Rank	1st	**.984**	.490	.211	**.994**	.333	.015
			2nd	.016	**.457**	.286	.006	**.662**	.288
			3rd	.000	.054	**.503**	.000	.005	**.697**
		Estimated Bias		–.02	.44	.71	–.01	.33	.32
		Est.CGR-Type II err.-Type III err.		.233	.700	.067	.389	.601	.010

TABLE 3.3

Simulation Results under Homogeneity, $C = 5$

Random Errors	Variance			$n = 6$ True Rank					$n = 12$ True Rank				
				1st	2nd	3rd	4th	5th	1st	2nd	3rd	4th	5th
Normal	$\sigma_{ij} = 0$	Estimated Rank	1st	**.972**	**.973**	**.974**	**.976**	**.980**	**.954**	**.957**	**.948**	**.948**	**.955**
			2nd	.016	.020	.016	.017	.011	.042	.020	.034	.021	.030
			3rd	.000	.001	.001	.002	.000	.002	.003	.003	.007	.004
			4th	.004	.002	.002	.001	.001	.001	.002	.006	.007	.000
			5th	.008	.004	.007	.005	.008	.001	.017	.009	.017	.011
		Estimated Bias		–.06	–.04	–.05	–.04	–.04	–.05	–.10	–.09	–.12	–.08
		Estimated CGR				.971					.946		
	$\sigma_{ij} = 0.4$	Estimated Rank	1st	**.972**	**.981**	**.977**	**.978**	**.981**	**.964**	**.963**	**.953**	**.959**	**.958**
			2nd	.010	.018	.015	.006	.011	.021	.022	.023	.024	.024
			3rd	.001	.000	.003	.001	.001	.001	.006	.007	.001	.003
			4th	.008	.001	.004	.002	.000	.001	.000	.010	.003	.009
			5th	.009	.000	.001	.013	.008	.013	.009	.007	.012	.006
		Estimated Bias		–.07	–.02	–.04	–.07	–.04	–.08	–.07	–.09	–.09	–.08
		Estimated CGR				.978					.953		
Student $t(3)$	$\sigma_{ij} = 0$	Estimated Rank	1st	**.973**	**.977**	**.979**	**.978**	**.976**	**.975**	**.967**	**.972**	**.966**	**.976**
			2nd	.017	.017	.015	.009	.017	.016	.022	.020	.024	.015
			3rd	.002	.002	.001	.001	.002	.001	.001	.002	.001	.001
			4th	.002	.002	.000	.003	.002	.000	.003	.001	.001	.003
			5th	.007	.002	.005	.010	.003	.009	.008	.005	.008	.006
		Estimated Bias		–.05	–.04	–.04	–.06	–.04	–.05	–.06	–.05	–.06	–.05
		Estimated CGR				.972					.964		

(Continued)

TABLE 3.3 (CONTINUED)

Simulation Results under Homogeneity, $C = 5$

Random Errors	Variance			$n = 6$					$n = 12$				
				True Rank					True Rank				
				1st	2nd	3rd	4th	5th	1st	2nd	3rd	4th	5th
	$\sigma_{ij} = 0.4$	Estimated Rank	1st	**.951**	**.964**	**.964**	**.966**	**.962**	**.968**	**.974**	**.966**	**.967**	**.972**
			2nd	.040	.019	.017	.025	.021	.020	.011	.024	.019	.018
			3rd	.000	.000	.000	.001	.004	.002	.002	.002	.002	.000
			4th	.005	.005	.000	.002	.004	.001	.005	.002	.003	.002
			5th	.005	.012	.018	.006	.008	.010	.009	.008	.010	.007
		Estimated Bias		–.07	–.08	–.09	–.06	–.08	–.06	–.06	–.06	–.07	–.05
		Estimated CGR				.949					.966		
Right–skewed	$\sigma_{ij} = 0$	Estimated Rank	1st	**.965**	**.967**	**.965**	**.983**	**.970**	**.960**	**.965**	**.970**	**.961**	**.963**
			2nd	.018	.017	.020	.012	.011	.023	.022	.007	.024	.006
			3rd	.088	.006	.004	.000	.009	.002	.006	.014	.005	.001
			4th	.003	.001	.003	.001	.004	.003	.006	.008	.001	.009
			5th	.007	.009	.008	.004	.006	.012	.002	.001	.008	.022
		Estimated Bias		–.07	–.07	–.07	–.03	–.06	–.08	–.06	–.06	–.07	–.12
		Estimated CGR				.962					.959		
	$\sigma_{ij} = 0.4$	Estimated Rank	1st	**.961**	**.948**	**.949**	**.944**	**.963**	**.954**	**.948**	**.964**	**.947**	**.952**
			2nd	.015	.028	.014	.035	.021	.030	.016	.019	.026	.037
			3rd	.002	.007	.001	.005	.004	.004	.008	.009	.006	.006
			4th	.002	.008	.007	.001	.005	.002	.001	.003	.003	.004
			5th	.021	.009	.002	.015	.006	.009	.028	.005	.019	.002
		Estimated Bias		–.11	–.10	–.07	–.11	–.07	–.08	–.14	–.07	–.12	–.07
		Estimated CGR				.944					.947		

TABLE 3.4

Simulation Results under Nonhomogeneity, $C = 5$

Random Errors	Variance			$n = 6$					$n = 12$				
				True Rank					True Rank				
				1st	2nd	3rd	4th	5th	1st	2nd	3rd	4th	5th
Normal	$\sigma_{ij} = 0$	Estimated Rank	1st	**.916**	.291	.025	.012	.007	**1.00**	.000	.000	.000	.000
			2nd	.074	**.679**	.138	.008	.006	.000	**1.00**	.013	.000	.000
			3rd	.010	.017	**.833**	.130	.004	.000	.000	**.987**	.009	.000
			4th	.000	.013	.004	**.845**	.174	.000	.000	.000	**.991**	.000
			5th	.000	.000	.000	.006	**.809**	.000	.000	.000	.000	**1.00**
		Estimated Bias		–.09	.25	.19	.18	.23	.00	.00	.01	.01	.00
		Est. CGR-Type II err.-Type III err.			.435	.465	.100			.988	.013	.000	
	$\sigma_{ij} = 0.4$	Estimated Rank	1st	**.907**	.437	.093	.019	.016	**1.00**	.119	.003	.001	.000
			2nd	.070	**.536**	.156	.013	.002	.000	**.881**	.129	.002	.000
			3rd	.023	.027	**.741**	.153	.024	.000	.000	**.868**	.148	.000
			4th	.000	.000	.010	**.809**	.269	.000	.000	.000	**.849**	.134
			5th	.000	.000	.000	.007	**.690**	.000	.000	.000	.000	**.866**
		Estimated Bias		–.12	.41	.33	.23	.38	.00	.12	.14	.16	.13
		Est. CGR-Type II err.-Type III err.			.235	.640	.125			.575	.425	.000	

(Continued)

TABLE 3.4 (CONTINUED)

Simulation Results under Nonhomogeneity, $C = 5$

Random Errors	Variance			$n = 6$					$n = 12$				
				True Rank					True Rank				
				1st	2nd	3rd	4th	5th	1st	2nd	3rd	4th	5th
Student $t(3)$	$\sigma_{ij} = 0$	Estimated Rank	1st	**.876**	.393	.082	.030	.015	**1.00**	.141	.000	.000	.000
			2nd	.093	**.552**	.194	.039	.006	.000	**.853**	.089	.000	.000
			3rd	.023	.050	**.690**	.179	.026	.000	.007	**.911**	.132	.000
			4th	.007	.006	.026	**.738**	.240	.000	.000	.000	**.868**	.119
			5th	.000	.000	.010	.014	**.713**	.000	.000	.000	.000	**.881**
		Estimated Bias		–.16	.33	.31	.33	.37	.00	.13	.09	.13	.12
		Est. CGR-Type II err.-Type III err.			.230	.572	.198			.599	.394	.007	
	$\sigma_{ij} = 0.4$	Estimated Rank	1st	.884	.471	.177	.089	.069	**.979**	.328	.046	.000	.000
			2nd	.089	**.463**	.236	.056	.030	.021	**.653**	.207	.024	.008
			3rd	.017	.061	**.508**	.220	.110	.000	.018	**.729**	.195	.057
			4th	.010	.005	.072	**.592**	.250	.000	.000	.018	**.767**	.294
			5th	.000	.000	.008	.044	**.542**	.000	.000	.000	.014	**.641**
		Estimated Bias		–.15	.40	.50	.55	.83	–.02	.31	.28	.23	.43
		Est. CGR-Type II err.- Type III err.			.235	.640	.125			.575	.363	.062	

(Continued)

TABLE 3.4 (CONTINUED)

Simulation Results under Nonhomogeneity, $C = 5$

Random Errors	Variance			n = 6					n = 12				
				True Rank					True Rank				
				1st	2nd	3rd	4th	5th	1st	2nd	3rd	4th	5th
Right-skewed	$\sigma_{ij} = 0$	Estimated Rank	1st	**.853**	.395	.061	.022	.009	**1.00**	.116	.000	.000	.000
			2nd	.113	**.590**	.186	.021	.000	.000	**.884**	.062	.000	.000
			3rd	.034	.016	**.739**	.160	.010	.000	.000	**.938**	.106	.000
			4th	.000	.000	.014	**.780**	.199	.000	.000	.000	**.894**	.097
			5th	.000	.000	.000	.017	**.781**	.000	.000	.000	.000	**.903**
		Estimated Bias		–.18	.38	.29	.25	.26	.00	.12	.06	.11	.10
		Est. CGR-Type II err.-Type III err.			.278	.544	.179			.680	.320	.000	
	$\sigma_{ij} = 0.4$	Estimated Rank	1st	**.901**	.562	.211	.079	.071	**1.00**	.287	.028	.000	.000
			2nd	.079	**.420**	.206	.061	.021	.000	**.713**	.190	.031	.000
			3rd	.012	.018	**.480**	.227	.100	.000	.001	**.757**	.210	.032
			4th	.007	.001	.104	**.537**	.281	.000	.000	.025	**.736**	.289
			5th	.000	.000	.000	.096	**.528**	.000	.000	.000	.023	**.679**
		Estimated Bias		–.13	.54	.52	.49	.83	.00	.29	.22	.25	.35
		Est. CGR-Type II err.-Type III err.			.235	.640	.125			.575	.377	.048	

small sample sizes ($n = 6, 12$) with respect to the assumed shifts ($\delta = 1$ for $C = 3$ and $\delta = 0.5$ for $C = 5$). Moreover, we may observe several interesting findings:

- Both correct individual and global ranking rates increase as the sample size increase, as suggested by the asymptotic estimator properties; both rates decrease in case of correlated errors and this confirms what we observed in different but related contexts (Corain and Salmaso, 2013, 2015): because correlation simply means less available information in the pairwise comparisons, the permutation test becomes less powerful and in turn the derived ranking estimator becomes less reliable.
- The estimated correct individual ranking rates are relatively smaller in the case of nonnormal random errors than the normal case; at any rate, the reduction in terms of lower probabilities for CGR (0.95) and CIR_j is reasonably contained.
- The estimated correct individual ranking rates are also larger for the best population than for the remaining ones, which in turn seem more or less similar.
- The estimated bias appears as relatively limited for the sample sizes we considered; it is worth noting that, as expected by the asymptotically unbiasedness of the ranking estimator, the bias decreases as the sample size increases as well and the rate of this decrease seems to be quite fast with respect to the actual finite sample sizes used in many real applications.

3.2.2 Simulation Study in Case of Unreplicated Designs

To validate the methodology presented in Chapter 1, Section 1.2.4 (see also Carrozzo et al., 2015), we carried out a Monte Carlo simulation study in the framework of the unreplicated design. The rationale of the simulation study was focussed on investigating the behaviour under the null hypothesis of equality of all populations and how the estimated global ranking is affected by the different strengths of dependence for random errors and by an increasing number of populations (Carrozzo et al., 2015). More specifically, the simulation study considered 1000 independent data generations of unreplicated multivariate normal samples and was designed to take into account 54 different settings, defined as combinations of the following configurations:

- Three values for the number of populations: $C = 3, 4, 5$, where the number of response variables was always kept fixed at $p = 10$.
- Three types of multivariate distributions: normal, heavy-tailed (Student's t with 28 d.f.) and skewed, where the latter has been

generated by using the method proposed by Vale and Maurelli (1983) and programmed in R by Zopluoglu (2011). The nonnormal heavy-tailed and skewed cases (with kurtosis and skewness parameters equal to 0.25 and 0.5 respectively) were considered to evaluate the possible robustness of the proposed methodology under the cases of moderate heavy-tailed or asymmetric multivariate distributions.

- Two types of variance/covariance matrices: (1) Σ_1 (heteroschedastic and independent errors, that is, $\sigma_k = k$ and $\sigma_{ks} = 0$, $\forall\ k,s = 1,\ldots,p$) and (2) Σ_2 (heteroschedastic and correlated errors, that is, $\sigma_k = k$ and $\sigma_{ks} = 0.75$, $\forall\ k,s = 1,\ldots,p$); at any rate, during the ranking estimation process, the variance/covariance will be assumed as unknown, so that empirical variance/covariance estimates will be used in the least significant difference (LSD) formula.
- Three situations for the true means $\mu_1, \mu_2, \ldots, \mu_C$:
  - Homogeneity of all populations: $\mu_1 = \mu_2 = \ldots = \mu_C$;
  - Nonhomogeneity and full ranked populations: $r(\Pi_1) = 1$, $r(\Pi_2) = 2, \ldots, r(\Pi_c) = C$;
  - Nonhomogeneity and some equally ranked populations, that is, $r(\Pi_1) = r(\Pi_2) = 1, \ldots, r(\Pi_{c-1}) = r(\Pi_c) = C$.

Under the nonhomogeneity settings, we adopted the rule "the lower the better" and we set the true means as $\mu_{jk} = (j - 1)2\sigma_k$, $k = 1,\ldots,p$.

Considering the α-level set as 0.05, the performance of the proposed method has been evaluated in terms of correct rank classification rates; more specifically we compute

- The overall –(CGR) rate, that is, the proportion of times the method simultaneously classifies *all populations* in the their correct ranking position
- The –(CIR) rate, that is, the proportion of times the method classifies *a given population* in its own correct ranking position
- Simulations results are displayed in Tables 3.5 through 3.7, where the true population ranks are reported along the columns while the estimated ranks are reported in the rows and the correct individual rank; CIR rates are highlighted in **bold**.

First, it is worth noting that, irrespective of the type of error distribution and of the possible correlation, under the homogeneity of all populations the proposed method has both CGR and CGI rates rather close or slightly greater than the nominal value 0.95 when $C > 3$. The reason why the nominal error rates are not respected in case of $C = 3$ is explained by the poor normal approximated distribution of θs estimator due to the estimates of correlation parameters. In fact, additional simulations (not reported here) have shown

TABLE 3.5

Simulations Results for $C = 3$ Populations ($\alpha = 0.05$)

Normal Errors					Heavy-Tailed Errors					Skewed Errors			
	True Rank			Estimated		True Rank			Estimated		True Rank		
CGR	1	2	3	Rank	CGR	1	2	3	Rank	CGR	1	2	3
Homogeneity of All Populations, Heteroscedastic and Independent Errors													
.903	**.941**	**.941**	**.943**	1	.891	**.937**	**.934**	**.932**	1	.909	**.945**	**.943**	**.945**
	.035	.031	.031	2		.034	.036	.037	2		.033	.031	.034
	.024	.028	.027	3		.029	.030	.031	3		.022	.026	.021
Homogeneity of All Populations, Heteroscedastic and Correlated Errors													
.924	**.955**	**.952**	**.954**	1	.905	**.943**	**.942**	**.943**	1	.916	**.948**	**.947**	**.953**
	.026	.026	.024	2		.031	.030	.032	2		.029	.028	.024
	.019	.022	.022	3		.026	.028	.025	3		.023	.025	.023
Nonhomogeneity, Full Ranked Populations and Independent Errors													
.976	**1.00**	.020	.017	1	.969	**1.00**	.024	.017	1	.978	**1.00**	.019	.015
	.000	**.980**	.003	2		.000	**.976**	.007	2		.000	**.981**	.003
	.000	.000	**.980**	3		.000	.000	**.976**	3		.000	.000	**.982**

(Continued)

TABLE 3.5 (CONTINUED)

Simulations Results for C = 3 Populations (α = 0.05)

Normal Errors					Heavy-Tailed Errors					Skewed Errors			
	True Rank			Estimated		True Rank			Estimated		True Rank		
CGR	1	2	3	Rank	CGR	1	2	3	Rank	CGR	1	2	3
Nonhomogeneity, Full Ranked Populations and Correlated Errors													
.927	**1.00**	.059	.046	1	.916	**1.00**	0.63	0.44	1	.929	**1.00**	.056	.045
	.000	**.941**	.013	2		.000	**.94**	.021	2		.000	**.944**	.015
	.000	.000	**.941**	3		.000	.000	**.93**	3		.000	.000	**.940**
Nonhomogeneity, Some Equally Ranked Populations and Independent Errors													
1.00	**1.00**	**1.00**	.000	1	1.00	**1.00**	**1.00**	.000	1	1.00	**1.00**	**1.00**	.000
	.000	.000	.000	2		.000	.000	.000	2		.000	.000	.000
	.000	.000	**1.00**	3		.000	.000	**1.00**	3		.000	.000	**1.00**
Nonhomogeneity, Some Equally Ranked Populations and Correlated Errors													
.009	**.959**	**.957**	.907	1	.015	**.942**	**.944**	.871	1	.013	**.956**	**.967**	.910
	.041	.043	.000	2		.058	.056	.000	2		.044	.033	.000
	.000	.000	**.093**	3		.000	.000	**.129**	3		.000	.000	**.090**

TABLE 3.6

Simulations Results for $C = 4$ Populations ($\alpha = 0.05$)

Normal Errors						Heavy-Tailed Errors						Skewed Errors				
	True Rank				Estimated		True Rank				Estimated		True Rank			
CGR	1	2	3	4	Rank	CGR	1	2	3	4	Rank	CGR	1	2	3	4
Homogeneity of All Populations, Heteroscedastic and Independent Errors																
.951	**.966**	**.967**	**.964**	**.966**	1	.941	**.959**	**.959**	**.960**	**.960**	1	.951	**.964**	**.969**	**.963**	**.968**
	.017	.018	.021	.019	2		.022	.020	.024	.021	2		.023	.019	.017	.017
	.006	.006	.005	.005	3		.007	.008	.006	.005	3		.008	.004	.004	.006
	.011	.009	.010	.011	4		.012	.014	.010	.014	4		.005	.008	.016	.009
Homogeneity of All Populations, Heteroscedastic and Correlated Errors																
.969	**.979**	**.977**	**.978**	**.980**	1	.955	**.967**	**.970**	**.969**	**.968**	1	.962	**.976**	**.975**	**.972**	**.972**
	.011	.012	.013	.011	2		.018	.016	.017	.018	2		.012	.013	.015	.017
	.003	.005	.004	.004	3		.005	.004	.004	.005	3		.004	.005	.003	.002
	.007	.007	.006	.005	4		.010	.010	.011	.009	4		.008	.007	.010	.009
Nonhomogeneity, Full Ranked Populations and Independent Errors																
.702	**1.00**	.136	.000	.000	1	.698	**1.00**	.139	.001	.000	1	.714	**1.00**	.139	.000	.000
	.000	**.864**	.022	.000	2		.000	**.861**	.022	.001	2		.000	**.861**	.022	.000
	.000	.000	**.977**	.139	3		.000	.000	**.978**	.142	3		.000	.000	**.978**	.125
	.000	.000	.000	**.861**	4		.000	.000	.000	**.857**	4		.000	.000	.000	**.875**

(Continued)

TABLE 3.6 (CONTINUED)

Simulations Results for C = 4 Populations ($\alpha = 0.05$)

Normal Errors						Heavy-Tailed Errors						Skewed Errors				
	True Rank						True Rank						True Rank			
CGR	1	2	3	4	Estimated Rank	CGR	1	2	3	4	Estimated Rank	CGR	1	2	3	4
Nonhomogeneity, Full Ranked Populations and Correlated Errors																
.604	**1.00**	.177	.001	.000	1	.615	**1.00**	.175	.003	.000	1	.605	**1.00**	.177	.000	.000
	.000	**.823**	.045	.002	2		.000	**.825**	.040	.004	2		.000	**.823**	.038	.004
	.000	.000	**.954**	.177	3		.000	.000	**.957**	.175	3		.000	.000	**.962**	.188
	.000	.000	.000	**.822**	4		.000	.000	.000	**.822**	4		.000	.000	.000	**.808**
Nonhomogeneity, Some Equally Ranked Populations and Independent Errors																
.998	**1.00**	**1.00**	.000	.000	1	.995	**1.00**	**1.00**	.000	.000	1	.996	**1.00**	**1.00**	.000	.000
	.001	.001	.000	.000	2		.003	.002	.000	.000	2		.002	.002	.000	.000
	.000	.000	**.999**	**.999**	3		.000	.000	**.998**	**.998**	3		.000	.000	**.997**	**.999**
	.000	.000	.001	.015	4		.000	.000	.002	.002	4		.000	.000	.003	.001
Nonhomogeneity, Some Equally Ranked Populations and Correlated Errors																
.000	**.977**	**.974**	.948	.948	1	.000	**.961**	**.959**	.914	.914	1	.000	**.962**	**.973**	.935	.935
	.023	.026	.020	.025	2		.040	.041	.038	.036	2		.038	.027	.030	.033
	.000	.000	**.006**	**.004**	3		.000	.000	**.008**	**.009**	3		.000	.000	**.002**	**.002**
	.000	.000	.026	.023	4		.000	.000	.040	.040	4		.000	.000	.033	.030

TABLE 3.7

Simulations Results for $C = 5$ Populations ($\alpha = 0.05$)

Normal Errors							Heavy-Tailed Errors							Skewed Errors					
	True Rank					Estimated		True Rank					Estimated		True Rank				
CGR	1	2	3	4	5	Rank	CGR	1	2	3	4	5	Rank	CGR	1	2	3	4	5
Homogeneity of All Populations, Heteroscedastic and Independent Errors																			
.975	**.981**	**.979**	**.981**	**.981**	**.982**	1	.960	**.969**	**.969**	**.967**	**.972**	**.970**	1	.974	**.980**	**.980**	**.981**	**.980**	**.980**
	.013	.013	.012	.011	.012	2		.020	.020	.020	.018	.019	2		.011	.011	.012	.010	.012
	.002	.002	.001	.002	.001	3		.002	.002	.003	.002	.003	3		.003	.002	.004	.004	.005
	.001	.002	.002	.001	.001	4		.002	.002	.002	.002	.002	4		.001	.001	.000	.002	.000
	.004	.004	.005	.005	.004	5		.007	.007	.007	.006	.007	5		.005	.006	.003	.004	.003
Homogeneity of All Populations, Heteroscedastic and Correlated Errors																			
.984	**.989**	**.989**	**.987**	**.988**	**.988**	1	.972	**.979**	**.978**	**.978**	**.979**	**.979**	1	.986	**.990**	**.990**	**.990**	**.986**	**.990**
	.007	.007	.009	.008	.006	2		.014	.015	.014	.012	.013	2		.006	.006	.009	.007	.006
	.001	.001	.001	.001	.001	3		.002	.002	.002	.002	.002	3		.001	.001	.000	.001	.001
	.001	.001	.001	.001	.001	4		.001	.001	.001	.002	.001	4		.001	.000	.000	.001	.002
	.002	.003	.003	.003	.003	5		.004	.005	.005	.006	.005	5		.002	.003	.001	.005	.001
Nonhomogeneity, Full Ranked Populations and Independent Errors																			
.362	**1.00**	.228	.000	.000	.000	1	.365	**1.00**	.219	.000	.000	.000	1	.343	**1.00**	.227	.000	.000	.000
	.000	**.772**	.106	.000	.000	2		.000	**.781**	.104	.000	.000	2		.000	**.773**	.111	.000	.000
	.000	.000	**.895**	.110	.000	3		.000	.000	**.896**	.110	.000	3		.000	.000	**.889**	.110	.000
	.000	.000	.000	**.890**	.228	4		.000	.000	.000	**.890**	.237	4		.000	.000	.000	**.889**	.242
	.000	.000	.000	.000	**.773**	5		.000	.000	.000	.000	**.763**	5		.000	.000	.000	.000	**.758**

(Continued)

TABLE 3.7 (CONTINUED)

Simulations Results for C = 5 Populations ($\alpha = 0.05$)

Normal Errors							Heavy-Tailed Errors							Skewed Errors					
	True Rank					Estimated		True Rank					Estimated		True Rank				
CGR	1	2	3	4	5	Rank	CGR	1	2	3	4	5	Rank	CGR	1	2	3	4	5
Nonhomogeneity, Full Ranked Populations and Correlated Errors																			
.327	**1.00**	.233	.000	.000	.000	1	.320	**1.00**	.251	.001	.000	.000	1	.333	**1.00**	.245	.001	.000	.000
	.000	**.767**	.116	.000	.000	2		.000	**.749**	.120	.000	.000	2		.000	**.755**	.108	.000	.000
	.000	.000	**.884**	.126	.000	3		.000	.000	**.879**	.122	.000	3		.000	.000	**.891**	.12	.003
	.000	.000	.000	**.874**	.243	4		.000	.000	.000	**.878**	.234	4		.000	.000	.000	**.879**	.238
	.000	.000	.000	.000	**.757**	5		.000	.000	.000	.000	**.766**	5		.000	.000	.000	.000	**.759**
Nonhomogeneity, Some Equally Ranked Populations and Independent Errors																			
.964	**.990**	**.988**	.000	.000	.000	1	.952	**.984**	**.988**	.000	.000	.000	1	.962	**.989**	**.986**	.000	.000	.000
	.010	.012	.000	.000	.000	2		.016	.013	.000	.000	.000	2		.011	.014	.000	.000	.000
	.000	.000	**.982**	**.983**	.001	3		.000	.000	**.979**	**.977**	.001	3		.000	.000	**.979**	**.985**	.000
	.000	.000	.018	.017	.000	4		.000	.000	.021	.024	.000	4		.000	.000	.021	.015	.000
	.000	.000	.000	.000	**1.00**	5		.000	.000	.000	.000	**.999**	5		.000	.000	.000	.000	**1.00**
Nonhomogeneity, Some Equally Ranked Populations and Correlated Errors																			
.693	**.879**	**.876**	.061	.061	.060	1	.696	**.875**	**.879**	.055	.055	.052	1	.696	**.860**	**.882**	.045	.046	.045
	.121	.124	.239	.239	.000	2		.125	.121	.235	.236	.000	2		.140	.118	.255	.257	.000
	.000	.000	**.696**	**.695**	.000	3		.000	.000	**.700**	**.702**	.000	3		.000	.000	**.696**	**.697**	.000
	.000	.000	.004	.005	.002	4		.000	.000	.010	.008	.004	4		.000	.000	.004	.000	.002
	.000	.000	.000	.000	**.939**	5		.000	.000	.000	.000	**.944**	5		.000	.000	.000	.000	**.953**

that if we replace the estimates with the true correlation values in the LSD formula, then the estimated rates become exactly matched with the nominal values also for $C = 3$. Under the nonhomogeneity settings the proposed method shows good behaviour in terms of detection of the true rank both partially and globally, especially recalling that we are considering an unreplicated design with a relatively small shift δ among populations ($\delta = 2\sigma$). As expected by the LSD formula, simulation results do confirm that the presence of correlation negatively affects the corrected classification rates while both estimated correct individual and global ranking rates are only slightly smaller in the case of nonnormal random errors than in the normal case. Note that the estimated correct individual ranking rates are larger for the best population than for the remaining ones, which in turn seem more or less similar. Finally, it is interesting to note the benefit provided by including a 'very worst' population, as highlighted by the case of nonhomogeneity and some equally ranked populations when errors are correlated. In fact, the inclusion of a 'worst' population we have when $C = 5$, with respect to the case when $C = 4$, allows us the obtain much higher corrected global as well as individual classification rates for the two tied populations with true rank equal to 3.

References

Arboretti Giancristofaro R., Bonnini S., Corain L., Solmi F. (2010). A comparison on FWE-type multiple comparison procedures. In *Proceedings of the 2010 JSM – Joint Statistical Meetings*, 31 July – 5 August, 2010, Vancouver, Canada, pp. 915–927.

Arboretti Giancristofaro R., Corain L., Ragazzi S. (2012a). The multivariate randomized complete block design: A novel permutation solution in case of ordered categorical variables. *Communications in Statistics – Theory and Methods*, 41(16–17), 3094–3109.

Arboretti Giancristofaro R., Bolzan M., Bonnini S., Corain L., Solmi F. (2012b). Advantages of closed testing method for multiple comparison procedures. *Communications in Statistics – Simulation and Computation*, 41(6), 746–763.

Arboretti Giancristofaro R., Bonnini S., Corain L., Salmaso L. (2014). A permutation approach for ranking of multivariate populations. *Journal of Multivariate Analysis*, 132, 39–57.

Barbiero A., Ferrari P. A. (2012). GenOrd: Simulation of ordinal and discrete variables with given correlation matrix and marginal distributions. R package version 1.0.1. http://CRAN.R-project.org/package=GenOrd.

Basso D., Pesarin F., Salmaso L., Solari A. (2009). *Permutation Tests for Stochastic Ordering and ANOVA: Theory and Applications with R*. Lecture Notes in Statistics. New York: Springer Science+Business Media.

Brombin C., Salmaso L. (2009). Multi-aspect permutation tests in shape analysis with small sample size. *Computational Statistics & Data Analysis*, 53(12), 3921–3931.

Carrozzo E., Corain L., Musci R., Salmaso L., Spadoni L. (2015). A new approach to rank several multivariate normal populations with application to life cycle assessment. *Communications in Statistics – Simulation and Computation*, doi: 10.1080/03610918.2014.925926.

Corain L., Salmaso L. (2013). Nonparametric permutation and combination-based multivariate control charts with applications in microelectronics. *Applied Stochastic Models in Business and Industry*, 24(4), 334–349.

Corain L., Salmaso L. (2015). Improving power of multivariate combination-based permutation tests. *Statistics and Computing*, 25(2), 203–214.

Ferrari P., Annoni P., Barbiero A., Manzi G. (2011). An imputation method for categorical variables with application to nonlinear principal component analysis. *Computational Statistics & Data Analysis*, 55, 2410–2420.

Finos L., Pesarin F., Salmaso L. (2003). Test combinati per il controllo della molteplicità mediante procedure di closed testing (Combined tests for controlling multiplicity by closed testing procedures). *Italian Journal of Applied Statistics*, 15(2), 301–329.

Folks J. L. (1984). 6 Combination of independent tests. *Handbook of Statistics*, 4, 113–121.

Gini C. (1912). "Italian: Variabilità e mutabilità" (Variability and mutability, C. Cuppini, Bologna, 156 pp. Reprinted in *Memorie di metodologica statistica* (Ed. Pizetti E, Salvemini, T). Rome: Libreria Eredi Virgilio Veschi (1955).

Holm S. (1979). A simple sequentially rejective multiple test procedure. *Scandinavian Journal of Statistics*, 6, 65–70.

Marcus R., Peritz E., Gabriel K. R. (1976). On closed testing procedures with special reference to ordered analysis of variance. *Biometrika*, 63, 655–660.

Montgomery D. C. (2012). *Design and Analysis of Experiments*, 8th ed. Hoboken, NJ: John Wiley & Sons.

Pesarin F., Salmaso L. (2010a). *Permutation Tests for Complex Data: Theory, Applications and Software*. Wiley Series in Probability and Statistics. Chichester, UK: John Wiley & Sons, pp. 159–163.

Pesarin F., Salmaso L. (2010b). Finite-sample consistency of combination-based permutation tests with application to repeated measures designs. *Journal of Nonparametric Statistics*, 22(5), 669–684.

Salmaso L., Solari A. (2005). Multiple aspect testing for case-control designs. *Metrika*, 12, 1–10.

Shaffer J. P. (1986). Modified sequentially rejective multiple test procedure. *Journal of the American Statistical Association*, 81, 826–831.

Shannon C. E. (1948). A mathematical theory of communication. *The Bell System Technical Journal*, 27, 379–423 and 623–656.

Vale C., Maurelli V. (1983). Simulating multivariate nonnormal distributions. *Psychometrika*, 48(3), 465–471.

Westfall P. H., Young S. S. (1993). *Resampling-Based Multiple Testing*. New York: John Wiley & Sons.

Westfall P. H., Tobias R. D., Wolfinger R. D. (2011). *Multiple Comparisons and Multiple Tests Using SAS*, 2nd ed. Cary, NC: SAS Institute Inc.

Zopluoglu C. (2011). *Applications in R: Generating Multivariate Non-normal Variables*. University of Minnesota.

Section II

Software Tools, Applications and Case Studies

4

Software Nonparametric Combination Global Ranking

The aim of this chapter is to give an overview to the software specifically projected for including all global ranking procedures described in the book. An important feature of the software is the possibility to be customized according to specific requests of the user. For any info or consulting on the software please contact: info.globalranking@gest.unipd.it.

4.1 Integrating R and PHP in a Web App Framework

Since Web applications have become more and more popular among an increasing number of users, one can predictably expect many more services to become accessible via your own browser. This trend is largely proved by the substantial migration from desktop applications to Web inside-my-browser applications, and pushed even more forward by the adoption of 'smart' mobile devices (smartphones or tablets).

As a result, statistical computation is one of the branches of services that could expect to increase the user base by making it easier to do the computation itself (Valero-Mora and Ledesma, 2012). R is doubtless widely used because of its power but its software environment is confined to a single-user paradigm, so far away from the typical client–server architecture of a Web application.

PHP is one of the most famous server-side scripting languages and one of the most used worldwide for the development of Web applications, so why not try to design a PHP Web app that uses R for statistical computation? As there are no language bindings for R and PHP, we have the issue of how to make them communicate.

The first idea is to run the R script by a system call from PHP, by using the function `exec`. The action is quite simple: when the Web app has collected the script's input data (dataset and parameters) from the user, it is ready to run the computation through the script; thus PHP make a system call to launch the R environment by redirecting the output to a file. Once the script has finished the computation, the Web app would read the file containing the script output and show it to the user. It's important to note that this 'naïve' behaviour allows the Web app to access a level so low in the server (a system call run as a normal user) that it opens a variety of security and workload management issues.

A more elegant way to handle this R-PHP communication is rApache (Horner, 2013), 'a project supporting web application development using the R statistical language and environment and the Apache web server'.

rApache runs on UNIX/Linux and Mac OS X operating systems and the only requirements are R and Apache (The Apache Software Foundation, 2014); the standard installation procedure follows the typical GNU/Linux source scheme (`configure`, `make` and `make install` steps); and it's quite simple.

Once installed, rApache provides an interpreter to the Web server Apache, by adding a `mod _ R` module, and transforms R de facto into a server-side scripting environment. It needs a few lines of configuration in Apache server configuration file(s):

```
# Load the R interpreter Apache module
LoadModule R_module/apache/module/path/mod_R.so
```

The previous line loads the Apache module that will handle every rApache directive. Now we need to locate the R script we are going to use for our Web application and enumerate any other files needed by the script that are normally called by the `source()` command.

```
<Location/my_rApache_project>
# Define the rApache handler
SetHandler r-handler
# Configure system with startup files
# equivalent to source () command in R
RSourceOnStartup "/var/www/lib/R/startup1.r"
RSourceOnStartup "/var/www/lib/R/startup2.r"
# Specify what file has to handle incoming web requests
RFileHandler/var/www/lib/R/myscript.r
</Location>
```

In the preceding example, we placed all the R files needed by the application in an Apache `Location` (http://httpd.apache.org/docs/2.2/mod/core.html#location) named `my _ rApache _ project`, assuming that the script needs the files `startup1.r and startup2.r`. We started to load them on the script startup and then we specified the full path of our script.

Now we can simply recall the script `myscript.r` using a Web request similar to `http://mysite/my _ rApache _ project` and rApache will handle the request by loading the `RSourceOnStartup` scripts before launching `myscript.r`.

The first line inside the Location block uses the Apache's SetHandler directive to force the URL to be parsed through the rApache handler. In other words, Apache lets the R environment parse and execute the script(s), in the same way it lets the PHP interpreter parse the php files.

As a proper Web request we can also pass a number of parameters by HTTP `GET` or `POST` methods; these variables will be injected into the environment of the R handler and they will be available in our R code via lexical scoping rules by the `GET` and `POST` list variables.

As for security, unlike the first idea there is no system call, and therefore all aspects are handled totally by the normal Apache configuration and directives, which allow us to consider the entire environment strictly confined into the Apache process.

The last thing to set is the connection of the page production in PHP and the output from the R script; in fact we would like to show to the user the result of R computation inside the page layout (HTML) generated by PHP. We can do it by calling the rApache URL in an *AJAX* (Garrett, 2005) call, hence by making an asynchronous request to the module, getting the result and showing it in the page content using JavaScript. The overall architecture is schematically described in Figure 4.1, where we can see the two different calls (*page request* and *AJAX request*) to the Web server. The first produces the page layout, and of course the logic within, and the second let rApache handles the computation of the R script.

4.2 User Registration and Input/Output Management

All the Nonparametric Combination (NPC) Web applications need to register yourself and hence acquire an individual username and password. This is necessary to control the use of the applications and to be able to manage the server's workload because every computation is potentially heavy enough to contribute to a general slowdown of the server resources.

Every interested person can simply click on the **Register** link on the top menu; a dialog then appears to show a brief explanation of the applications. At the bottom of the dialog, you can insert your email address and click **Confirm email** to confirm the intention to register to the NPC Web Apps, as shown in Figure 4.2.

If you type a valid email address and this address is not already registered in the application, you will soon receive an email from **NPC Web Apps mailer** with the instructions to register (see Figure 4.3); otherwise an error message will show up. Do not reply to the email, but simply follow the link within the email body (click on it or copy and paste in your browser) and you'll be asked to enter a password.

If you click on the link or paste the correct URL on the address bar in your browser you will be prompted with a green success message, as shown on Figure 4.4.

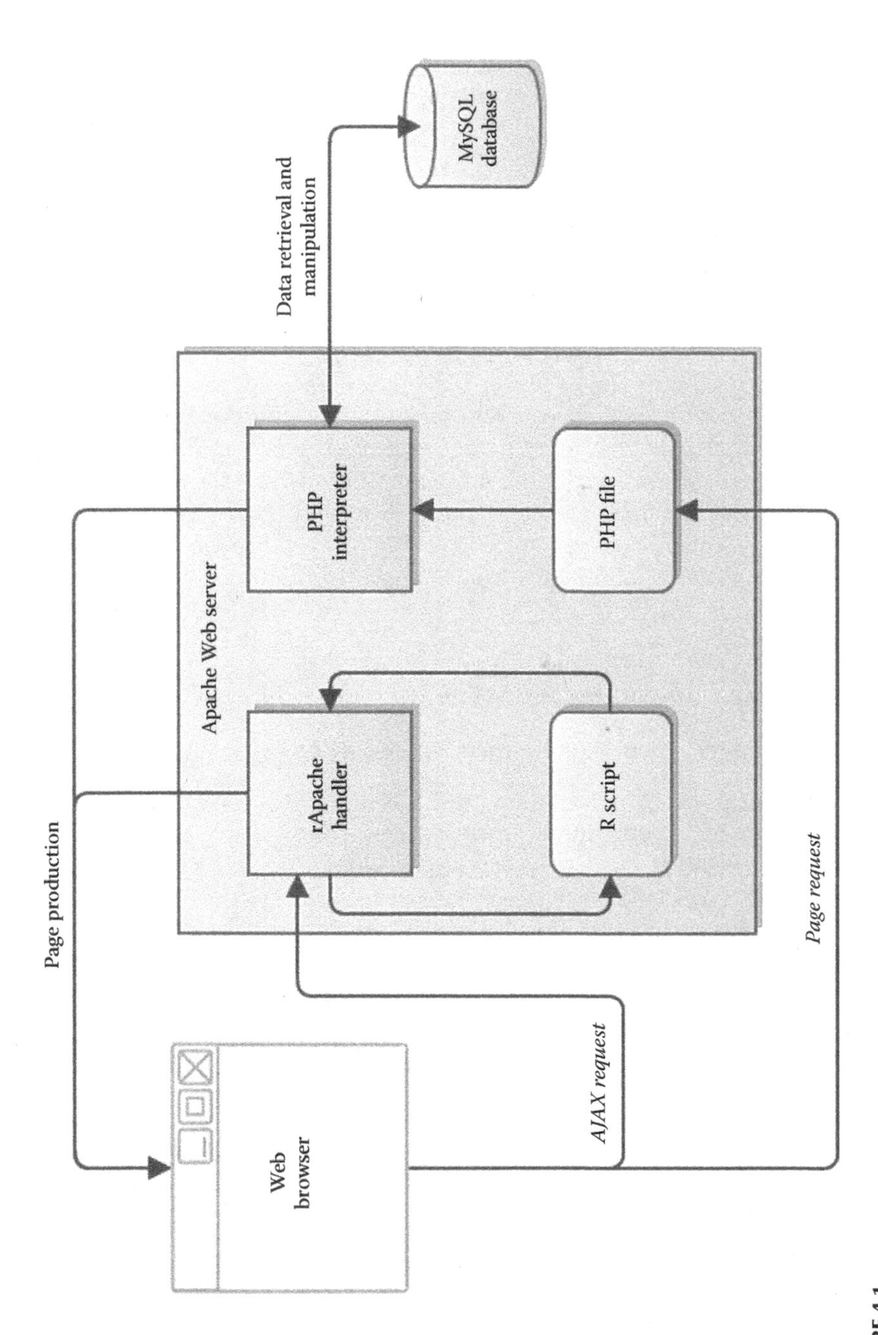

FIGURE 4.1
Overall R-PHP architecture.

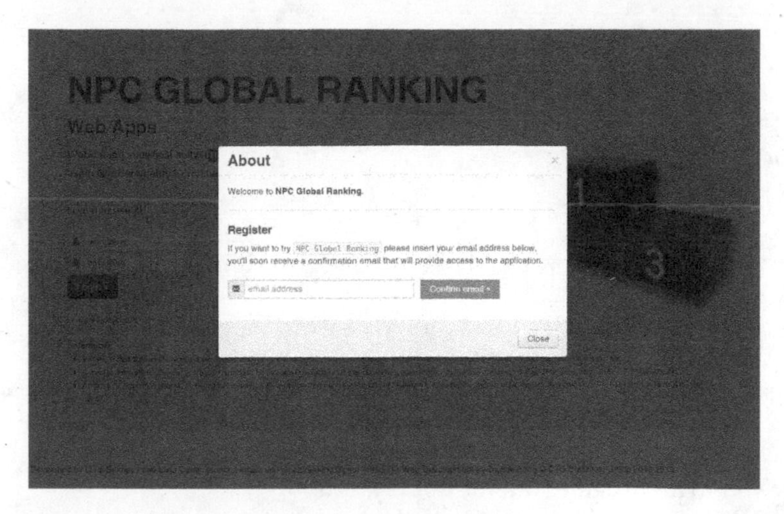

FIGURE 4.2
User registration.

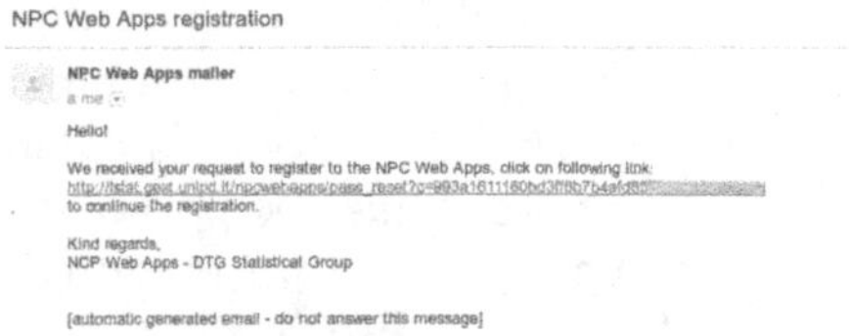

NPC Web Apps registration

NPC Web Apps mailer

Hello!

We received your request to register to the NPC Web Apps, click on following link:

to continue the registration.

Kind regards,
NCP Web Apps - DTG Statistical Group

[automatic generated email - do not answer this message]

FIGURE 4.3
Example of email to request a password setting.

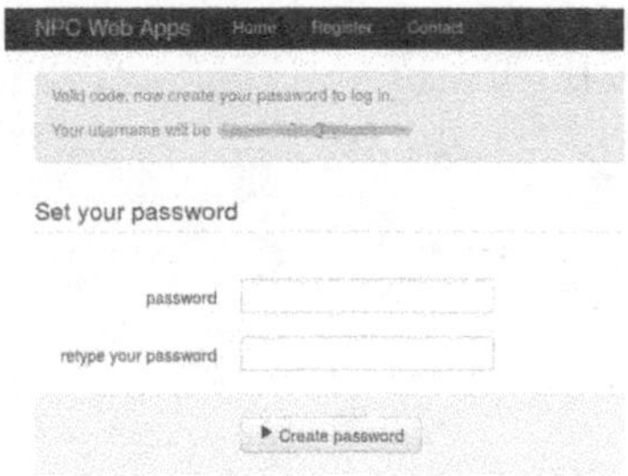

NPC Web Apps Home Register Contact

Valid code, now create your password to log in.
Your username will be

Set your password

password

retype your password

Create password

FIGURE 4.4
Setting your password.

Now you can set your password and type it twice to check it is correct. For security reasons please be sure to enter a password at least eight characters long or a message will warn you to retype a new password.

When you click on **Create password** and everything is fine, you should receive a message that informs you that your credentials have been created and you can go back to the home page to log in, as shown on Figure 4.5.

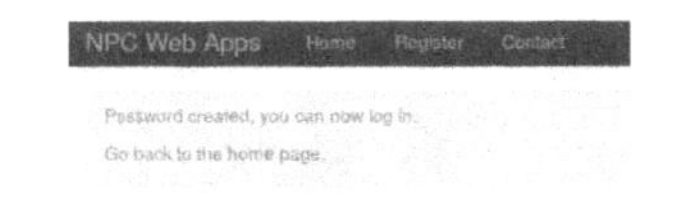

FIGURE 4.5
Password created.

Otherwise, you will be warned about a password mismatch or a password too short to continue.

From now on the email address and password pair will let you login to the NPC Web applications.

In case you forget your password, you can click the **Lost your password?** link to start a password reset procedure: an email will be sent to your registered email address with the instructions for changing your password and allow you to login again into the application.

The NPC Web applications share the same function scheme: basically you have the chance to load your own dataset, view the graphical version of the data within, choose the input parameters to set the preferred computation and launch the computation.

By design policy you are not allowed to store different datasets; hence for every computation session you can upload only one dataset. Once the computation has finished and your output has been printed you can simply start a new session, choose a new dataset and start over with the new computation.

To prevent an overload of the server, for each compute session you'll be notified about the server's workload and availability. Every session obviously takes a variable amount of load on the server computation capability; hence to prevent a general slowdown every user is warned if the server is temporarily unavailable. If the server workload is above a certain bound, a red warning will notify the temporary unavailability to launch a new computation session and will ask the user to try again in a certain number of minutes.

4.3 NPC Global Ranking

To provide an easy-to-use interface implementing the permutation-based multivariate ranking methodology proposed in this book, we developed a companion book Web-based software named NPC Global Ranking. Within a set of Web-based apps, the access to NPC Global Ranking software is available at http://lstat.gest.unipd.it/npcwebapps/ (Figure 4.6).

Once you've logged in using your username and password (see the previous section), you would move inside the NPC Global Ranking's home page (Figure 4.7).

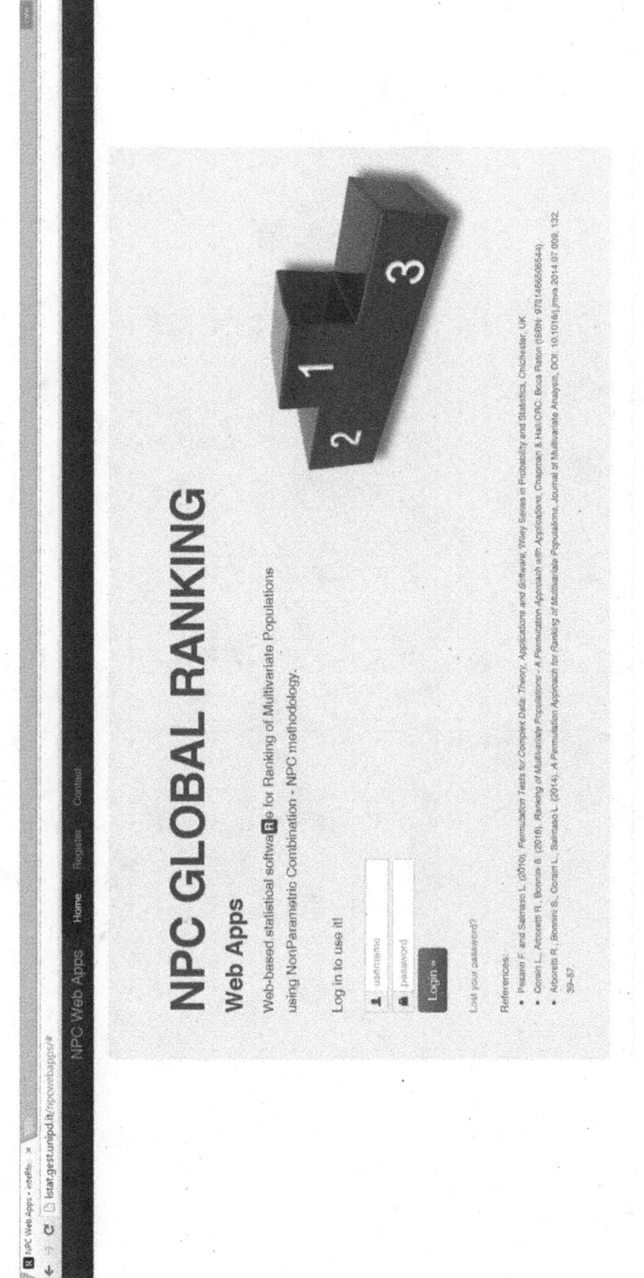

FIGURE 4.6
NPC Global Ranking access.

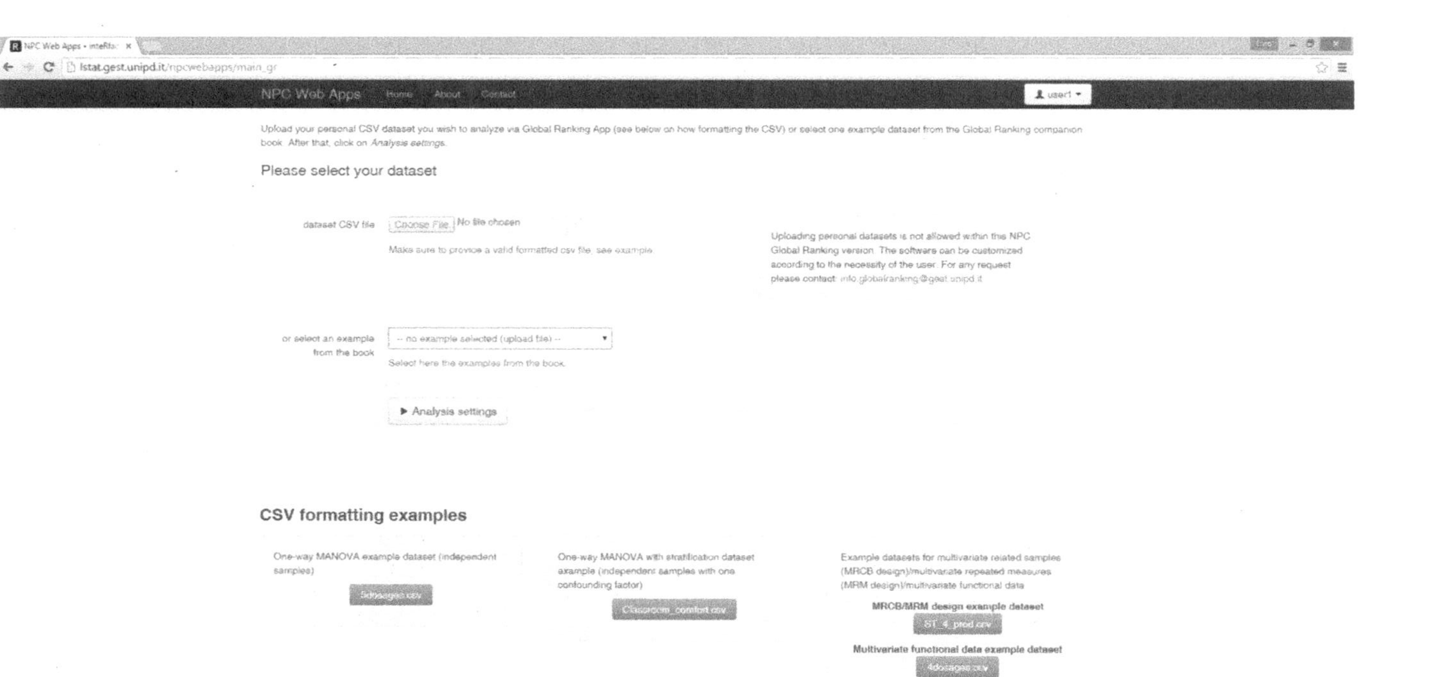

FIGURE 4.7
NPC Global Ranking's home page.

First you need to load the dataset you wish to analyze via NPC Global Ranking App. For this purpose you can choose between two alternatives: upload your personal CSV dataset or select one example dataset from the Global Ranking companion book. Uploading personal datasets is not allowed within this NPC Global Ranking version. The software can be customized according to the necessity of the user. For any request please contact: info .globalranking@gest.unipd.it. Please note that selecting an example from the drop-down list will forbid you to upload your personal dataset; moreover if you have already selected the dataset file and then you choose from the examples drop-down list, the dataset will not be considered and the analysis will take place on the example.

After the dataset selection, you must click on the **Analysis settings** button.

4.3.1 Dataset Formatting

For suitable formatting of your CSV dataset, a set of simple rules must be followed.

- The first column of the dataset must always be used to define the multivariate populations to be ranked; the number of populations must be at least equal to three, but due to possible computational problems consider that no more that six or seven populations are advised.
- In case the design of interest involves a stratification or a blocking factor (for details see Chapter 3, Sections 3.1.1 and 3.1.2), the second column of the dataset must always be used to define the stratification or blocking factor.
- Both population and stratification/blocking labels can be defined by numeric or alphanumeric symbols but in the latter case the labels will be converted in numeric values according to their order of appearance.
- All the remaining columns, from the second to the last (in case of the simple one-way MANOVA layout) or from the third to the last (for all the other more complex designs), must always be used to define the response variables you want to use to rank the populations; actually there is no limit to the number of response variables. As a reminder, within the NPC testing approach the number of observed variables may be larger than the number of observations (Corain and Salmaso, 2013, 2015).

Finally, when creating the dataset the user must keep in mind an additional list of suggestions and hints.

- As usual, the first row of the dataset must contain the labels with the name for each variable.
- When formatting the CSV file, as the delimiter symbol and decimal mark you must use the semicolon, ';' and the dot, i.e. '.'; as a reminder, simply by using the option **Save as** available in Microsoft Excel® you can create a suitable CSV file according to the requests just mentioned.
- In the dataset no one missing value is allowed at all so that each cell of the dataset must contain a valid entry.
- All the response variables must be of the same type, that is, all continuous/binary or all ordered categorical; moreover, no nominal categorical entries are allowed for the response variables.
- For any given treatment, that is, combination of population group and stratification/blocking factor, the number of replicates must be equal to at least one.

4.3.2 Analysis Setting

After loading the dataset you wish to analyze and clicking on the **Analysis** settings button, the NPC Global Ranking's analysis setting interface will appear as shown in Figure 4.8. Note that a data preview sheet is shown at the top to allow the user to check whether the CSV file has been properly uploaded and read. You can also drag the bottom line of this table to increase or decrease its height.

Obviously, the specific setting choices depend either on the particular features of the multivariate ranking problem at hand (see Chapter 3, Section 3.1) or on the user preferences. The first five setting items (from Type of design to Combining function) must be set up according to the actual ranking problem at hand, and the last four setting items are related mainly to the user preferences on computational issues and output visualization.

To provide a basic descriptive sketch of the data, at the bottom of the page you can select the way you would like to represent the dataset graphically. The selectable function-to-plot options are sensitive to the type of response variables you previously chose: the sample mean is the only available choice in case of continuous or binary response variables while in the case of ordered categorical responses the additional top-box percentage option is allowed.

Finally, to run the multivariate ranking analysis you must simply click on the **Launch R analysis** button located at the lower bottom of the page.

If a domain analysis is required (see Chapter 3, Section 3.1.3), you need to select the number of desired domains and then assign each response variable to a given domain. To make easier the procedure, an automatic domain assistant will guide you throughout this process (Figure 4.9). As you select 'Yes' on the Domain analysis setting, you will be asked to set the number of

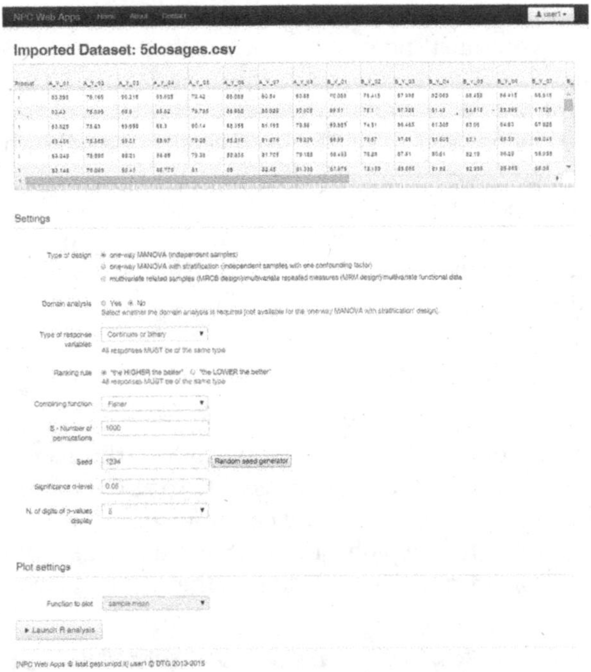

FIGURE 4.8
NPC Global Ranking's analysis setting interface.

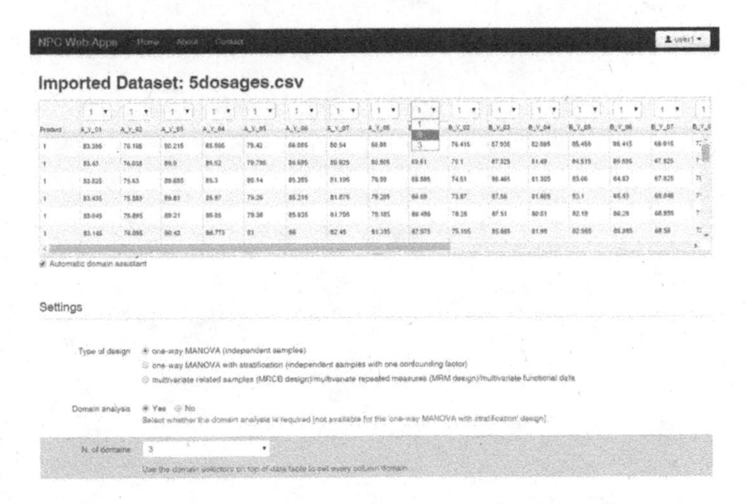

FIGURE 4.9
Domain analysis setting.

domains in a drop-down list; this number will automatically set the maxim domain number you must indicate for each column in the selectors on the top the dataset table. The **Automatic domain assistant**, enabled by default on the bottom of the dataset table, will guide you to easily set the domain for each column: when you select the domain for one column, every other column on the right side will be set with the same number, to increasingly set the next columns domains.

Note that the domain analysis option is available for two out of three types of designs, that is, this choice is not allowed for the 'one-way MANOVA with stratification' design.

4.3.3 Global Ranking Results

Finally, to run the multivariate ranking analysis you have to click on the **Launch R analysis** button located at the bottom of the page. When this is done, the user is transferred to a new page where at the top there is a work-in-progress bar (Figure 4.10). Below you can find some plots that can be hidden by clicking on the toggle plots button. Plots can be also downloaded (as images or PDF files) or printed by clicking on the chart context menu located

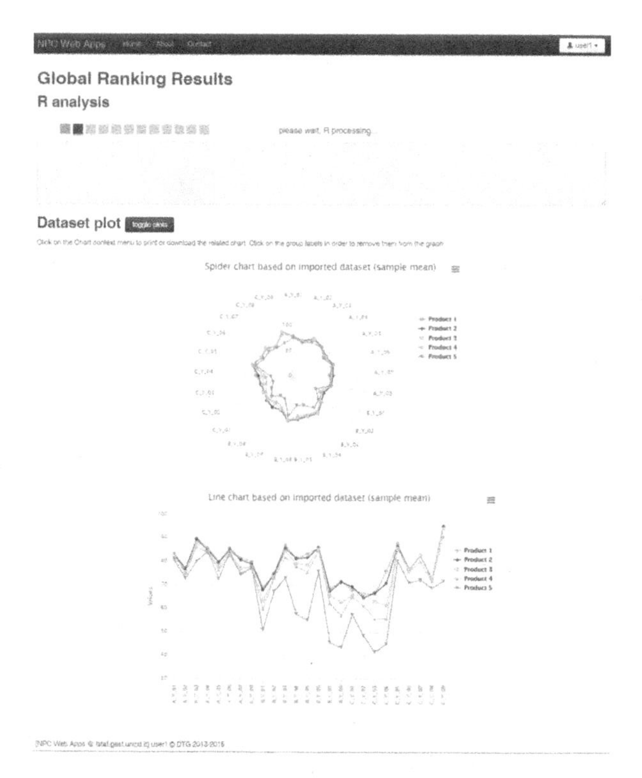

FIGURE 4.10
Global ranking results: Main interface and dataset plots.

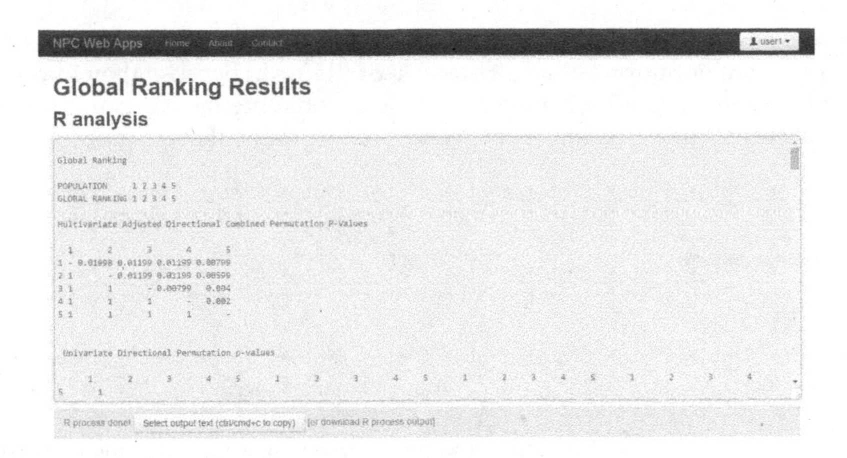

FIGURE 4.11
Global ranking results: Output performed by R.

at the upper-right corner of the graph. Moreover, when clicking on the group labels some series can be switched off or on from the graph's plotting area.

Once the computational calculations performed in background by R are completed, the output of the multivariate ranking analysis will appear within the R analysis frame (Figure 4.11).

To export the R analysis results, simply click on **Select output text** and then paste the selection in a text editor or click on **Download R process output**. In the latter case, a new Web page will be created providing the output you can save in a text file.

4.3.4 Warning and Error Messages

The NPC Global Ranking Web app operates with a limited number of warning and error messages. Basically, the messages are concerned with the issues described in Section 4.3.1, as reported in the following list.

- 'The column(s) defining the type of design has/have been forced to be numeric'.
- 'The Global Ranking analysis cannot be performed because there is not at least one observation for each treatment. Please check and modify your dataset and try again'.
- 'The Global Ranking analysis cannot be performed because the response variables are not numeric/binary/ordered categorical. Please check and modify your dataset and try again'.
- 'The Global Ranking analysis cannot be performed because there are some missing values in the dataset. Please check and modify your dataset and try again'.

As a common suggestion, while you are using the NPC Web apps do not use the browser buttons 'Back', 'Forward' or 'Refresh', because they interfere with the proper behaviour of the applications and therefore can cause unpredictable errors that will cause you to start the process all over again.

References

Corain L., Salmaso L. (2013). Nonparametric permutation and combination-based multivariate control charts with applications in microelectronics. *Applied Stochastic Models in Business and Industry*, 29(4), 334–349.

Corain L., Salmaso L. (2015). Improving power of multivariate combination-based permutation tests. *Statistics and Computing*, 25(2), 203–214.

Garrett J. J. (2005). AJAX: A new approach to Web applications. http://www.adaptivepath.com/ideas/ajax-new-approach-web-applications/.

Horner J. (2013). rApache: Web application development with R and Apache. http://www.rapache.net/.

The Apache Software Foundation, http://httpd.apache.org/.

Valero-Mora P. M., Ledesma R. D. (2012). Graphical user interfaces for R. *Journal of Statistical Software*, 49(1), 1–8.

5

Applications to New Product Development in Industry

In this chapter we present several real case studies aimed at demonstrating the usefulness of the proposed ranking method in the field of new product development (NPD) and industrial research. It is worth noting that the first two applications concern a typical problem in the context of benchmarking new products where a dose–response industrial experiment is carried out and the 'true' underlying ranking can be *a priori* assumed as 'almost' known. From our point of view, this is very important because the multivariate ranking results should be viewed as 'toy-examples', meaning that the actual goal was to illustrate with real data the theoretical validity and the practical relevance of the multivariate ranking method we propose in this book.

In the field of NPD research is often aimed at evaluating the product/service performances from a multivariate aspect, that is, in connection with more than one aspect/dimension and/or under several conditions. In this framework the main goal of statistical data analysis is to effectively compare and rank the products/services from the 'best' to the 'worst', usually by means of the calculation of a composite indicator that represents the global performance evaluation and that generates a synthesis of the information given by whole performance data. This chapter illustrates several in-depth applications on the topics of ranking multivariate populations specifically devoted to support the development process of new products in industry.

The application of the ranking methodology for multivariate populations in the field of NPD in industry is particularly advisable because the emphasis of the process is to take decisions under uncertainty regarding treatments with multivariate responses, usually in the case of very low sample size and where decisions under the alternative hypothesis go in the direction of identifying the best product/treatment within a reference set.

In general, the use of suitable statistical methods for the analysis of experimental data is a key factor in obtaining an effective and fast NPD process. The process may potentially slow down because of changes in the middle of design project when engineers lack information on the product features and technology they are using. To avoid future change the designers and engineers may plan the future task, though this takes time without completely avoiding possible changes during the process. Stockstrom and Herstatt (2008) show this kind of planning have a great deal of influence in external and especially internal success. Calantone and Di Benedetto (2000) and Davis et al. (2002) have demonstrated a positive correlation between NPD success and speed, that is, fast NPD processes. The firms that rapidly develop new products enjoy substantial competitive advantages while at the same time maintaining quality. A faster development process offers many advantages, such as rapid market introduction, delayed start of the design, frequent introduction in the market, and more opportunities for innovation.

5.1 One-Way MANOVA Dose–Response Experiment

With the aim of validating the multivariate ranking procedure on real data we propose in this book, we consider a typical problem in the context of benchmarking new products. The main objective here is to quantitatively

TABLE 5.1
Response Classification

Response Type	Response Label
A	A_Y_01
A	A_Y_02
A	A_Y_03
A	A_Y_04
A	A_Y_05
A	A_Y_06
A	A_Y_07
A	A_Y_08
B	B_Y_01
B	B_Y_02
B	B_Y_03
B	B_Y_04
B	B_Y_05
B	B_Y_06
B	B_Y_07
B	B_Y_08
C	C_Y_01
C	C_Y_02
C	C_Y_03
C	C_Y_04
C	C_Y_05
C	C_Y_06
C	C_Y_07
C	C_Y_08
C	C_Y_09

evaluate the possible relative benefits of five products and rank them from a multivariate aspect. Here the populations of interest are five dosages, i.e., concentrations: P1 = 100%, P2 = 80%, P3 = 60%, P4 = 40% and finally, as control treatment, we considered also P5 = 0%. Of course, as usual a higher dosage is intended to improve product performance. As multivariate response variables we considered the performance of 25 outputs. As dosage reduction actually means decreasing the effectiveness of the product, it is worth noting that in this situation, a priori we may deduce an important indication of the true ranking, that is, P1 ≥ P2 ≥ P3 ≥ P4 ≥ P5. The experiment was randomly and independently replicated six times for each treatment.

Each output can be classified by its main properties (labelled A, B, C), as reported in Table 5.1; hence a domain analysis can be useful to evaluate the separate effect on the same type of response.

A first look at the experimental results is provided in the radar graph in Figure 5.1, where response-by-response sample means by product are reported.

The setting of the ranking problem (see Chapter 3) can be summarized as follows.

- Type of design: one-way MANOVA (five independent samples, i.e., five multivariate populations to be ranked)
- Domain analysis: yes (three domains, see Table 5.1)
- Type of response variables: continuous (25 continuous responses)
- Ranking rule: 'the higher the better'

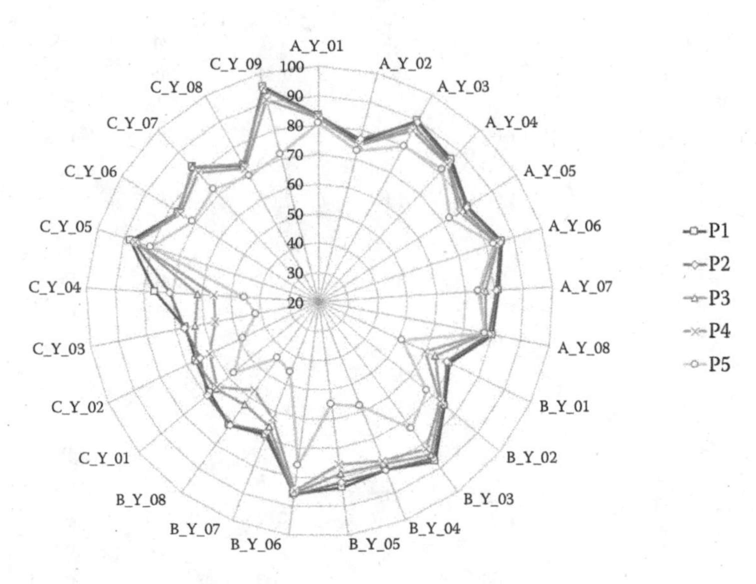

FIGURE 5.1
Radar graph of five dosages (response-by-response sample means by product).

- Combining function: Fisher
- B (number of permutations): 2000
- Significance α-level: 0.05

When considering the multivariate ranking analysis (significance α-level = 0.05), performed using 2000 permutations and the difference of sample means as the test statistic and finally the Fisher combining function, we obtain the results shown in Table 5.2, where directional multivariate permutation p-values have been adjusted by using the Bonferroni–Holm–Shaffer

TABLE 5.2

Global Ranking and Domain Analysis Results for the Dose–Response Experiment on Five Dosages

Global ranking	P1	P2	P3	P4	P5
All responses	1	2	3	4	5
Type A	1	2	3	4	5
Type B	1	2	3	4	5
Type C	1	2	3	4	5
All responses	**P1**	**P2**	**P3**	**P4**	**P5**
P1		**.010**	**.006**	**.006**	**.004**
P2	1.00		**.006**	**.006**	**.003**
P3	1.00	1.00		**.004**	**.002**
P4	1.00	1.00	1.00		**.001**
P5	1.00	1.00	1.00	1.00	
Type A	**P1**	**P2**	**P3**	**P4**	**P5**
P1		**.010**	**.006**	**.006**	**.004**
P2	1.00		**.006**	**.006**	**.004**
P3	1.00	1.00		**.003**	**.003**
P4	1.00	1.00	1.00		**.002**
P5	1.00	1.00	1.00	1.00	
Type B	**P1**	**P2**	**P3**	**P4**	**P5**
P1		**.010**	**.010**	**.006**	**.004**
P2	1.00		**.006**	**.006**	**.004**
P3	1.00	1.00		**.006**	**.003**
P4	1.00	1.00	1.00		**.002**
P5	1.00	1.00	1.00	1.00	
Type C	**P1**	**P2**	**P3**	**P4**	**P5**
P1		**.023**	**.010**	**.006**	**.006**
P2	1.00		**.012**	**.006**	**.004**
P3	1.00	1.00		**.006**	**.004**
P4	1.00	1.00	1.00		**.003**
P5	1.00	1.00	1.00	1.00	

method (Shaffer, 1986). As a reminder, the proper significance level to take into account for the ranking analysis is actually $\alpha/2$, so in this case it is 0.025.

All ranking analysis results can be obtained by using the book Web-based software (see Chapter 4) within the NPC Web Apps (lstat.gest.unipd.it /npcwebapps/). The reference dataset is labelled 5dosages.csv. To replicate the permutation p-values reported in Tables 5.2 and 5.3 exactly, a seed value equal to 1234 must be set up.

Note that the domain analysis always provides the same ranking we obtain considering all 25 responses. Details on the univariate pairwise comparisons are displayed in Table 5.3 (directional permutation p-values are not adjusted by multiplicity), where p-values significant at 5% are highlighted in bold.

As a final conclusion, we may say that as the dosage increases the global performance also increases in a way that the multivariate ranking procedure is properly able to rank the dosages following their 'natural' ordering, which in this case was known. Remember that the experiment included only six replicates and a relatively very high number of response variables (25 different responses).

This ranking study should be viewed as a 'toy-example', meaning that the actual goal was to illustrate the multivariate ranking method we propose in this book within a typical problem in the context of benchmarking new products. In fact, only in case of knowing the true underlying multivariate ranking were we able to understand if the application of the ranking method would provide the 'true' result.

5.2 Multivariate Repeated Measures Experiments for NPD

When developing new products, the performances are occasionally observed as either longitudinal data vectors from repeated measures on the same units/subjects (Diggle et al., 2002) or observations in the form of multivariate curves or trajectories, called functional data, lying to infinite dimensional spaces commonly called functional spaces (Ferraty and Vieu, 2006). In this framework data are typically observed as a sequence of time occasions, where time means any sequentially ordered entity including also space, lexicographic ordering, and so forth.

The need to take into account the presence of the time effect suggests considering as a reference statistical multivariate model a repeated measure design in which the effect of the population under investigation (products) can be modelled by a discrete or discretized stochastic process for which at most a countable set of data points is observed. In this way an observed curve is nothing more than a set of *repeated measures* in which each subject/unit (a given curve) is observed on a finite or at most a countable number of occasions so that successive responses are dependent (Pesarin and Salmaso, 2010). Within this framework we apply the ranking approach for multivariate population where multivariate directional

TABLE 5.3

Details of Univariate Directional Permutation p-Values for the Dose–Response Experiment on Five Dosages

	P1	P2	P3	P4	P5	P1	P2	P3	P4	P5	P1	P2	P3	P4	P5
			A_Y_01					**A_Y_02**					**A_Y_03**		
P1		**.004**	**.001**	**.001**	**.001**		.569	**.001**	**.002**	**.001**		.034	**.001**	**.001**	**.001**
P2	.998		**.003**	**.001**	**.001**	.432		**.001**	**.001**	**.001**	.967		**.001**	**.001**	**.001**
P3	1.00	.999		.077	**.001**	1.00	1.00		.194	.001	**1.00**	1.00		**.001**	**.001**
P4	1.00	1.00	.926		**.001**	1.00	1.00	.809		.001	**1.00**	1.00	1.00		**.001**
P5	1.00	1.00	1.00	1.00		1.00	1.00	1.00	1.00		1.00	1.00	1.00	1.00	
			A_Y_04					**A_Y_05**					**A_Y_06**		
P1		**.007**	**.004**	**.001**	**.001**		.309	**.001**	**.001**	**.001**		**.004**	**.001**	**.001**	**.001**
P2	.995		.104	**.001**	**.001**	.693		**.003**	**.001**	**.001**	.997		**.001**	**.001**	**.001**
P3	.997	.897		**.016**	**.001**	1.00	.998		**.002**	**.001**	1.00	1.00		.184	**.001**
P4	1.00	1.00	.985		**.002**	1.00	1.00	1.000		**.001**	1.00	1.00	.820		**.005**
P5	1.00	1.00	1.00	.999		1.00	1.00	1.00	1.00		1.00	1.00	1.00	.997	
			A_Y_07					**A_Y_08**					**B_Y_01**		
P1		.044	**.001**	**.001**	**.001**		.033	**.001**	**.001**	**.001**		.125	**.001**	**.001**	**.001**
P2	.957		**.001**	**.001**	**.001**	.968		**.004**	**.001**	**.001**	.876		**.001**	**.001**	**.001**
P3	1.00	1.00		**.008**	**.001**	1.00	.998		.057	.103	1.00	1.00		**.003**	**.001**
P4	1.00	1.00	.994		.070	1.00	1.00	.945		.800	1.00	1.00	.999		**.001**
P5	1.00	1.00	1.00	.931		1.00	1.00	.900	.202		1.00	1.00	1.00	1.00	

(*Continued*)

TABLE 5.3 (CONTINUED)

Details of Univariate Directional Permutation p-Values for the Dose–Response Experiment on Five Dosages

	P1	P2	P3	P4	P5	P1	P2	P3	P4	P5	P1	P2	P3	P4	P5
			B_Y_02					**B_Y_03**					**B_Y_04**		
P1		.252	**.005**	**.001**	**.001**		.009	**.001**	**.001**	**.001**		.407	**.009**	**.001**	**.001**
P2	.749		.071	.026	**.001**	.992		**.009**	**.001**	**.001**	.600		**.019**	**.001**	**.001**
P3	.996	.930		.095	**.001**	1.00	.993		**.003**	**.001**	.993	.982		.110	**.001**
P4	1.00	.975	.906		**.001**	1.00	1.00	1.00		.001	1.00	1.00	.893		**.001**
P5	1.00	1.00	1.00	1.00		1.00	1.00	1.00	1.00		1.00	1.00	1.00	1.00	
			B_Y_05					**B_Y_06**					**B_Y_07**		
P1		**.016**	**.001**	**.001**	**.001**		.908	.212	**.007**	**.001**		.026	**.001**	**.001**	**.001**
P2	.985		**.005**	**.001**	**.001**	.098		.050	**.001**	**.001**	.978		.033	**.001**	**.001**
P3	1.00	.997		**.005**	**.001**	.789	.951		.070	**.001**	1.00	.968		**.003**	**.001**
P4	1.00	1.00	.996		**.001**	.99	1.00	.931		**.001**	1.00	1.00	.999		**.001**
P5	1.00	1.00	1.00	1.00		1.00	1.00	1.00	1.00		1.00	1.00	1.00	1.00	
			B_Y_08					**C_Y_01**					**C_Y_02**		
P1		.330	**.001**	**.001**	**.001**		.935	**.011**	**.001**	**.001**		.122	.522	**.001**	**.001**
P2	.673		**.001**	**.001**	**.001**	.066		**.001**	**.001**	**.001**	.879		.877	**.003**	**.001**
P3	1.00	1.00		**.001**	**.001**	.990	1.00		.194	**.001**	.479	.124		**.001**	**.001**
P4	1.00	1.00	1.00		**.001**	1.00	1.00	.808		**.001**	1.00	1.00	1.00		**.001**
P5	1.00	1.00	1.00	1.00		1.00	1.00	1.00	1.00		1.00	1.00	1.00	1.00	

(Continued)

TABLE 5.3 (CONTINUED)

Details of Univariate Directional Permutation p-Values for the Dose–Response Experiment on Five Dosages

	P1	P2	P3	P4	P5	P1	P2	P3	P4	P5	P1	P2	P3	P4	P5
			C_Y_03					C_Y_04					C_Y_05		
P1		.644	**.010**	**.001**	**.001**		**.001**	**.001**	**.001**	**.001**		.041	**.009**	**.004**	**.001**
P2	.357		**.008**	**.001**	**.001**	1.00		**.001**	**.001**	**.001**	.962		.148	.122	**.001**
P3	.991	.993		**.001**	**.001**	1.00	1.00		**.001**	**.001**	.993	.854		.612	**.001**
P4	1.00	1.00	1.00		**.001**	1.00	1.00	1.00		**.001**	1.00	.879	.389		**.001**
P5	1.00	1.00	1.00	1.00		1.00	1.00	1.00	1.00		1.00	1.00	1.00	1.00	
			C_Y_06					C_Y_07					C_Y_08		
P1		.803	.678	.044	**.001**		.154	.400	**.003**	**.001**		.210	.775	**.003**	**.001**
P2	.201		.337	**.023**	**.001**	.847		.629	**.003**	**.001**	.792		.909	**.001**	**.001**
P3	.323	.670		.040	**.001**	.603	.373		**.008**	**.001**	.227	.092		**.004**	**.001**
P4	.957	.978	.961		**.001**	1.00	1.00	.995		**.001**	1.00	1.00	1.00		**.001**
P5	1.00	1.00	1.00	1.00		1.00	1.00	1.00	1.00		1.00	1.00	1.00	1.00	
			C_Y_09												
P1		**.013**	**.001**	**.001**	**.001**										
P2	.989		.087	**.001**	**.001**										
P3	1.00	.914		**.001**	**.001**										
P4	1.00	1.00	1.00		**.001**										
P5	1.00	1.00	1.00	1.00											

permutation tests are obtained via nonparametric combination by the so-called time-to-time permutation solution we described in Chapter 2, Section 2.3.2.

5.2.1 Multivariate Repeated Measures Dose–Response Experiment on Four Dosages

Let us consider the following real case study in the field of the industrial development process of a new product. The treatments of interest are four dosages (P1: 100%, P2: 80%, P3: 60%, P4: 50%) intended to increase the product efficacy; note that, because dosage reduction means actually decreasing the effectiveness of the product, we may deduce an important indication of the true ranking: P1 ≥ P2 ≥ P3 ≥ P4. Product performances are assessed by measuring a multivariate response labelled W, and where the univariate responses are labelled W_1, W_2, W_3, W_4, W_5, W_6, W_7. Owing to the technical specifications in which the industrial experiment is performed, the W's were measured on the same experimental units at increasing time measures, namely (NM): 5, 10, 15, 20, 25 minutes. Two independent replicates of each treatment were performed. Figure 5.2 represents all seven observed W series (continuous line), along with their sample mean values (dashed line).

The setting of the ranking problem (see Chapter 3) can be summarized as follows.

- Type of design: multivariate functional data (four samples, i.e., four multivariate functional populations to be ranked)
- Domain analysis: no
- Type of response variables: continuous (seven continuous functional responses)
- Ranking rule: 'the higher the better'
- Combining function: Fisher
- B (number of permutations): 5000
- Significance α-level: 0.05

When applying the ranking method on experimental data, performed using the difference of sample means as a univariate test statistic and finally the Fisher combining function, we obtain the results shown in Table 5.4, where directional multivariate permutation p-values have been adjusted by using the Bonferroni–Holm–Shaffer method (1986).

All ranking analysis results can be obtained by using the book Web-based software (see Chapter 4) within the NPC Web Apps (lstat.gest.unipd.it/npcwebapps/). The reference dataset is labelled 4dosages.csv. To replicate the permutation p-values in Tables 5.4 and 5.5 exactly, a seed value equal to 1234 must be set up.

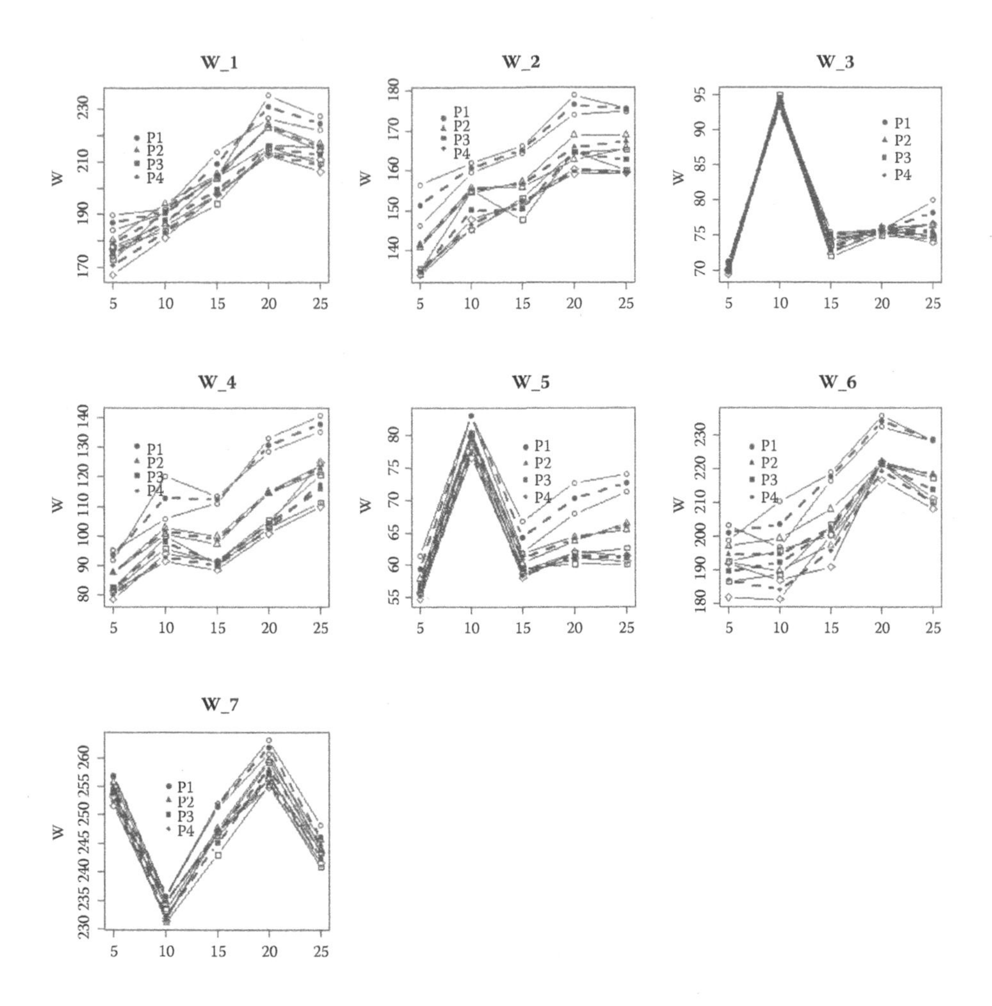

FIGURE 5.2
Observed W data for each univariate response and by the four products (dosages). The numbers of minutes are reported on the x-axis.

TABLE 5.4
Global Ranking Results for the W Ranking Analysis of Four Dosages ($\alpha = 5\%$)

Product	P1	P2	P3	P4
Global ranking	**1**	**2**	**3**	**3**
	P1	P2	P3	P4
P1		**.0012**	**.0006**	**.0006**
P2	1.000		**.0006**	**.0004**
P3	1.000	1.000		.0378
P4	1.000	1.000	1.000	

TABLE 5.5

Details of Univariate Directional p-Values for the W Ranking Analysis of Four Dosages

	P1	P2	P3	P4	P1	P2	P3	P4
		W_1				W_2		
P1		**.0018**	**.0004**	**.0002**		**.0002**	**.0002**	**.0002**
P2	.9988		**.0064**	**.0002**	1.000		**.0024**	**.0002**
P3	.9998	.9940		.0270	1.000	.9978		.0796
P4	1.000	1.000	.9742		1.000	1.000	.9238	
		W_3				W_4		
P1		**.0006**	**.0060**	**.0016**		**.0002**	**.0002**	**.0002**
P2	.9998		.3155	.6019	1.000		**.0002**	**.0002**
P3	.9942	.7081		.7425	1.000	1.000		.1788
P4	.9994	.4167	.2769		1.000	1.000	.8248	
		W_5				W_6		
P1		**.0006**	**.0002**	**.0002**		**.0002**	**.0002**	**.0002**
P2	1.000		**.0006**	**.0002**	1.000		.0754	**.0026**
P3	1.000	1.000		.0606	1.000	.9252		**.0132**
P4	1.000	1.000	.9460		1.000	.9976	.9884	
		W_7						
P1		**.0010**	**.0002**	**.0002**				
P2	.9992		**.0186**	**.0024**				
P3	1.000	.9828		.3687				
P4	1.000	.9982	.6427					

Note that P3 and P4, that is, the 60% and 50% dosages, do not differ at the 5% α-level because the combined directional permutation p-value is greater than 0.025. Accordingly, P3 and P4 are both tied at the third global ranking position.

As far as the comparisons for each univariate response, Table 5.5 displays the set of unadjusted univariate directional pairwise p-values, where significant p-values at 5% are highlighted in bold.

Like the first case study described previously, this ranking study should be viewed as a 'toy-example' as well, meaning that the actual goal was to illustrate the multivariate ranking method we propose in this book within a typical problem in the context of benchmarking new products in the case of a multivariate repeated measures dose–response experiment.

5.2.2 Multivariate Repeated Measures Experiment on Four Products

The treatments of interest for this study are four types of products (P1, P2, P3, P4). Product performances are assessed by measuring a multivariate

14-dimension response labelled CT. In contrast to what happened in the first two case studies, the CT measures must be interpreted in the opposite direction, that is, higher CT values mean a worse product. In practice, in this case the rule 'the lower the better' holds.

Owing to the technical specifications in which the industrial experiment is performed, the CT's (labelled CT_01, CT_02,..., CT_14) were sequentially measured on the same experimental units at different numbers of minutes spent in the experimental device (NM: 5, 10, 15, 20). Three independent replicates of each treatments were performed. Experimental data for each univariate response are displayed in Figure 5.3, where the continuous lines represent the individual CT's series while their sample mean values are displayed by a dashed line.

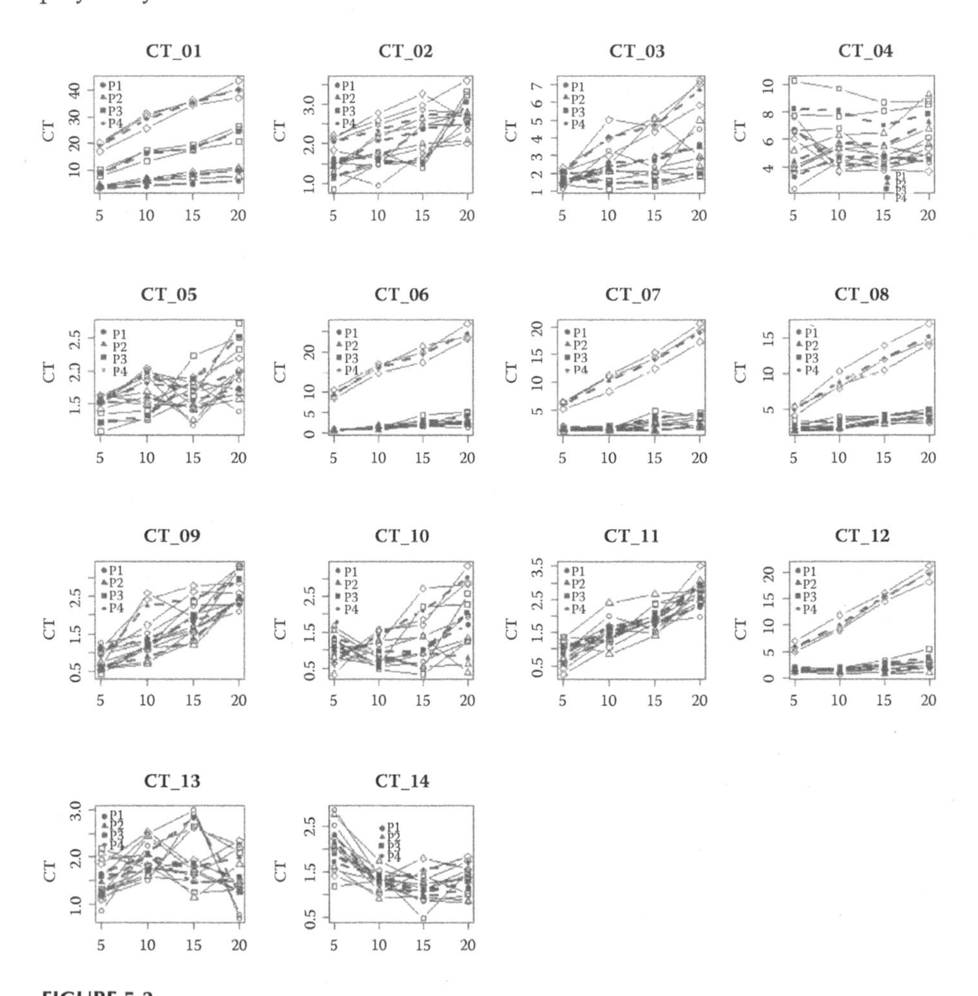

FIGURE 5.3
Observed CT data for each univariate response and by the four products. The numbers of minutes are reported on the x-axis.

TABLE 5.6

Global Ranking Results for the CT's Ranking Analysis of Four Products

Product	P1	P2	P3	P4
Global ranking	**1**	**2**	**3**	**4**
	P1	P2	P3	P4
P1		1.000	1.000	.9970
P2	**.0226**		1.000	1.000
P3	**.0012**	**.0006**		1.000
P4	**.0012**	**.0012**	**.0008**	

The setting of the analysis ranking problem (see Chapter 3) can be summarized as follows.

- Type of design: multivariate functional data (four samples, i.e., four multivariate functional populations to be ranked)
- Domain analysis: no
- Type of response variables: continuous (14 continuous functional responses)
- Ranking rule: 'the lower the better'
- Combining function: Fisher
- *B* (number of permutations): 5000
- Significance α-level: 0.05

When applying the ranking method on experimental data using directional parametric *p*-values performed via 5000 permutations we obtain the results shown in Tables 5.6 and 5.7, where directional multivariate permutation *p*-values have been adjusted by using the Bonferroni–Holm–Shaffer method (1986).

All results can be performed by using the book Web-based software (see Chapter 4) within the NPC Web Apps (lstat.gest.unipd.it/npcwebapps/). The reference dataset is labelled CTs.csv. To replicate the permutation *p*-values in Tables 5.6 and 5.7 exactly, a seed value equal to 1234 must be set up.

Table 5.7 displays the set of the univariate directional pairwise *p*-values, related to the multiple comparison procedures for the problem at hand.

5.2.3 An Unreplicated Multivariate Repeated Measures Experiment on Three Products

The treatments of interest for this study are three products (P1, P2, P3). Performance is assessed by measuring a multivariate 14-dimension response labelled CF, which measures the responses (labelled CF_01, CF_02,..., CF_14) after repeated measures. The CF's were measured on the same

TABLE 5.7

Details of Univariate Directional p-Values for the CT's Ranking Analysis of Four Products

	P1	P2	P3	P4	P1	P2	P3	P4	P1	P2	P3	P4
		CT_01				CT_02				CT_03		
P1		1.000	1.000	1.000		.6309	.1434	.9620		.4565	**.0006**	1.000
P2	**.0002**		1.000	1.000	.3701		.0488	.9704	.5441		**.0020**	1.000
P3	**.0002**	**.0002**		1.000	.8574	.9518		.9942	.9996	.9982		.9998
P4	**.0002**	**.0002**	**.0002**		.0384	.0300	**.0060**		**.0002**	**.0002**	**.0004**	
		CT_04				CT_05				CT_06		
P1		1.000	1.000	.8894		.0326	.5569	.7295		.7317	.9124	1.000
P2	**.0002**		.9978	.0720	.9686		.9092	.9954	.2694		.8798	1.000
P3	**.0002**	**.0024**		**.0004**	.4445	.0918		.6221	.0878	.1206		1.000
P4	.1108	.9282	.9998		.2726	**.0048**	.3795		**.0002**	**.0002**	**.0002**	
		CT_07				CT_08				CT_09		
P1		.7149	.9972	1.000		.4629	1.000	1.000		**.0026**	.5289	.9972
P2	.2861		.9988	1.000	.5385		1.000	1.000	.9976		.9916	1.000
P3	**.0030**	**.0014**		1.000	**.0002**	**.0002**		1.000	.4719	**.0086**		.9966
P4	**.0002**	**.0002**	**.0002**		**.0002**	**.0002**	**.0002**		**.0030**	**.0002**	**.0038**	
		CT_10				CT_11				CT_12		
P1		.3039	.6755	.9990		.8996	.7918	.5855		.6261	1.000	1.000
P2	.6967		.7697	.9948	.1008		.2320	.1378	.3745		.9998	1.000
P3	.3255	.2308		.9912	.2092	.7691		.3235	**.0002**	**.0004**		1.000
P4	**.0012**	**.0054**	**.0090**		.4157	.8628	.6783		**.0002**	**.0002**	**.0002**	
		CT_13				CT_14						
P1		.1738	.2486	.5501		.5035	.3321	.8680				
P2	.8284		.6411	.9420	.4983		.3187	.8834				
P3	.7525	.3591		.8500	.6693	.6819		.9704				
P4	.4515	.0582	.1506		.1324	.1176	.0298					

experimental units at different numbers of minutes spent in the experimental device (NM: 1, 5, 10, 20, 25). Just one replicate of each treatment was performed. Experimental data are displayed in Figure 5.4, where the continuous lines represent the individual CF series.

The setting of the ranking problem (see Chapter 3) can be summarized as follows.

- Type of design: multivariate functional data (three samples, that is, three multivariate functional populations to be ranked)
- Domain analysis: no
- Type of response variables: continuous (14 continuous functional responses)

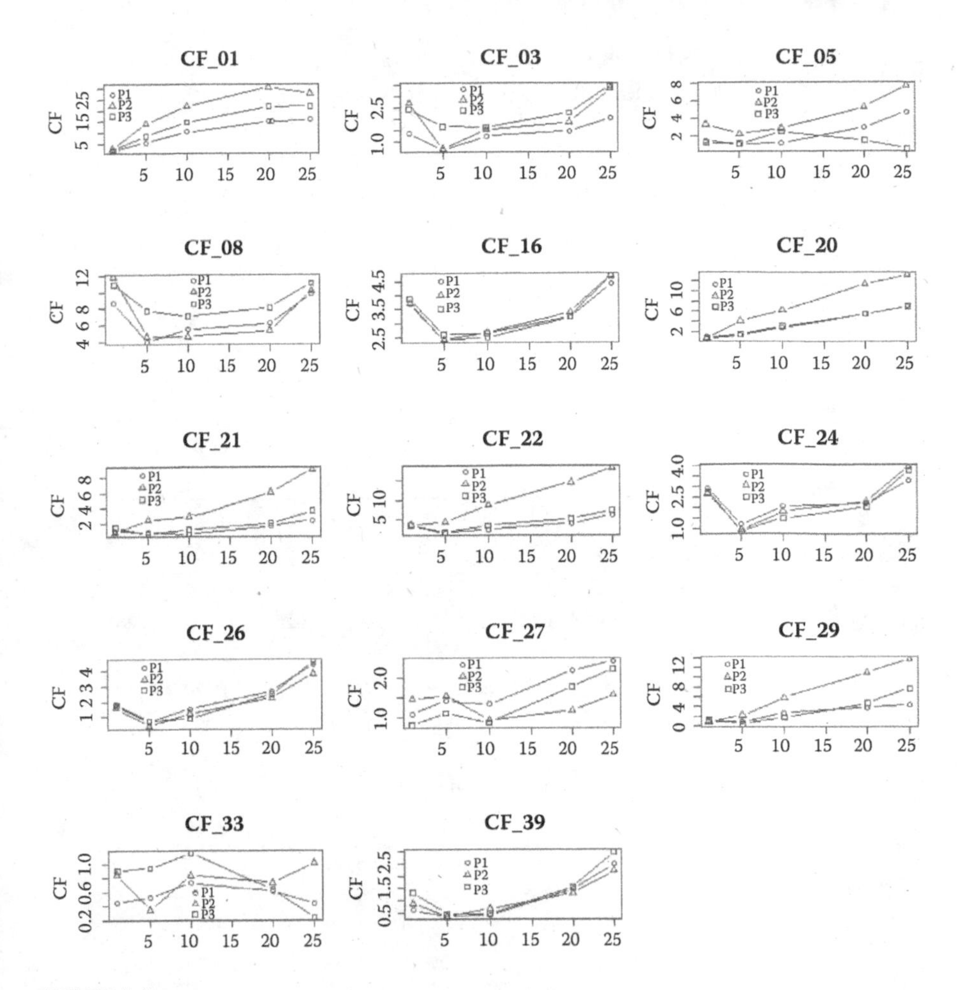

FIGURE 5.4
Observed CF data for each univariate response and by the three products. The number of minutes are reported on the x-axis.

- Ranking rule: 'the lower the better'
- Combining function: Fisher
- B (number of permutations): 2000
- Significance α-level: 0.1

When applying the ranking method on experimental data using directional parametric p-values performed via 5000 permutations we obtain the shown in Tables 5.8 and 5.9, where directional multivariate permutation p-values have been adjusted by using the Bonferroni–Holm–Shaffer method (1986).

All ranking analysis results can be obtained by using the book Web-based software (see Chapter 4) within the NPC Web Apps (lstat.gest.unipd.it

TABLE 5.8

Global Ranking Results for the CF Ranking Analysis of Three Products ($\alpha = 10\%$)

Product	P1	P2	P3
Global ranking	**1**	**2**	**3**
	P1	P2	P4
P1		.848	.744
P2	.078		.078
P3	**.026**	.658	

TABLE 5.9

Details of Univariate Directional p-Values for the CF Ranking Analysis of Three Products

	P1	P2	P3	P1	P2	P3	P1	P2	P3
		CF_01			**CF_03**			**CF_05**	
P1		1.00	1.00		1.00	1.00		1.00	.182
P2	**.026**		**.026**	**.026**		.874	**.026**		**.026**
P3	**.026**	1.00		**.026**	.152		.844	1.00	
		CF_08			**CF_16**			**CF_20**	
P1		.683	1.00		1.00	1.00		1.00	.900
P2	.343		.939	**.026**		**.626**	.026		**.026**
P3	**.026**	.087		**.026**	.400		.126	1.00	
		CF_21			**CF_22**			**CF_24**	
P1		.965	.970		1.00	.965		.496	.180
P2	.061		.061	**.026**		**.026**	.530		.061
P3	.056	.965		.061	1.00		.846	.965	
		CF_26			**CF_27**			**CF_29**	
P1	**.026**	.142		.160	**.026**		1.00	.744	.744
P2		.88	.866		.566	**.026**		.061	.062
P3	.145		1.00	.460		.282	.965		
		CF_33			**CF_39**				
P1	.913	.933		.641	1.00	1.00			
P2		.566	.385		.944	.937			
P3	.460		**.026**	.082					

/npcwebapps/). The reference dataset is labelled CFs.csv. To exactly replicate the permutation p-values in Tables 5.8 and 5.9 exactly, a seed value equal to 1234 must be set up.

It is worth noting that despite the fact that the CF experiments had only one replicate per treatment, that is, it was an unreplicated design, the multivariate ranking method is able to determine an α-level set as 10% which is the best, the second and the worst product.

Table 5.9 displays the set of univariate directional pairwise p-values, related to the multiple comparison procedures for the problem at hand.

5.3 Sensory Test: ST Analysis

Sensory analysis is a quantitative statistical subject aimed at using human senses (sight, smell, taste, touch and hearing) for the purposes of evaluating consumer products. The discipline requires a panel of human assessors, by whom the products are evaluated (Meilgaard et al., 2006). These kinds of measures are in fact thought to be comparable with the assessment that final consumers will make when using the new product.

By applying suitable statistical techniques to the results of a sensory test it is possible to make inferences and gain insights about the products being tested.

Within this framework the experimental design typically handles human assessors as blocks. If the evaluations of panellists are multivariate in nature the reference design is a multivariate Randomized Complete Block (RCB) and the related statistical inferences become hard to cope with within the traditional parametric approach, and a suitable inferential, possibly non-parametric, approach should be developed to build up an effective global performance comparison.

In general, the requirement to take into consideration an RCB design occurs when the experimental units are heterogeneous; hence the notion of blocking is used to control the extraneous sources of variability. The major criteria of blocking are characteristics associated with the experimental material and the experimental setting. The purpose of blocking is to sort experimental units into blocks, so that the variation within a block is minimized while that among blocks is maximized. An effective blocking not only yields more precise results than an experimental design of comparable size without blocking, but also increases the range of validity of the experimental results.

5.3.1 ST of Four Products

A company is studying three possible new products (labelled P2, P3, P4) to compare with their own presently marketed product (labelled P1). The experiment is designed as follows: after blind testing of a given product using the sense of smell, taste or touch, the human assessor assigned three different scores to it, describing the three most important aspects of the product, labelled S (1–5 points), P (1–5 points) and A (Yes = 2, No = 1). The same experiment is replicated under different assessment conditions (labelled B, D, L, N and W), which should represent the situations in which the final customers will make use of the product. Note that the experiment has a set of mixed-type response variables and considers a total of 15 variables.

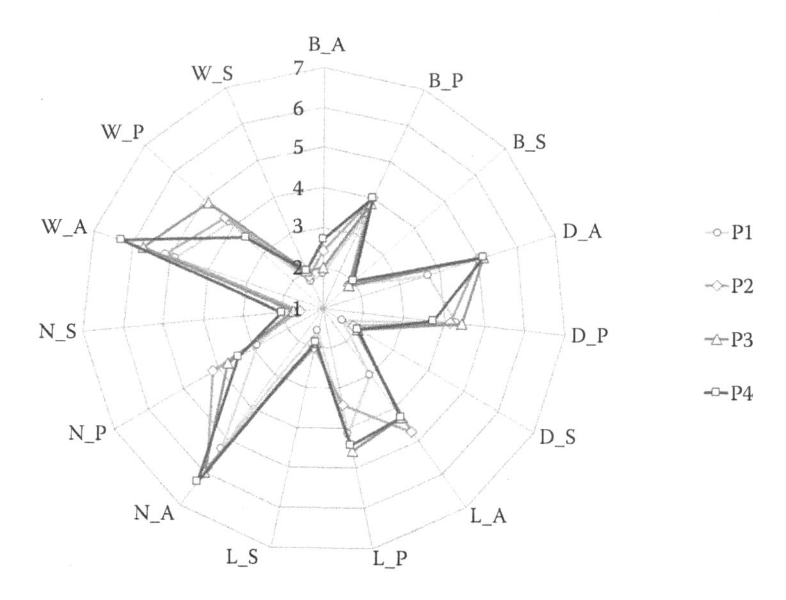

FIGURE 5.5
Radar graph for ST of four products (sample score means by product).

A first look at the experimental results is provided in the radar graph in Figure 5.5, where variable-by-variable sample score means by product are reported.

Note that, from the descriptive point of view, sample score means appear generally similar and the differences between products rather mild.

The setting of the ST analysis ranking problem (see Chapter 3) can be summarized as follows.

- Type of design: multivariate randomized complete block design (four related samples, i.e., four multivariate populations to be ranked);
- Domain analysis: yes (three domains, see Table 5.10)
- Type of response variables: ordered categorical (15 ordered categorical responses; 10 on a 1–5 rating scale and 5 on a 1–2 binary scale)
- Ranking rule: 'the higher the better'
- Combining function: Fisher
- B (number of permutations): 1000
- Significance α-level: 0.1

When applying the ranking method to sensory test data using directional parametric p-values performed via 1000 permutations we obtain results reported in Tables 5.10 and 5.11, where directional multivariate permutation p-values in Table 5.11 have been adjusted by using the Bonferroni–Holm–Shaffer method (1986).

TABLE 5.10
Global Ranking for the ST Analysis of Four Products ($\alpha = 10\%$)

Product	P1	P2	P3	P4
Global ranking	**4**	**3**	**1**	**1**
S	4	3	1	1
P	1	1	1	1
A	1	1	1	1
All aspects	P1	P2	P3	P4
P1		1.00	1.00	1.00
P2	.264		1.00	1.00
P3	**.036**	.281		1.00
P4	**.018**	.249	.210	
S	P1	P2	P3	P4
P1		1.00	1.00	1.00
P2	.186		1.00	1.00
P3	**.036**	.278		1.00
P4	**.018**	.164	.054	
P	P1	P2	P3	P4
P1		1.00	1.00	1.00
P2	.590		1.00	1.00
P3	.390	.351		1.00
P4	.195	.369	.458	
A	P1	P2	P3	P4
P1		1.00	.928	1.00
P2	1.00		1.00	1.00
P3	1.00	.944		1.00
P4	1.00	.909	.810	

All ranking analysis results can be obtained by using the book Web-based software (see Chapter 4) within the NPC Web Apps (lstat.gest.unipd.it/npcwebapps/). The reference dataset is labelled ST_4_prod.csv. To replicate the permutation p-values in Tables 5.10 and 5.11 exactly, a seed value equal to 1234 must be set up.

The results allow us to show the slight but significantly better multivariate performance of products P4 and P3 over all the remaining products, where the global ranking is mainly explained by their superiority with respect to the S aspect.

Table 5.11 displays the set of the univariate permutation directional pairwise p-values, related to the set of the multiple comparison procedures for the problem at hand.

Univariate directional p-values do confirm the importance of the S aspect in discriminating the four products under investigation.

TABLE 5.11

Details on Univariate Directional p-Values for the ST Analysis of Four Products

	P1	P2	P3	P4	P1	P2	P3	P4	P1	P2	P3	P4
		B_S				**D_S**				**L_S**		
P1		.758	.532	.991		.807	.968	.986		.975	.984	.952
P2	.241		.078	.790	.192		.676	.556	**.024**		.336	.239
P3	.467	.921		.975	**.031**	.323		.511	**.015**	.663		.559
P4	**.008**	.209	**.024**		**.013**	.443	.488		**.047**	.760	.440	
		N_S				**W_S**				**B_P**		
P1		.935	.967	.967		.642	.975	.987		.874	.605	.936
P2	.064		.793	.864	.357		.957	.992	.125		.325	.573
P3	**.032**	.206		.878	**.024**	**.042**		.968	.394	.674		.564
P4	**.032**	.135	.121		**.012**	**.007**	**.031**		.063	.426	.435	
		D_P				**L_P**				**N_P**		
P1		.482	.620	.227		.118	.759	.802		.830	.751	.736
P2	.517		.794	.277	.881		.928	.955	.169		.163	.184
P3	.379	.205		.123	.240	.071		.429	.249	.836		.408
P4	.772	.722	.876		.197	**.044**	.570		.263	.815	.591	
		W_P				**B_A**				**D_A**		
P1		.558	.808	.130		.500	.486	.744		.805	.939	.939
P2	.441		.717	.248	.500		.486	.744	.194		.757	.757
P3	.191	.282		.145	.513	.513		.757	.060	.242		.500
P4	.869	.751	.854		.255	.255	.242		.060	.242	.500	
		L_A				**N_A**				**W_A**		
P1		.935	.871	.871		.345	.486	.879		.690	.879	.879
P2	.064		.255	.255	.654		.683	.938	.309		.750	.750
P3	.128	.744		.500	.513	.316		.885	.120	.250		.500
P4	.128	.744	.500		.120	.061	.114		.120	.250	.500	

5.3.2 ST of Three Products

A company is studying three possible new products (labelled P1, P2, P3) and for this goal a suitable sensory test has been designed. The experiment is designed as follows: after blind testing a given product using a sense such as smell or touch, each one of the nine enrolled human assessors assigns three different scores to it, describing the three most important aspects of the product, labelled St, Pl and Sf. The rating scale used for all evaluations is the 1–9 scale, where 1 means the worst and 9 the best score. The same experiment was independently replicated three times, and under different assessment conditions (labelled N, W, D, LL), which should represent the situations in which the final customers will make use of the product. In the case of the first two conditions, that is, N and W, the Sf score was not requested because it is not an admissible evaluation under those conditions.

A first look at the experimental results is provided in the radar graph in Figure 5.6, where variable-by-variable sample score means by product are reported.

Note that, from the descriptive point of view, sample score means appear to support the somewhat better performances of products 2 and 3 although product 1 highlights two response variables (N_St and W_St), where it seems to be better than both of the remaining products. The setting of the PT analysis ranking problem (see Chapter 3) can be summarized as follows.

- Type of design: multivariate randomized complete block design (three related samples, i.e., three multivariate populations to be ranked)
- Domain analysis: yes (three domains; see Table 5.12)
- Type of response variables: ordered categorical (five ordered categorical responses on a 1–9 rating scale)
- Ranking rule: 'the higher the better'
- Combining function: Fisher
- B (number of permutations): 1000
- Significance α-level: 0.05

When applying the ranking analysis using directional permutation p-values ($\alpha = 5\%$), calculated via Anderson–Darling test statistic and performed with 1000 permutations and using the Fisher combining function, we obtain the

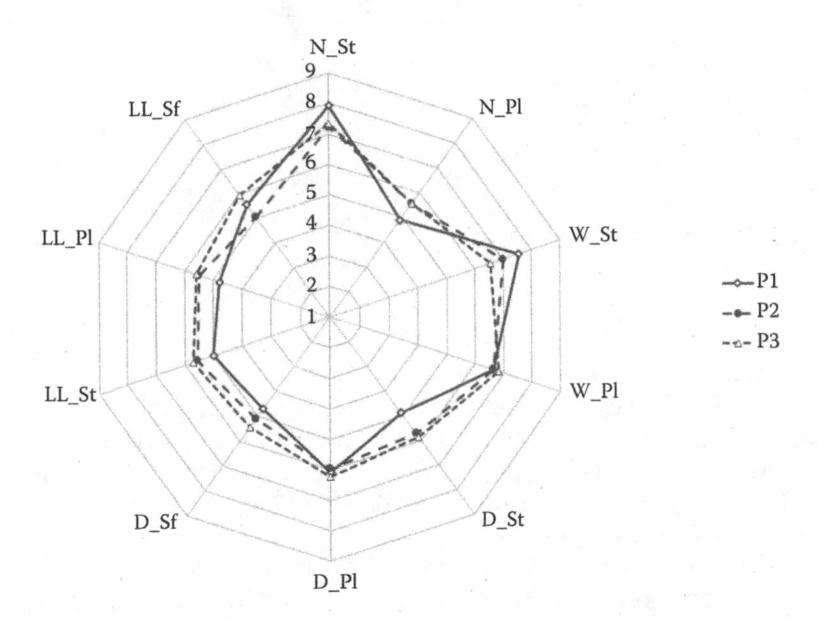

FIGURE 5.6
Radar graph for ST of three products (sample score means by product).

TABLE 5.12

Details of Univariate Directional p-Values for the ST Analysis of Three Products

Product	P1	P2	P3
Global ranking	**2**	**2**	**1**
N	1	2	2
W	1	2	3
D	3	2	1
LL	2	2	1
All aspects	P1	P2	P3
P1		.039	.084
P2	.098		.602
P3	**.003**	.022	
N	P1	P2	P3
P1		**.006**	.011
P2	.413		.913
P3	.212	.39	
W	P1	P2	P3
P1		**.012**	.005
P2	.871		**.013**
P3	.702	1.00	
D	P1	P2	P3
P1		.801	.949
P2	.040		1.00
P3	**.024**	.327	
LL	P1	P2	P3
P1		1.00	1.00
P2	.124		.983
P3	**.006**	.002	

results shown in Tables 5.12 and 5.13, where the domain by assessment condition has been performed as well.

All ranking analysis results can be obtained by using the book Web-based software (see Chapter 4) within the NPC Web Apps (lstat.gest.unipd.it /npcwebapps/). The reference dataset is labelled ST_3_prod.csv. To exactly replicate the permutation p-values in Tables 5.12 and 5.13 exactly, a seed value equal to 1234 must be set up.

Table 5.13 displays the set of the univariate permutation directional pairwise p-values, related to the set of the multiple comparison procedures for the problem at hand.

TABLE 5.13

Global Ranking for the ST Analysis of Three Products (α = 5%)

	P1	P2	P3	P1	P2	P3	P1	P2	P3
		N_St			**N_Pl**			**W_St**	
P1		**.001**	**.003**		.865	.938		**.001**	**.001**
P2	.999		.917	.134		.675	.998		**.003**
P3	.997	.082		.061	.324		.999	.997	
		W_Pl			**D_St**			**D_Pl**	
P1		.479	.657		.994	.998		.344	.565
P2	.520		.676	**.005**		.472	.655		.698
P3	.342	.323		**.002**	.527		.434	.301	
		D_Sf			**LL_St**			**LL_Pl**	
P1		.786	.956		.938	.974		.876	.984
P2	.213		.814	.061		.792	.123		.803
P3	.043	.185		.025	.207		**.015**	.196	
		LL_Sf							
P1		.060	.897						
P2	.939		.998						
P3	.102	**.001**							

It is worth noting that despite the fact that P1 has a better significance performance over P2 and P3 in two out of ten response variables (N_St and W_St) and two out of four conditions/domains (N and W), the ranking analysis supports the decision to classify P1 as the second ranked product and P3 as the best product. This result is explained mainly by the better performance of P3 under the LL condition where P3 is clearly much better than P1 and P2 (see Table 5.12, last row of the long-lasting domain table).

5.4 Ranking Approach on the Issue of Consumer Relevance in NPD

In the context of intensifying competition both in products and services markets, the question of properly taking into account consumer relevance is certainly one of the main challenges that firms have to face today during the development process of new products. Product innovations that can directly incorporate the customer's point of view are intended to be more successful than innovations that are emphasized from the technical and technological aspects, but are not of real interest to the target market. Managerial and organizational practices during the NPD process frequently suffer in the contrast between objective performance, as measured by laboratory instrumental tests, and perceptual evaluation by the end customers, measured by surveys such as consumer tests. Very often the objective performances that are used to determine the effectiveness of a new product are not directly comparable between the products or with the subjective perceptions of the end users of the product.

The main goal of this section is to validate that the joint use of well-designed experiments and consumer surveys, along with suitable statistical techniques based on the ranking methods, may allow one to demonstrate the possible causal relationship between objective instrumental product performance and subjective assessments of consumers. This would confirm that a 'truly' better product can also be perceived by consumers as the more effective and more acceptable option.

A case study is provided in which several products are evaluated from a multivariate point of view and a final global ranking is obtained. The statistical analyses confirm the direct correlation between results obtained via instrumental measures with the results obtained from the corresponding consumer surveys conducted on the same products. Better performances and a higher position in the instrumental global ranking are associated with better assessments on product efficacy, better general evaluations (overall opinion, satisfaction, etc.) and a higher position in the consumer global ranking.

This section is organized as follows. In the first subsection we review the managerial literature around the subject of the Voice of Customers (VOC). We also discuss how to directly capture real needs of the target market in the NPD process. In the second and third subsections, we present a detailed ranking analysis from the instrumental laboratory tests and from a large consumer test that has been performed on the same products. Finally, we critically compare the instrumental versus the consumer results to demonstrate that in the case study under consideration a causal relation between objective instrumental product performance and subjective assessments of consumers does exist.

5.4.1 VOC in the NPD Process

Innovation in the form of developing or improving new products is a fundamental activity for the survival of a business in almost all industries. The product, the technology to produce it and the organization are components that help a company gain a competitive advantage in the long run (Foreman, 2011). Innovation is the result of a combination of external factors and social influences, such as cultural changes, while market conditions facilitate or constrain the extent of firm innovation activity. Original products are the result of unique individuals; there is always someone who has a more individualistic innovation. Scientific and technological developments lead to knowledge input while societal changes and market needs lead to demands and opportunities; together they drive a firm to develop knowledge, processes and products. The recent product development literature indicates that there are numerous methods to achieve innovation, no method is the best and the results of methods are attributable to many factors. In the text that follows we briefly review such methods.

All stakeholders should participate in the innovation process so that their direct involvement will reduce the risk of developing products that will experience market failure. A company can achieve significant improvements by observing or involving external parties in addition to internal employees. Collecting customer feedback can be a massive undertaking that companies typically do more than once a year; response rates are usually very low and the responses obtained are generally supportive of the company's interests. Cooper and Dreher (2010) identify eight methods to capture the VOC: ethnographic research, customer visit teams, customer focus groups, lead-user analysis, customer or user designs, customer brainstorming, customer advisory board or panel and community of enthusiasts. Some VOC methods are used very extensively, notably customer visit teams and focus groups that identify customer problems using the lead-user method. Readers can find more information about the lead-user method in Cooper and Dreher (2010).

The traditional market research is the first step toward an external link with the market, trying to understand early the needs of the customer. There are companies that are looking for new ideas or talents and companies that involve the customer in some phases of product development. A person's creative abilities are a function of his or her intellectual ability, knowledge, personality and environment. We interpret intellectual ability as the ability to see problems from unconventional perspectives, to recognize and formulate ideas and effectively communicate these ideas to others. To progress creative ideas, the internal environment of the company should support and recognize the creative ideas. Innovation requires the integration of creative idea with resources and expertise that can give the idea a 'useful' form. Finding solutions that meet the needs of consumers has pushed companies to collaborate with others. Companies can form alliances with customers, suppliers, producers and final consumers.

NPD practices may also create problems for the company. Conflicts can arise from partner selection, determining the timing and intensity of customer involvement, the ability and willingness to provide knowledge and the nature and extent of that knowledge. Other risks include negative publicity due to the untimely circulation of test results, product performance, inaccurate data, or unrepresentative feedback. Innovative products, undertaken in partnership, could incite more trouble than they are worth because of the complexity of managing a partnership for the development of a new product. In addition, if we involve users at all stages we may find an increase in costs that is not justified by the revenue obtained. It is better to prioritize the stages of product development and concentrate on the most important tasks identified by the company for a successful outcome.

The strategy adopted to collect data on the VOC is very important to obtain data with minimal uncertainty. The choice depends on many factors but the most important factor, which contains all the others, is the kind of industry to be involved in. The strategies to collect VOC data can be grouped into two categories: one approach concentrates on efforts in IT technologies, where a disruptive technology can decrease the process time and eliminate some tasks but is a great investment. The other category is a deliberate strategic exercise to assess the external world and identify trends and threats. This is less expensive than disruptive technology but its effect is in the short term (Cooper and Edgett, 2010). Gibbons (2012) identifies six steps in the VOC strategy: build the team, gather the right insight, make it relevant, communicate, close the loop to join all tasks and demonstrate the benefit to the company.

A study by Hsieh and Chen (2005) found that interaction with the customer is vital to achieving a good performance in NPD. In this study, authors identified two main variables that describe the performance of the NPD: the intensity of customer integration in the NPD process and knowledge management capabilities. For the first variable, they identified five additional variables that describe it, where the five variables correspond to hypothetical five-step product development and each variable indicates the strength of interaction with the customer in the specification phase. They divided the second variable, knowledge management competence, into capacity of the structure and the process capability. For the survey they considered a product sold on the market not customized and created interviews on it to be submitted to the companies. From the analysis it was found that each variable was significant; it was also shown that the interaction between the two main variables is equally significant.

An important aspect for the success of the project is to have an effective knowledge management system. Integrating market information in the NPD process has been recognized as a prior step to get good performance from the new product. Many companies implement a knowledge management system, but many knowledge management projects are information projects. Although many companies are using customer data, few are turning the data into knowledge. This aspect is crucial but often collides with a lack of expertise of knowledge management competence from the interaction with the user.

A knowledge management system is based on two types of organizational capabilities of a company: the infrastructure capability and the process capability. The infrastructure capability is improved by removing barriers to communication that naturally occur between the different parties within the company. The birth and development of the Internet along with new IT technology has helped customers and prospects to interact and communicate with businesses. The process capability includes obtaining information about users during interactions with them and transforming the information into a useful format for all business functions.

5.4.2 Instrumental Performance Ranking Analysis

At a laboratory of a company, two instrumental tests (labeled as IT_01 and IT_02) were performed to compare five different products, labeled as P1, P2,…,P5.

Four independent test replications were performed in a random order, and a multivariate response was measured where each univariate output was classified as A, B and C. A descriptive look at the experimental results is provided by radar graphs in Figure 5.7, where output-by-output sample means by product are reported. At first sight we can certainly note that no dramatic

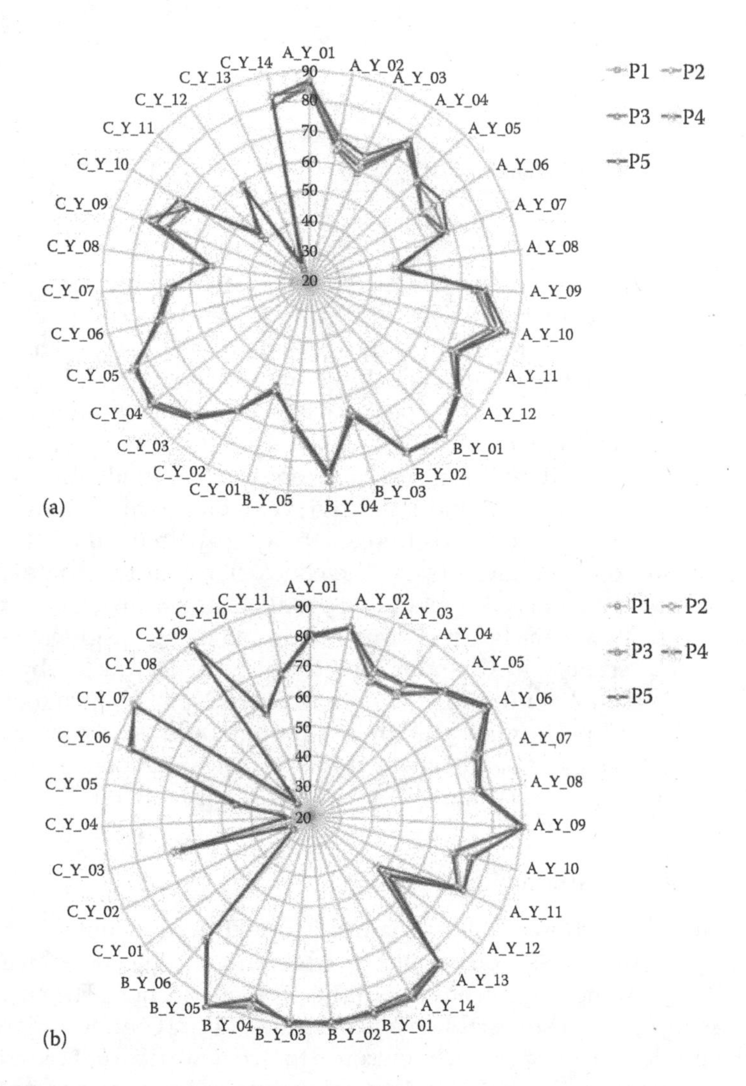

FIGURE 5.7
Radar graph for (a) IT_01 and (b) IT_02 (output-by-output sample means by product).

TABLE 5.14

Global Ranking for the Instrumental Analysis of Five Products (α = 10%)

IT_01					
Ranking	P1	P2	P3	P4	P5
A (12)	2	2	2	5	1
B (5)	1	1	1	1	1
C (14)	2	4	4	2	1
Global ranking (31)	**2**	**4**	**2**	**4**	**1**
IT_02					
Ranking	P1	P2	P3	P4	P5
A (14)	2	2	2	5	1
B (6)	1	1	1	1	1
C (11)	2	5	4	3	1
Global ranking (31)	**2**	**4**	**2**	**4**	**1**

performance difference among products does exist and only for a few subsets of outputs could any possible significant differences ascertained.

For each instrumental tests, Table 5.14 reports a summary of the overall and the by-output category global ranking analysis we obtained setting the α-level as 10% (we did not use 5% because of the very low number of replicates of these experiments).

It is worth noting both ranking results strongly agree in identifying P5 as the best product and both P3 and P4 as the worst ones while P1 and P3 are tied and ranked in the second position. Globally, instrumental performance ranking results convincingly support the conclusion that P5 can be considered as the best one among the set of five products under investigation.

Statistical analysis of the two tests was performed using the global ranking methodology presented in Chapter 2, so that univariate output-by-output comparisons and global ranking allow us to establish which products have the best and worst performances, either from the multivariate (overall and by category; see Tables 5.15 and 5.16) or from the univariate point of view (for the sake of simplicity, univariate results are not reported here).

5.4.3 Consumer Test Ranking Analysis

A marketing research institute was commissioned to carry out a large consumer survey enrolling several subjects, where the products under investigation were the same five considered in the previous section. The consumer survey's main goal was to establish a full diagnostic and better understanding of the product labelled P4 with reference to two new different customized versions (labelled P4.Lf and P4.Hf). Actually, the consumer survey focussed

TABLE 5.15

Global Ranking Results for the IT_01 Analysis of Five Products (α = 10%)

Global ranking	P1	P2	P3	P4	P5
All outputs	**2**	**4**	**2**	**4**	**1**
A	2	2	2	5	1
B	1	1	1	1	1
C	2	4	4	2	1
All outputs	P1	P2	P3	P4	P5
P1		.774	.516	1.00	1.00
P2	1.00		1.00	.660	.794
P3	1.00	1.00		1.00	1.00
P4	.856	1.00	1.00		1.00
P5	.129	**.020**	.168	**.048**	
A	P1	P2	P3	P4	P5
P1		1.00	.492	1.00	1.00
P2	1.00		.468	.055	.966
P3	.808	1.00		1.00	1.00
P4	1.00	1.00	1.00		1.00
P5	.126	.144	.171	**.005**	
B	P1	P2	P3	P4	P5
P1		.555	.805	.870	.612
P2	.805		1.00	.722	.825
P3	1.00	1.00		1.00	1.00
P4	1.00	.195	.450		1.00
P5	1.00	.305	1.00	.609	
C	P1	P2	P3	P4	P5
P1		.115	.498	.687	.594
P2	1.00		.848	1.00	1.00
P3	1.00	1.00		1.00	.761
P4	.798	.870	.682		1.00
P5	.057	**.010**	**.036**	.150	

on six products: P1, P2, P3, P4.Lf, P4.Hf, P5. Note that P4 was submitted in two versions.

Statistical analysis of the consumer survey was performed using the global ranking methodology presented in Chapter 2, separately for each one of the two aspects of interest (product performance and product opinions), so that univariate and multivariate comparisons, along with global ranking, allow us to establish which products have the best and worst performances, either

TABLE 5.16

Global Ranking Results for the IT_02 Analysis of Five Products ($\alpha = 10\%$)

Global ranking	P1	P2	P3	P4	P5
All outputs	**2**	**4**	**2**	**4**	**1**
A	2	2	2	5	1
B	1	1	1	1	1
C	2	5	4	3	1
All outputs	P1	P2	P3	P4	P5
P1		.810	.555	1.00	1.00
P2	1.00		1.00	.700	.803
P3	1.00	1.00		1.00	1.00
P4	.858	1.00	1.00		1.00
P5	.222	**.040**	.144	**.036**	
A	P1	P2	P3	P4	P5
P1		1.00	.303	1.00	1.00
P2	1.00		.438	.435	1.00
P3	1.00	.811		1.00	1.00
P4	1.00	1.00	1.00		.997
P5	.282	.147	.237	**.030**	
B	P1	P2	P3	P4	P5
P1		1.00	.804	.819	.807
P2	1.00		.770	.759	.792
P3	1.00	1.00		.570	1.00
P4	.774	1.00	1.00		.919
P5	1.00	.405	.219	.290	
C	P1	P2	P3	P4	P5
P1		**.035**	.441	.369	.132
P2	1.00		1.00	1.00	1.00
P3	1.00	1.00		1.00	.708
P4	.855	.390	.632		.393
P5	.063	**.010**	**.033**	.345	

from the univariate or from the multivariate aspect. Note that unlike what happened for the instrumental performance analysis, where the measures were numeric and continuous, the response variables of interest here are ordered categorical measures as usually occur in sample consumer surveys. The flexibility of the global ranking methodology allows us to easily rank the products at hand simply by changing the permutation test statistic, which in this case is the Anderson–Darling statistic.

For both consumer test aspects, we identified several sub-domains, labelled Perf_A and Perf_B, while with reference to product opinions analysis, we defined three main sub-domains: GO, Per and Sat. Table 5.17 reports the global ranking analysis for each one of the two consumer test aspects and the related sub-domains.

Accordingly, with the ranking analysis results from instrumental data, it is worth noting that P5 always appears as the best product. Note also that P4's first version P4.Lf is always assessed as the worst product, which is also not tied with P4's second version P4.Hf. Finally, it is worth noting that global rankings estimated from two separate marketing research questionnaire sections, that is, PER and GO, match each other perfectly.

A first look at the product performance data is provided in the radar graphs in Figure 5.8, where score sample means and top-two-box (T2B) percentages by product are reported. Note that products P5 and P4.Lf seem to be the products with the best and the worst performance assessments, respectively.

Details on multivariate product comparisons via directional permutation tests are presented in Table 5.18 ($\alpha = 10\%$), whereas for sake of simplicity, univariate results are not reported here.

It is interesting that, as mentioned previously, the same product P4 in two different versions is ranked at different positions.

A first look at product opinion data is provided in the radar graphs in Figure 5.9, where score sample means and top-two-box percentages by product are reported. Note that P5 seems also in this case to marginally be the product with the best performance assessments.

Details on multivariate product comparisons via directional permutation tests ($\alpha = 10\%$) are presented in Table 5.19 (for the sake of simplicity, univariate results are not reported here).

With reference to the general product opinions aspect, there are no significant differences in the consumer evaluations on the first and second versions of P4.

TABLE 5.17

Global Ranking Results from Consumer Survey on Six Products ($\alpha = 5\%$)

			Product Performance			
Ranking	P1	P2	P3	P4.Lf	P4.Hf	P5
Perf_A (15)	2	2	2	6	2	1
Perf_B (12)	2	2	2	6	2	1
Global ranking (27)	**2**	**2**	**2**	**6**	**2**	**1**
			Product Opinions			
Ranking	P1	P2	P3	P4.Lf	P4.Hf	P5
GO (8)	1	1	1	1	1	1
PER (5)	4	2	2	4	4	1
SAT (5)	2	2	2	6	2	1
Global ranking (18)	**2**	**2**	**2**	**6**	**2**	**1**

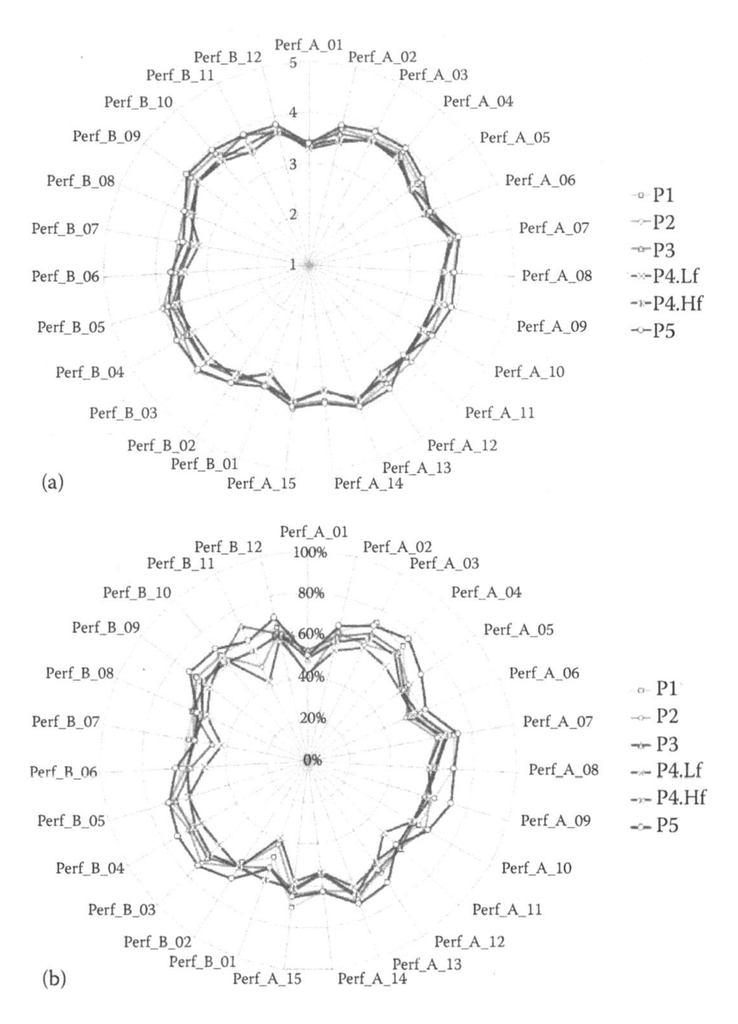

FIGURE 5.8
Radar graphs for product performances from consumer survey. (a) Performance-by-performance sample means by product. (b) Performance-by-performance T2B percentages by product.

5.4.4 Objective Instrumental Product Performance and Subjective Assessments of Consumers

The main goal of this work was to look for a possible positive causal relation between objective instrumental product performance and subjective assessments of consumers, possibly to argue that a 'truly' better product may also be perceived by consumers as being more effective and as more acceptable as well. From two ranking analyses on both objective instrumental performances and subjective consumer assessments we obtained two global

TABLE 5.18

Details of Global Ranking Results for Product Performance Analysis from Consumer Survey (α = 10%)

Global ranking	P1	P2	P3	P4.Lf	P4.Hf	P5
All performances	**2**	**2**	**2**	**6**	**2**	**1**
Perf_A	2	2	2	6	2	1
Perf_B	2	2	2	6	2	1
All performances	P1	P2	P3	P4.Lf	P4.Hf	P5
P1		1.00	1.00	.712	.135	.392
P2	.545		1.00	1.00	**.045**	.170
P3	1.00	1.00		1.00	.205	1.00
P4.Lf	1.00	1.00	1.00		1.00	1.00
P4.Hf	1.00	1.00	1.00	1.00		.485
P5	.637	.745	.890	**.007**	**.035**	
Perf_A	P1	P2	P3	P4.Lf	P4.Hf	P5
P1		1.00	.993	.190	.340	1.00
P2	1.00		.989	.240	.220	1.00
P3	1.00	1.00		.70	.83	1.00
P4.Lf	1.00	.97	1.00		1.00	1.00
P4.Hf	1.00	1.00	1.00	1.00		1.00
P5	.699	1.000	.340	**.020**	.049	
Perf_B	P1	P2	P3	P4.Lf	P4.Hf	P5
P1		1.00	.839	.050	.769	1.00
P2	1.00		.440	**.030**	.160	1.00
P3	1.00	1.00		.17	1.00	1.00
P4.Lf	1.00	.985	1.00		1.00	1.00
P4.Hf	1.00	1.00	1.00	.090		1.00
P5	1.00	.779	.280	**.015**	.190	

ranking reported in Table 5.20, where for product P4 we included only the P4.Lf version. Recall that the two global rankings we estimated from the two separate marketing research questionnaire sections, that is product performance and product opinions, matched each other perfectly.

The two rankings estimated from the two data sources are obviously quite similar, though a more formal comparison between rankings can be performed by calculating suitable measures of agreement between rankings such as Cohen's kappa and Spearman rho (Hollander et al., 2014). We obtained the values of $\kappa = 0.600$ (95% bootstrap confidence: [0.286,0.917]) and $\rho = 0.775$ (95% bootstrap confidence: [0.707,1]). The formal agreement analysis proves that the two data sources do provide consistent rankings that are strongly in accord with each other.

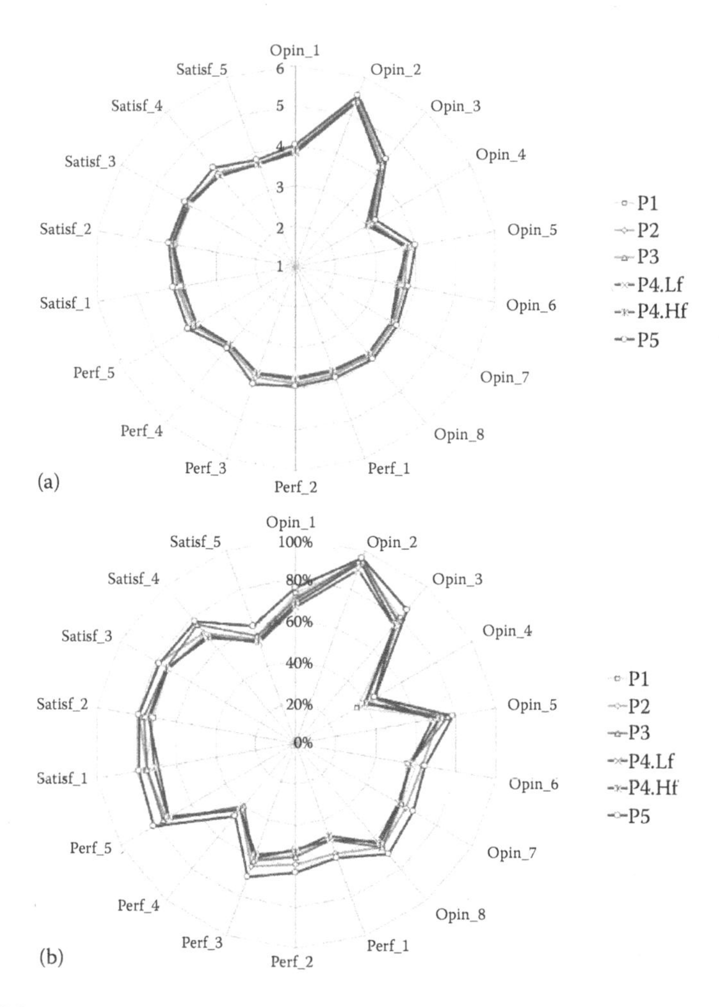

FIGURE 5.9
Radar graphs for product opinions from consumer survey. (a) Opinion-by-opinion sample means by product. (b) Opinion-by-opinion T2B percentages by product.

5.5 Conclusion

The competitive context affirmed over recent years has highlighted the critical state of the process of definition of new products and services. From a company's point of view one of the conditions of competitive success lies in a high level of correspondence between the products and services offered and the more and more varying needs of clients (Barney, 2002). Such a condition is even more critical, given that a growing tendency toward market segmentation and product customization is clear in many sectors. Therefore, to

TABLE 5.19

Details of Global Ranking Results for Product Opinion Analysis from Consumer Survey (α = 10%)

Global ranking	P1	P2	P3	P4.Lf	P4.Hf	P5
All opinions	**2**	**2**	**2**	**6**	**2**	**1**
GO	1	1	1	1	1	1
PER	4	2	2	4	4	1
SAT	2	2	2	6	2	1
All opinions	P1	P2	P3	P4.Lf	P4.Hf	P5
P1		1.00	1.00	1.00	1.00	1.00
P2	1.00		1.00	1.00	1.00	1.00
P3	1.00	1.00		1.00	1.00	1.00
P4.Lf	1.00	.941	1.00		1.00	1.00
P4.Hf	1.00	1.00	1.00	1.00		1.00
P5	.070	.420	.280	**.023**	.049	
GO	P1	P2	P3	P4.Lf	P4.Hf	P5
P1		1.00	1.00	1.00	1.00	1.00
P2	1.00		1.00	1.00	1.00	1.00
P3	1.00	1.00		1.00	1.00	1.00
P4.Lf	1.00	.906	1.00		1.00	1.00
P4.Hf	1.00	1.00	1.00	1.00		.992
P5	.250	.869	.719	.165	.150	
PER	P1	P2	P3	P4.Lf	P4.Hf	P5
P1		1.00	1.00	1.00	1.00	1.00
P2	1.00		1.00	.480	1.00	1.00
P3	1.00	1.00		1.00	1.00	1.00
P4.Lf	1.00	.988	1.00		1.00	1.00
P4.Hf	1.00	1.00	1.00	1.000		1.00
P5	**.020**	.340	.049	**.015**	**.020**	
SAT	P1	P2	P3	P4.Lf	P4.Hf	P5
P1		1.00	1.00	1.00	1.00	1.00
P2	1.00		1.00	1.00	1.00	1.00
P3	1.00	1.00		1.00	1.00	1.00
P4.Lf	1.00	.873	1.00		1.00	1.00
P4.Hf	1.00	1.00	1.00	1.00		1.00
P5	.220	.390	.370	**.042**	**.022**	

achieve the important strategic objective of satisfying the differing needs of clients, successful companies invest greater and greater resources and competences in planning and development (Campbell and Cooper, 1999). This leads to the search for constant client satisfaction without, however, forgoing the control of production and service provision costs. One of the reasons for

TABLE 5.20

Global Ranking Results from Instrumental Performances and Consumer Assessments of Five Products

Data Source	P1	P2	P3	P4	P5
Instrumental data	2	4	2	4	1
Consumer data	2	2	2	2	1

this change in strategy is linked not only to changes in the competitive and technological context but also to the evolution of the demand: companies must in fact satisfy clients who are no longer content with the basic features of the product/service that today are considered prerequisites. Clients are looking for the satisfaction of their own specific needs. As a consequence, the product/service planning stages are more and more prominently aimed at identifying the complementary attributes of the product/service, over and above the fundamental basic ones, to obtain greater satisfaction of clients' needs.

A strategic factor that can lead to the success of a new product lies in the ability to include the voice of customers and a strong association of the wording in the actual survey question given to customers and the recorded data. In addition, strategic factors include a well-designed experiment and analysing with suitable statistical techniques such as the ranking methods. If the consumer test surveys are expressly provided for suitable questions that allow us to directly connect some aspects of perceptual assessments of clients with objective assessments of instrumental experiments, we have an opportunity to demonstrate the possible causal relation between objective instrumental product performance and subjective assessments of consumers.

Referring to a real case study on a set of products, we highlighted in this specific case the presence of a strong direct correlation between instrumental test results and results of the corresponding consumer survey: better performances and higher position in the instrumental global ranking are associated with better assessments, better general evaluations (overall opinion, satisfaction, etc.) and higher position in the consumer global ranking. The work allowed us to establish a methodology capable of defining a relationship between an objective instrumental performance evaluation and consumer perceptions obtained from the consumer survey. Taking into account some of the obvious limitations given that we evaluated a single product of a particular industry, the results obtained suggest that it is possible to think of a possible extension of the results of instrumental measures to estimate the potential outcome of a corresponding consumer survey.

References

Barney J. B. (2002). *Gaining and Sustaining Competitive Advantage*. Upper Saddle River, NJ: Prentice Hall.

Calantone R. J., Di Benedetto C. A. (2000). Performance and time to market: Accelerating cycle time with overlapping stages. *IEEE Transactions on Engineering Management*, 47(2), 232–244.

Campbell A. J., Cooper R. G. (1999). Do customer partnerships improve new product success rates? *Industrial Marketing Management*, 28(5), 507–519.

Cooper G. R., Dreher A. (2010). Voice-of-customer methods: What is the best source of new-product ideas? *Marketing Management Magazine*, 19(4), 38–43.

Cooper G. R., Edgett S. J. (2010). Developing a product innovation and technology strategy for your business. *Research Technology Management*, 53(3), 33–40.

Davis P. S., Dibrell C. C., Janz B. D. (2002). The impact of time on the strategy–performance relationship: Implications for managers. *Industrial Marketing Management*, 31(4), 339–347.

Ferraty F., Vieu P. (2006). *Nonparametric Functional Data Analysis*. New York: Springer Science+Business Media.

Foreman L. (2011). Edison nation. *Research Technology Management*, 54(6), 37–41.

Gibbons P. (2012). Don't doom your VoC strategies: A six-step plan to overcoming common obstacles. *Customer Relationship Management Magazine*, 16(11), 52.

Hollander M., Wolfe D. A., Chicken E. (2014). *Nonparametric Statistical Methods*, 3rd ed. Hoboken, NJ: John Wiley & Sons.

Hsieh L., Chen S. (2005). Incorporating voice of the consumer: Does it really work? *Industrial Management & Data Systems*, 105(6), 769–785.

Meilgaard M., Civille G., Carr B. (2006). *Sensory Evaluation Techniques*, 4th ed. Boca Raton, FL: CRC Press.

Pesarin F., Salmaso L. (2010). *Permutation Tests for Complex Data: Theory, Applications and Software*. Wiley Series in Probability and Statistics. Chichester, UK: John Wiley & Sons.

Shaffer J. P. (1986). Modified sequentially rejective multiple test procedure. *Journal of the American Statistical Association*, 81, 826–831.

Stockstrom C., Herstatt C. (2008). Planning and uncertainty in new product development. *R&D Management*, 38(5), 480–490.

6

Subjective Evaluations and Perceived Quality of the Indoor Environment

Questionnaire-based surveys and observational studies are nonexperimental designs where implicitly the need to define a global ranking of multivariate populations from a set of ordered categorical response variables occurs many times. This chapter presents three real case studies in the field of perceived quality and well-being in which survey-based subjective evaluations are used to perform multivariate ranking analysis to rank, from the 'best' to the 'worst', several indoor environments in terms of perceived indoor environmental quality. The need to provide a high level of indoor quality is of particular importance, as it has been widely demonstrated that working, studying and recovering in a comfortable environment enhance productivity, learning and even the process of healing (De Carli et al., 2008). In this connection, the application of multivariate ranking methods proves to be an insightful tool to assess the perceived well-being of indoor environments to establish priorities in operation, maintenance and refurbishment actions.

6.1 Evaluation of Indoor Environmental Conditions in Educational Buildings

This study focuses on three Italian educational buildings, monitored from February to June 2011, involving about 160 pupils 9 to 11 years old (De Giuli et al., 2014a). It is a development of a pilot study, based on spot measurements and surveys previously conducted in seven primary schools (De Giuli et al., 2012). The survey, distributed three times, to assess the actual thermal sensation in different seasons has been somewhat improved together with the statistical analysis using the multivariate ranking approach. On the basis of the collected data from the questionnaire, this work aims to provide, by means of innovative ranking methods, a way to rank schools in terms of perceived indoor environmental quality with reference to microclimatic conditions and building-related factors, psychological factors and interaction between occupants and environment. The multivariate ranking method has proved to be a possible tool to assess building stocks in order to establish priorities in operation, maintenance and refurbishment actions. The main problems observed concern thermal comfort in the hot season and solar penetration, corresponding as well to pupils' complaints.

6.1.1 Introduction

As mentioned previously, the need to provide a high level of indoor quality is of particular importance; in fact, standards suggest, for each microclimatic parameter, a range of acceptable values, but human comfort is, first, subjective and, second, depends on many factors that refer to different aspects, not only to the actual indoor conditions that can be easily recorded with instruments available today (Lundquist et al., 2002). Environmental comfort analysis therefore becomes particularly tricky and requires a multidisciplinary approach with careful investigation from many different research fields, such as engineering, psychology, statistics, medicine, educational science, and so on (Frontczak and Wargocki, 2011).

The panorama of Italian schools is constituted by old buildings, lack of proper thermal insulation (concrete or masonry envelope without any kind of insulation); not energy efficient; classrooms that are undersized for the number of students; insufficient space for break times; and acoustic problems, especially in lunchrooms and gyms, as well as in classrooms. In general they are equipped with neither mechanical ventilation nor a cooling system; further, a room thermostat is not available and in many cases indoor temperatures in winter are extremely high. Venetian blinds are the common shading device, but often they are broken and remain unusable. Comfort inside educational buildings is of particular interest because children spend much of their time at school. Moreover, in contrast with adult who can often make adjustments in their workplaces, they cannot interact with the environment and have to accept indoor conditions passively.

Children are particularly sensitive to environmental factors (Mendell and Heath, 2005); for example, absenteeism has been found to be related to air pollution (Chen et al., 2000) and even has been used as a proxy in air quality investigation (Houghton et al., 2003). Classroom lighting may be important for pupils' learning and teachers' instruction: teacher and pupil preferences about the classroom environment have been studied (Schneider, 2003); glare and imperceptible 100-Hz flicker from fluorescent lamps were examined in a sample of UK schools (Winterbottom and Wilkins, 2009); teachers' preferences for daylight emerged in Hathaway (1983) and for control over lighting levels in Lang (2002).

To investigate the well-being of building occupants, a combination of measurements and survey submissions provides a more complete overview of environmental quality analysis, as occupants' satisfaction usually does not correspond to indoor conditions, even though standards are met. Several studies adopted this method of investigation, looking for a correlation between physical measurements and subjective response, such as another Italian study, in high school and university classrooms, that focussed only on thermal sensation (Corgnati et al., 2007, 2009). We would like to go further, aiming at summing all the collected information, to rank school buildings according to different fields: indoor environmental quality, satisfaction with building-related factors and interaction between occupants and the environment. This study therefore validates a method, while also providing innovative statistical techniques, that allows a global ranking of the investigated indoor environments.

One of the main problems is to establish the weight of each indoor parameter relative to the total environmental evaluation (Humphreys, 2005). Environmental aspects cannot be considered independent of each other; moreover it has been demonstrated that satisfaction with only one aspect can influence a positive evaluation of the whole (i.e., personal control over the environment has been suggested to be a 'forgiveness factor'; see Leaman and Bordass, 1999). The weight that each indoor microclimatic parameter has compared to the overall comfort is subjective and can even change over time and from culture to culture. It has in fact been demonstrated that the attitude towards environmental features changes from one country to another; it would therefore be wrong as well as impossible to define a unique multifactor index for all countries. In addition, individuals have their own priority ranking for various environmental aspects, and hence a unique index for all kinds of evaluation in different kinds of buildings cannot be established. Humphreys (2005) proposed an index of overall comfort, obtained by using an equation in which a different coefficient was applied to each aspect, depending on the relative importance that respondents assigned to them. A comparison between the overall comfort index obtained thereby and the overall comfort described by the occupants showed no correlation. According to Humphreys (2005), as the ranking of buildings by using a combined index seems difficult or even wrong, it would be more prudent to continue to consider each aspect separately.

6.1.2 Description of the Case Study

The educational buildings involved in this research are located in Padua, a town in the north-east of Italy. The primary school X is near the historical centre, in a residential area (Figure 6.1a), while schools Y and Z (Figure 6.1b), two buildings that are part of the same school, are located in the suburbs, facing a busy street.

School X and Z were built in the 1970s, and they have traditional classrooms (Figure 6.2a and c), undersized for the number of students, as often occurs in Italian schools. On the contrary, school Y dates from the early 20th century and has large classrooms (around 70 m^2), with high ceilings and high

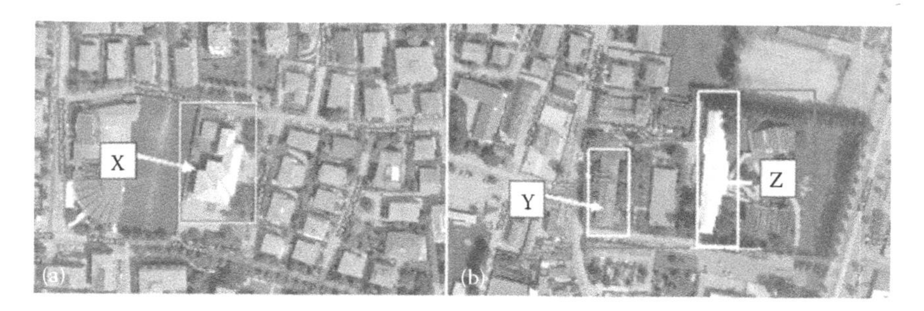

FIGURE 6.1
Aerial pictures of the three educational buildings.

FIGURE 6.2
Pictures of representative classrooms: (a) School X, (b) School Y and (c) School Z.

windowsills and with the windows in front of the blackboard (Figure 6.2b), whereas in the other schools the windows are on the side. All the schools have fluorescent lamps, venetian blinds and radiator heating systems. No mechanical ventilation or cooling system is present.

This research involved a total of 160 9- to 11-year-old students across eight classrooms: four were east-facing (School X, Figure 6.3a); two west-facing (School Y, Figure 6.3b); and two east-facing (School Z, Figure 6.3c). Classes take place from 8 am to 1 pm, Monday through Saturday at Schools X and Y, and from 8 am to 4 pm, Monday through Friday at School Z.

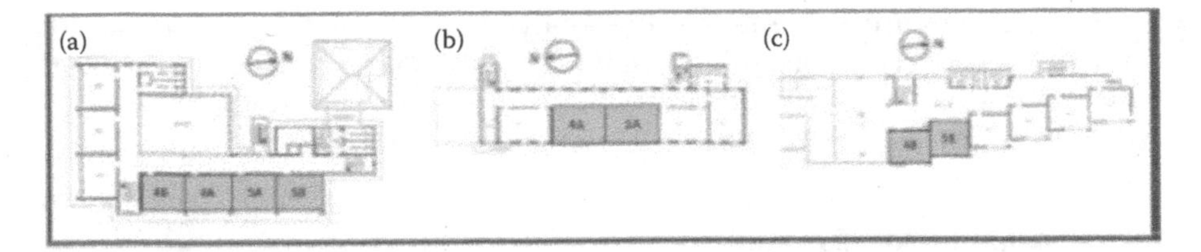

FIGURE 6.3
Schools plants: (a) X, (b) Y and (c) Z.

6.1.3 Materials and Methods

The questionnaire was administered three times: February, April and June. The reason behind the multi-time-submission strategy was not merely to investigate the thermal sensation in three different thermal conditions, but also to overcome the uncertainty of the children's memories, as many of the questions required the children to think over an extended period (i.e., the current school year). Moreover, administering the questionnaire on a number of occasions may help the children to understand the questions better, thus making them more disposed to completing it.

The survey focussed on many aspects that can be summarized as follows: general information (gender, age, health status and psychological state), information about the type of clothing worn at the time the questionnaire was filled out and indoor environmental discomfort (thermal, air quality, visual and acoustical). In the present work we focussed on multivariate ranking analysis using the pupils' assessments on seven discomfort items reported in Table 6.1.

Pupils were asked how often they were annoyed by each of these items and the response options ranged from 1 (never) to 4 (always). For all the questions concerning frequency of discomfort, a pair of four-point scales was chosen to remove the central neutral answer so as to allow a clear distinction between positive and negative answers.

TABLE 6.1
Definitions of Discomfort Items Used to Rank Educational Buildings

Item	Definition
Winter temperature	Frequency of winter thermal discomfort
Summer temperature	Frequency of summer thermal discomfort
Poor air	Frequency of poor air quality
Draughts	Frequency of inconvenient draughts
Artificial light	Frequency of visual discomfort due to artificial light
Sunlight	Frequency of visual discomfort due to sunlight
Noise	Frequency of noisy acoustic discomfort

The questionnaire was made more enjoyable and easier for students to complete by associating a smiley face with the most positive and a frowning face with the most negative answers; the full questionnaire is available as appendix in De Giuli et al. (2014a).

6.1.4 Ranking Analysis Results

A first look at discomfort data is provided by radar graphs in Figures 6.4 and 6.5, where score sample means and top-two-box percentages-T2B (i.e.,

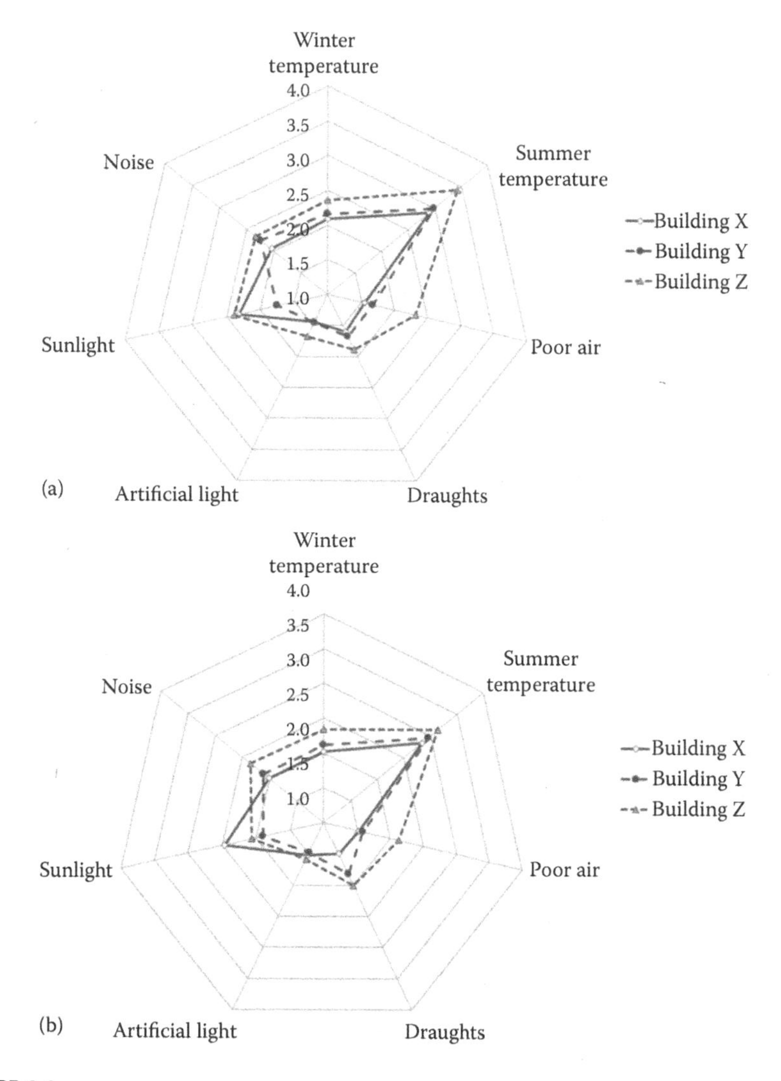

FIGURE 6.4
Radar graphs of score means by time of questionnaire submission. (a) Item-by-item sample score means (all times). (b) Item-by-item sample score means (time 1). *(Continued)*

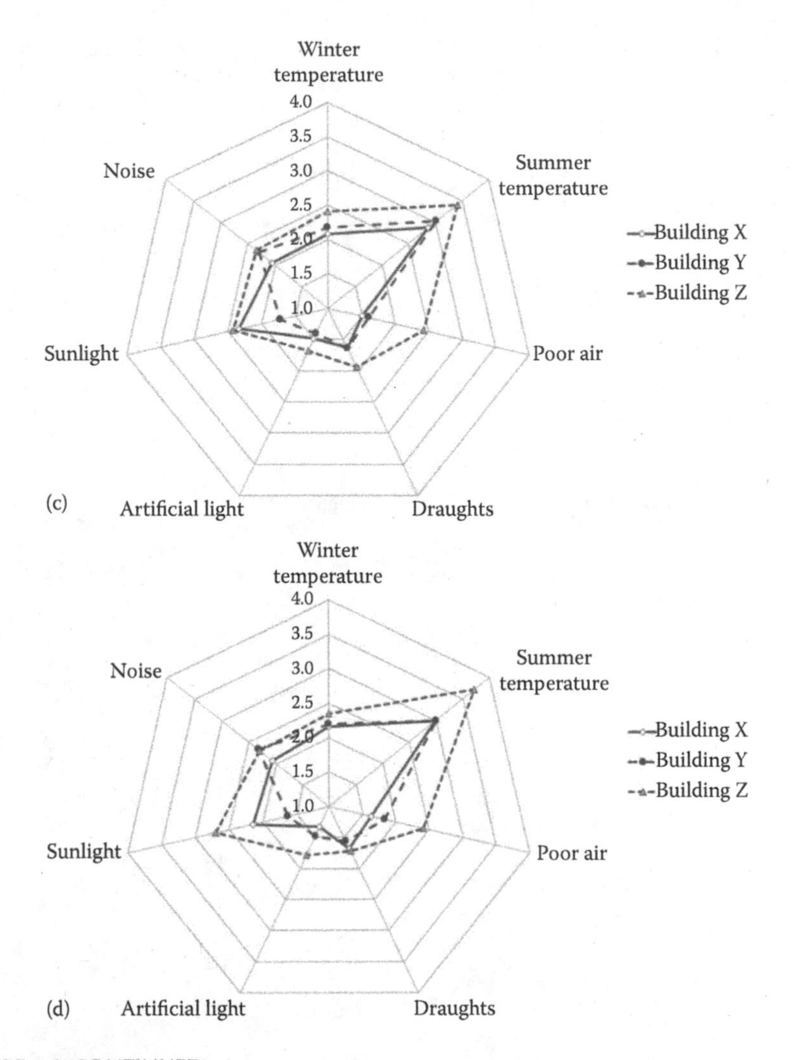

FIGURE 6.4 (CONTINUED)
Radar graphs of score means by time of questionnaire submission. (c) Item-by-item sample score means (time 2). (d) Item-by-item sample score means (time 3).

the percentage of scores greater than 2) by buildings and times are reported. According to De Giuli et al. (2014a), from a descriptive point of view it appears that the most uncomfortable building is Z because the related score means and top-two-box percentages tend to take generally greater values. This remark seems to be more or less valid for each observation time.

The main objective of this work was to rank the three investigated educational buildings from the pupils' point of view using their assessments on discomfort items. For this goal, because the questionnaire was submitted several times, a stratified global ranking analysis is here advised. We recall that by

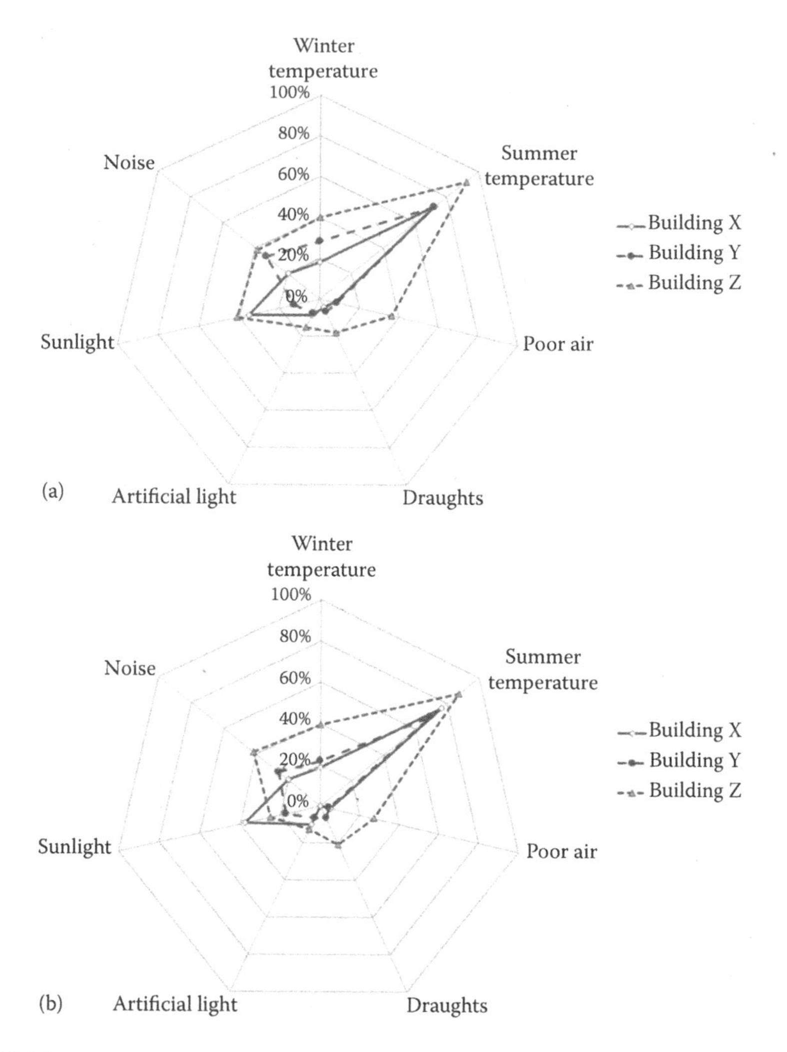

FIGURE 6.5
Radar graphs of T2B percentage by time of questionnaire submission. (a) Item-by-item sample T2B percentages (all times). (b) Item-by-item T2B percentages (time 1). *(Continued)*

'stratified analysis' we mean that several separated ranking analyses for each stratum (time) along with an overall ranking analysis have been performed.

The setting of the indoor environmental conditions ranking problem (see Chapter 3) can be summarized as follows.

- Type of design: one-way MANOVA with stratification (three independent samples, i.e., three multivariate populations to be ranked, and one confounding factor to take into account, i.e., the specific time period of questionnaire submission)

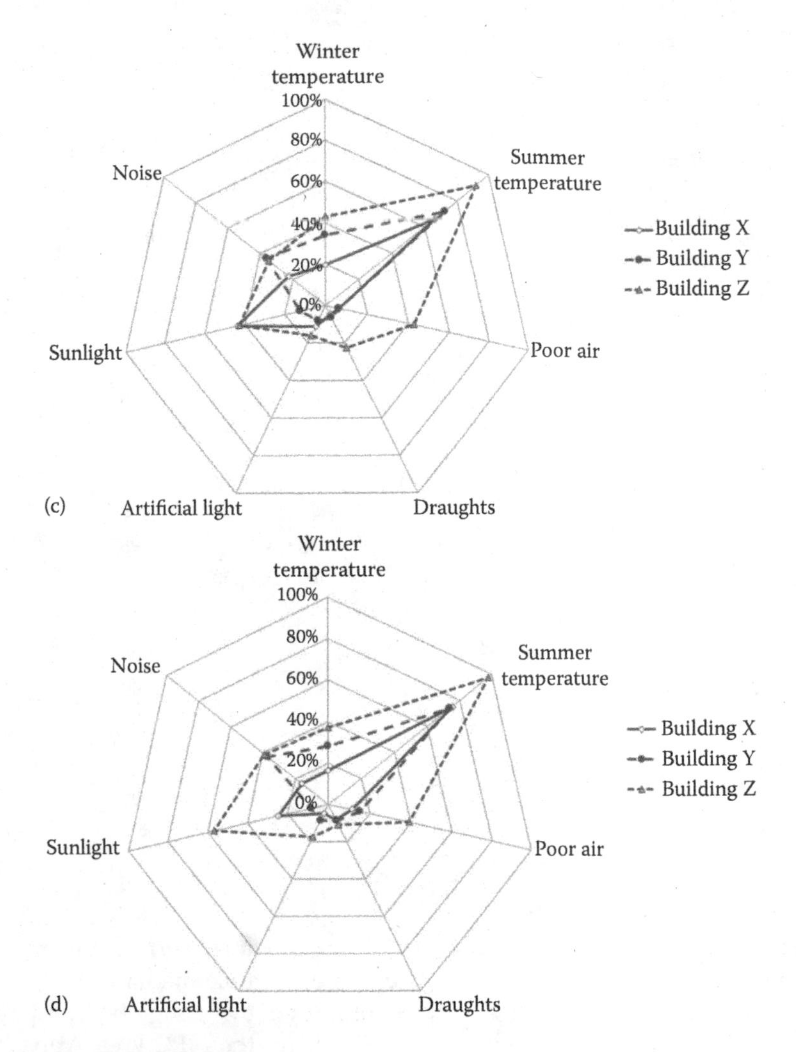

FIGURE 6.5 (CONTINUED)
Radar graphs of T2B percentage by time of questionnaire submission. (c) Item-by-item sample T2B percentages (time 2). (d) Item-by-item T2B percentages (time 3).

- Domain analysis: no
- Type of response variables: ordered categorical (seven ordinal 1–4 scores)
- Ranking rule: 'the lower the better'
- Combining function: Fisher
- *B* (number of permutations): 2000

Details on multivariate and univariate educational building comparisons via directional permutation tests are presented in Tables 6.2 and 6.3. Because

TABLE 6.2

Details of Stratified Global Ranking Results for Satisfaction Items ($\alpha = 5\%$)

Educ. Building	**X**	**Y**	**Z**
All Times	**1**	**1**	**3**
Time 1	1	1	3
Time 2	1	1	3
Time 3	1	1	3
All Times	X	Y	Z
X	–	.565	1.00
Y	.133	–	1.00
Z	**.002**	**.001**	–
Time 1	X	Y	Z
X	–	.477	.494
Y	.063	–	.996
Z	**.002**	**.007**	–
Time 2	X	Y	Z
X	–	.511	.999
Y	.370	–	1.00
Z	**.002**	**.001**	–
Time 3	X	Y	Z
X	–	.412	.998
Y	.375	–	.992
Z	**.002**	**.001**	–

the survey presented several missing values, before performing ranking analysis we replaced them by using the related sample median.

All ranking analysis results can be obtained by using the companion book Web-based software (see Chapter 4) within the NPC Web Apps (lstat .gest.unipd.it/npcwebapps/). The reference dataset is labelled Building _discomfort.csv. To replicate the permutation p-values reported in Tables 6.2 and 6.3 exactly, a seed value equal to 1234 must be set up.

Ranking analysis results allow us to easily rank the three educational buildings under investigation using the pupils' assessments on discomfort items. The buildings perceived as most comfortable are X and Y, both tied for classification as the less uncomfortable, while Z had the most unfavourable results. It is also worth noting that the stratified ranking analysis provided several interesting insights into how pupils evaluate the building under different climate times. In fact, the within-strata estimated rankings are exactly the same although the univariate buildings are not identical time-by-time. For example, a significantly greater frequency of visual discomfort due to artificial light is found in building Z at time 3, but not at times 1 and 2.

TABLE 6.3

Overall and Stratified Item-by-Item Univariate Directional Permutation p-Values

	All Times											
	X	Y	Z	X	Y	Z	X	Y	Z	X	Y	Z
	Winter Temperature			*Summer Temperature*			*Poor Air*			*Draughts*		
X	–	.827	.999	–	.803	.999	–	.818	.999	–	.786	.998
Y	.172	–	.987	.196	–	.999	.181	–	.999	.214	–	.992
Z	**.000**	**.012**	–	**.000**	**.000**	–	**.000**	**.000**	–	**.002**	**.008**	–
	Artificial Light			*Sunlight*			*Noise*					
X	–	.452	.993	–	**.000**	.809	–	.978	.997			
Y	.547	–	.996	.999	–	.999	**.022**	–	.786			
Z	**.006**	**.004**	–	.191	**.000**	–	**.003**	.213	–			
	Time 1											
	X	Y	Z	X	Y	Z	X	Y	Z	X	Y	Z
	Winter Temperature			*Summer Temperature*			*Poor Air*			*Draughts*		
X		.805	.995		.792	.962		.624	.999		.998	.999
Y	.195		.953	.207		.801	.375		.999	**.001**		.767
Z	.004	.047		.038	.198		**.000**	**.000**		**.000**	.232	
	Artificial Light			*Sunlight*			*Noise*					
X	–	.395	.697	–	**.000**	**.002**	–	.713	.971			
Y	.605	–	.781	.999	–	.851	.287	–	.901			
Z	.302	.219	–	.997	.149	–	.028	.099	–			
	Time 2											
	X	Y	Z	X	Y	Z	X	Y	Z	X	Y	Z
	Winter Temperature			*Summer Temperature*			*Poor Air*			*Draughts*		
X	–	.633	.984	–	.780	.999	–	.789	.999	–	.431	.956
Y	.367	–	.942	.220	–	.977	.211	–	.999	.568	–	.984
Z	**.016**	.058	–	**.001**	.023	–	**.000**	**.000**	–	.043	**.016**	–
	Artificial Light			*Sunlight*			*Noise*					
X	–	.270	.850	–	**.001**	.751	–	.931	.970			
Y	.730	–	.947	.999	–	.999	.069	–	.620			
Z	.150	.053	–	.249	**.001**	–	.029	.379	–			

(Continued)

TABLE 6.3 (CONTINUED)

Overall and Stratified Item-by-Item Univariate Directional Permutation p-Values

	Time 3											
	X	Y	Z	X	Y	Z	X	Y	Z	X	Y	Z
	Winter Temperature			*Summer Temperature*			*Poor Air*			*Draughts*		
X	–	.552	.849	–	.364	.999	–	.754	.999	–	.212	.625
Y	.448	–	.802	.635	–	.999	.245	–	.999	.788	–	.876
Z	.151	.198	–	**.000**	**.000**	–	**.000**	**.001**	–	.374	.123	–
	Artificial Light			*Sunlight*			*Noise*					
X	–	.880	.997	–	**.001**	.999	–	.914	.871			
Y	.120	–	.975	.999	–	.999	.086	–	.433			
Z	**.003**	.025	–	**.000**	**.000**	–	.128	.566	–			

6.1.5 Conclusions

The purpose of this research was not merely to present a case study, but also to draw up a roadmap that could be applied when a building has to be evaluated for occupant–building interactions or needs repairs. The approach presented here may be useful for designing human-centred schools and for finding a solution to the problems brought to light by both objective measurements and human perception.

This work presents a survey-based method to investigate indoor environmental conditions in school buildings: the analysis consists of a subjective approach, which allows an exhaustive evaluation. An innovative statistical approach has been used to evaluate the items of the questionnaire: a multivariate ranking of the schools has been established, considering different aspects referring to the same domain, that is, the frequency of discomfort. Buildings X and Y turn out to be the best schools from the pupils' perceived environmental comfort point of view: this fact confirms that these buildings require the fewest environmental changes because of the more comfortable indoor conditions.

6.2 Evaluation of Indoor Environment Conditions in Classrooms

In this work students' sensations perceived in the classroom during school activities were investigated by means of two surveys submitted on several occasions, one related to actual microclimatic conditions and one to overall satisfaction, occupant–building interaction and reactions to discomfort

(De Giuli et al., 2014b). Multivariate nonparametric ranking methods were then applied to compare and rank the classrooms in accordance with students' subjective perceptions; a global ranking analysis has also been performed, jointly considering thermal and visual comfort and air quality.

6.2.1 Introduction

Living in a pleasant, comfortable environment enhances people's well-being and satisfaction. These conditions are particularly important in commercial, educational and healthcare buildings, where good Indoor Environmental Quality (IEQ) influences productivity, learning and convalescence (Frontczak and Wargocki, 2011). This study focuses on Italian children (9–11 years old) and their schools, where they usually spend about 8 hours a day, that is, one third of their time. Children are very sensitive to indoor conditions because they are at a growing age and may develop hypersensitivity if they are exposed to poor indoor conditions (Mendell and Heath, 2005). Bakó-Biró et al. (2012) investigated the influence of poor air quality on pupils' attention and vigilance, and results reveal that poor ventilation lowers memory and concentration.

Many Italian schools have structural problems, such as insufficient space and safety, poor indoor air quality, low thermal comfort and bad acoustics; they are often energy inefficient due to obsolete systems, the facades are poorly, insulated if at all and in some cases windows are single glazed.

Many studies have evaluated the indoor conditions of educational buildings, taking into account single aspects (Bakó-Biró et al., 2012; Corgnati et al., 2009; Mendell and Heath, 2005), or various aspects of overall comfort perception (Astolfi and Pellerey, 2008; Astolfi et al., 2003; De Giuli et al., 2012; Mumovic, 2009).

Four different classrooms and a total of 62 students were involved in this research. The school is attended by pupils from different countries and many of them belong to families with economic difficulties; therefore the sample represents a range of cultures, habits and social classes, which may influence perception of comfort and expectations. Some studies have analysed the role of ethnicity, socioeconomic status, teacher quality and school size in learning achievements (Nichols, 2003; Toutkoushian and Curtis, 2005). One study surveyed New York City public schools to clarify the role that the condition of school facilities plays in educational outcomes (Duran-Narucki, 2008), starting from the fact that poor living conditions are related to a higher risk of social and emotional difficulties (Evans and English, 2002).

The IEQ analysis presented here is a development of the methodology explained and applied in De Giuli et al. (2012), where environmental snapshots were taken in seven primary schools near Venice; spot measurements were taken and pupils filled in a survey about their general satisfaction with indoor conditions. In that study, no specific reference was made to actual conditions, except for thermal sensation; moreover, spot measurements were recorded only once and no continuous measurements were carried out. The present analysis, however, overcomes these limitations with a specific 'spot survey' designed to compare recorded parameters with actual personal feelings.

6.2.2 Description of the Case Study

In this study the level of children's satisfaction was investigated in an educational building by means of a subjective approach, that is, via survey administration.

The school itself is in a residential area near the centre of Padua (Figure 6.6a), north-east Italy. In Italy, children begin primary school when they are 6 and finish when they are 11. School starts in mid-September and ends in early June. This research studied four of the school's classrooms: two were west-facing (A, B) and two east-facing (C, D). These classrooms were used by two fourth-year classes and two fifth-year classes, the oldest years at the school.

The research lasted three months, from March to June; it therefore covered one month in heating conditions (until April 16) and two months in free-running conditions. The four classrooms are the same size (around 45 m^2); their façades are singled glazed, shaded with internal venetian blinds (Figure 6.7) and the blackboard is opposite the windows. Classrooms B and C have two facades, but classrooms A and D only one. All of the classrooms are heated by radiators, and lit by four fluorescent lamps;

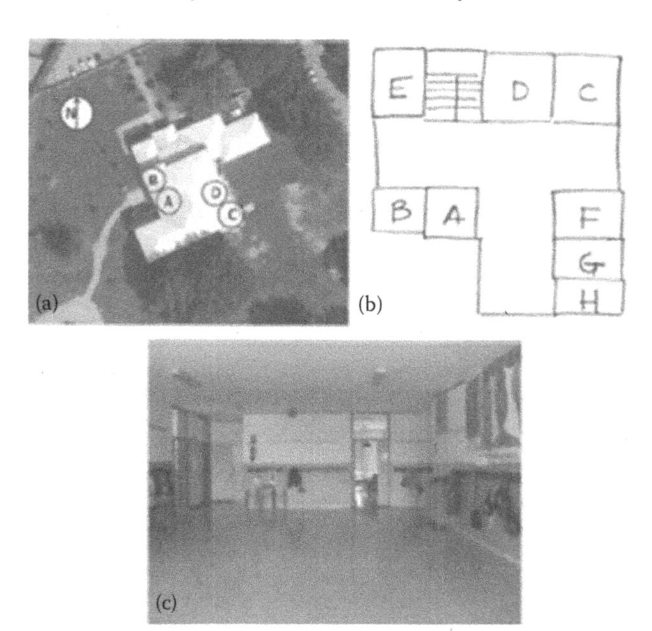

FIGURE 6.6
(a) Aerial picture of the school, (b) classroom map and (c) classroom representative picture.

FIGURE 6.7
Classrooms A (a), B (b), C (c) and D (d).

they have no cooling or mechanical ventilation systems. Pupils used two classrooms during the monitoring period: the fourth-year classes moved between classrooms A and B, while the fifth-year classes moved between classrooms C and D.

From the answers given in the survey, the multivariate ranking methodology has been applied to provide a global ranking of the four classrooms, jointly considering thermal comfort, air quality and visual comfort. Considering classrooms' orientation and position inside the building, it could be expected that, as regards thermal aspect, classroom B should be the least comfortable (north-west oriented and two external façades).

6.2.3 Materials and Methods

Two different surveys were administered to pupils in a subjective approach; one referred to actual conditions (i.e., the spot survey) and one (i.e., the general survey) to satisfaction with building-related aspects (e.g., building, space, furniture, etc.), occupant–building interaction (raising/lowering blinds, switching lights on/off, opening/closing windows) and reactions to discomfort. The spot questionnaire was distributed four times during the recording of spot measurements. The general survey was administered only twice, at the beginning and at the end of the research.

Although it was completed anonymously, each questionnaire had an identification number on the front page so that all the surveys for each pupil could be grouped. Subjective scales have been simplified and changed from

TABLE 6.4
Definitions of Comfort Items Used to Rank the Classrooms

Item	Definition
Thermal comfort	How do you describe your thermal comfort at this time?
Air quality	How good do you feel the air quality is at this time?
Visual comfort	Can you see well at your desk?

the ones suggested in ISO 28802 because of the tender age of the respondents. Perceptual and affective scales were used and not preference ones. As regards the thermal sensation scale, the traditional seven-point scale has been reduced to a five-point one: this scale has been used in the spot survey, ranking actual thermal sensation. In the long survey, thermal sensation was investigated in terms of frequency of thermal discomfort during both cold and warm seasons, by means of a four-point scale. Questions about air quality and visual quality have a four-point scale, like the one suggested in ISO 28802. In the long survey, all the questions about satisfaction of the overall building have an affective scale of four point and even the ones about the use of shading, window and lighting.

In the present work we focussed on multivariate ranking analysis using the pupils' assessments on three comfort four-point scale items reported in Table 6.4.

6.2.4 Ranking Analysis Results

A first look at comfort data is provided in the radar graphs in Figure 6.8, where score sample means by classroom and questionnaire submission time are reported. Accordingly to De Giuli et al. (2014b), from a descriptive point of view it appears that the most comfortable room seems to be A, especially in times 3 and 4, because the related score means tend to take generally large values. Conversely, classroom D appears to be the least comfortable one because its score means are generally lower than those of the other classrooms, especially in time 3. In terms of which comfort dimension is the one that most discriminates the perception of well-being in the classroom, the radar graphs seem to suggest that thermal comfort should be considered as the most valuable feature whereas for visual comfort and air quality the score means by classroom look quite similar to each other.

The main objective of this work was to rank the four investigated classrooms from the pupils' point of view using their assessments on comfort items. For this goal, because the questionnaire was submitted several times, a stratified global ranking analysis is advised. We recall that with 'stratified analysis' we mean that several separated ranking analyses for each stratum (time) along with an overall ranking analysis should be performed.

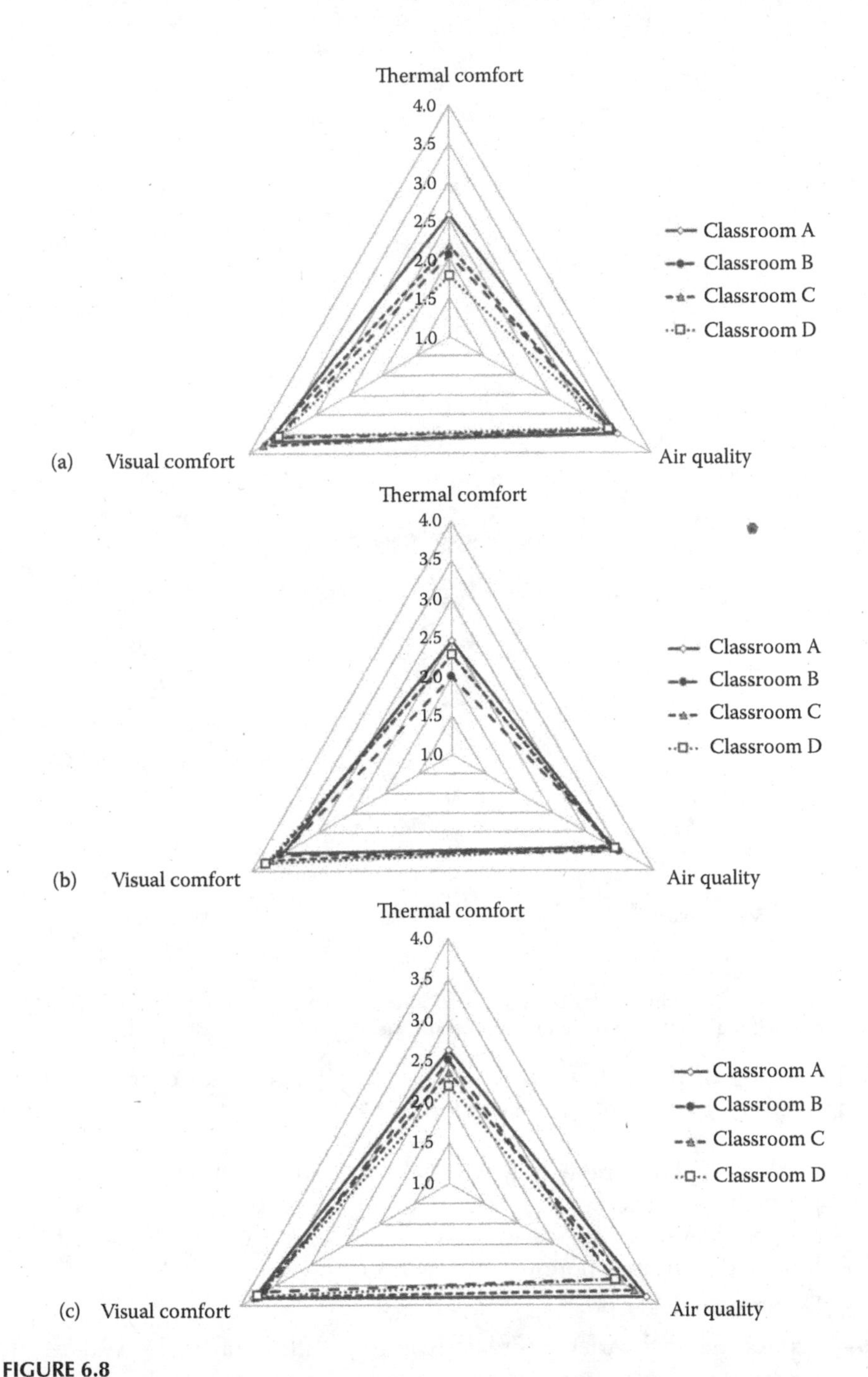

FIGURE 6.8
Radar graphs of score means by time of questionnaire submission. (a) Item-by-item sample score means (all times). (b) Item-by-item sample score means (time 1). (c) Item-by-item sample score means (time 2).

(Continued)

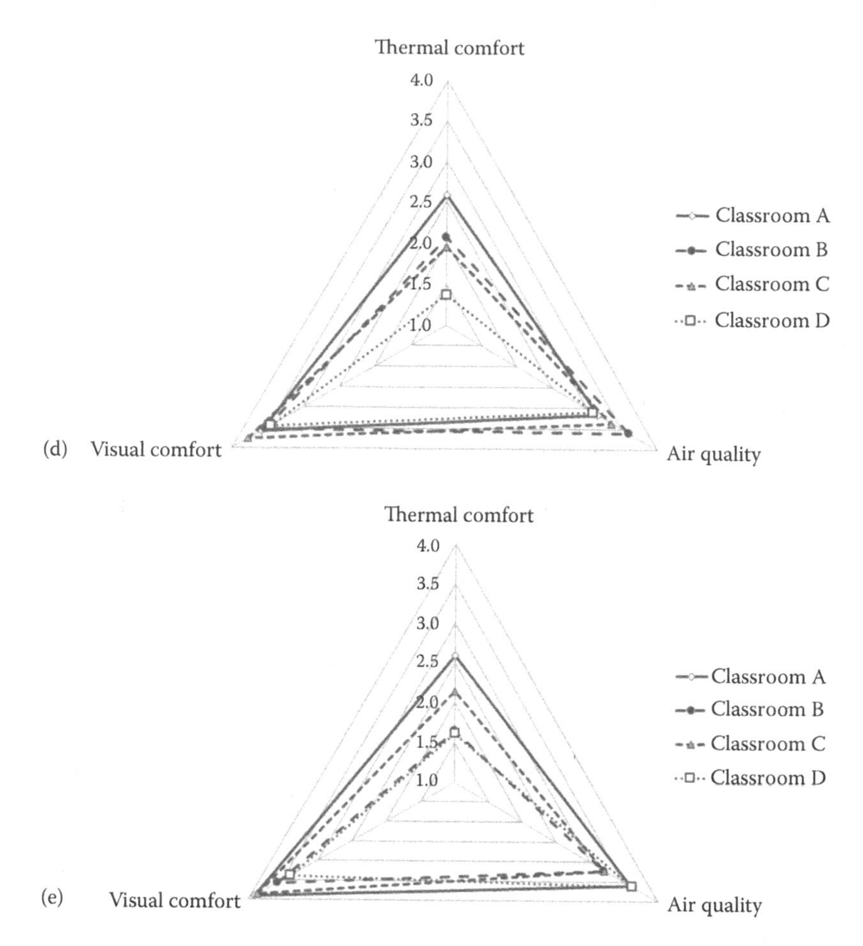

FIGURE 6.8 (CONTINUED)
Radar graphs of score means by time of questionnaire submission. (d) Item-by-item sample score means (time 3). (e) Item-by-item sample score means (time 4).

The setting of the indoor environment conditions ranking problem (see Chapter 3) can be summarized as follows.

- Type of design: one-way MANOVA with stratification (four independent samples, i.e., four multivariate populations to be ranked, and one confounding factor to take into account, i.e., the specific time period of questionnaire submission)
- Domain analysis: no
- Type of response variables: ordered categorical (three ordinal 1–4 scores)
- Ranking rule: 'the higher the better'
- Combining function: Fisher
- B (number of permutations): 2000

Details on multivariate and univariate classroom comparison via directional permutation tests are presented in Tables 6.5 and 6.6. As the survey presented several missing values, before performing ranking analysis we replaced them by using the related sample median.

All ranking analysis results can be obtained by using the companion book Web-based software (see Chapter 4) within the NPC Web Apps (lstat.gest.unipd.it/npcwebapps/). The reference dataset is labelled Classroom_comfort.csv. To replicate the permutation *p*-values reported in Tables 6.5 and 6.6 exactly, a seed value equal to 1234 must be set up.

TABLE 6.5

Details of Stratified Global Ranking Results on Four Classrooms (α = 10%)

Classroom	**A**	**B**	**C**	**D**
All Times	**1**	**3**	**2**	**4**
Time 1	1	1	1	1
Time 2	1	2	2	4
Time 3	1	2	2	4
Time 4	1	3	2	3
All Times	*A*	*B*	*C*	*D*
A	–	**.006**	**.024**	**.006**
B	1.00	–	.894	.575
C	1.00	.525	–	**.037**
D	.987	1.00	1.00	–
Time 1	*A*	*B*	*C*	*D*
A	–	1.00	1.00	1.00
B	.582	–	1.00	.908
C	1.00	.904	–	1.00
D	.936	.780	1.00	–
Time 2	*A*	*B*	*C*	*D*
A	–	.166	.357	**.036**
B	1.00	–	.819	.857
C	1.00	1.00	–	.462
D	.992	1.00	1.00	–
Time 3	*A*	*B*	*C*	*D*
A	–	.478	.447	**.030**
B	1.00	–	.307	.082
C	1.00	1.00	–	.109
D	1.00	1.00	.991	–
Time 4	*A*	*B*	*C*	*D*
A	–	.027	.157	**.039**
B	.992	–	.953	1.00
C	1.00	.363	–	.072
D	1.00	1.00	1.00	–

TABLE 6.6

Overall and Stratified Item-by-Item Univariate Directional Permutation *p*-Values

	A	*B*	*C*	*D*	*A*	*B*	*C*	*D*	*A*	*B*	*C*	*D*
						All Times						
	Thermal Comfort				*Air Quality*				*Visual Comfort*			
A	–	**.000**	**.001**	**.000**	–	.261	.175	.125	–	.145	.808	.246
B	.999	–	.821	.089	.738	–	.446	.428	.855	–	.965	.574
C	.999	.179	–	**.014**	.825	.553	–	.458	.191	**.035**	–	**.043**
D	.999	.910	.986	–	.875	.571	.542	–	.754	.426	.957	–
						Time 1						
	Thermal Comfort				*Air Quality*				*Visual Comfort*			
A	–	**.047**	.252	.274	–	.731	.393	.645	–	660	.810	.874
B	.953	–	.841	.875	.268	–	.233	.360	.339	–	.774	.875
C	.748	.159	–	.557	606	.766	–	.699	.189	.226	–	.651
D	.726	.124	.443	–	.354	.639	.300	–	.125	.124	.349	–
						Time 2						
	Thermal Comfort				*Air Quality*				*Visual Comfort*			
A	–	.225	.086	**.018**	–	**.024**	.185	**.024**	–	.185	.440	.359
B	.774	–	.497	.126	.976	–	.846	.552	.315	–	.764	.739
C	.914	.503	–	.177	.815	.154	–	.160	.560	.236	–	.405
D	.982	.874	.822	–	.976	.448	.840	–	.640	.261	.594	–
						Time 3						
	Thermal Comfort				*Air Quality*				*Visual Comfort*			
A	–	**.049**	**.019**	**.001**	–	.960	.728	.335	–	.343	.606	.249
B	.950	–	.341	**.016**	**.039**	–	.104	**.048**	.657	–	.815	.394
C	.980	C	–	**.040**	.272	.895	–	.181	.393	.184	–	.137
D	.999	.984	.959	–	.664	.951	.819	–	.751	.606	.863	–
						Time 4						
	Thermal Comfort				*Air Quality*				*Visual Comfort*			
A	–	**.006**	.088	**.006**	–	.163	.118	.445	–	.082	.418	**.020**
B	.993	–	.928	.492	.837	–	.532	.880	.917	–	.918	.260
C	.912	.071	–	.076	.881	.467	–	.881	.581	.082	–	**.014**
D	.993	.507	.923	–	.555	.119	.119	–	.979	.740	.985	–

Ranking analysis results allow us to easily rank the four classrooms under investigation using the pupils' assessments on comfort items. The classroom perceived as most comfortable is A, while the worst results are for D. It is also worth noting that the stratified ranking analysis provided several interesting insights into how pupils evaluate the classrooms under different climate times. In fact, the within-strata estimated rankings by time are not always the same; in particular, at time 1 all classrooms are tied and ranked in first position.

6.2.5 Conclusions

A comprehensive analysis in four classrooms, under different indoor conditions, was conducted in which two different kinds of surveys were administered. The global ranking analysis performed considering subjective responses about air quality and thermal and visual comfort allows us to state that classroom A, the classroom with only one external façade, which faces west, is the most comfortable with reference to all monitoring times. The two classrooms with two external facades (B and C) almost always have the same ranking, while classroom D, an east-facing corner room, is almost always ranked as the worst. Accordingly, to improve the pupils' well-being, classroom D should be considered at the top of the list for actions aimed at improving the indoor climatic conditions.

6.3 Measured and Perceived Indoor Environmental Quality at a Hospital

This section presents the results of a field monitoring of three medical wards, Orthopaedics, Internal Medicine and Paediatrics, located at Padua General Hospital in north-east Italy (De Giuli et al., 2013). The perceived indoor environmental quality was investigated by means of a survey administered to both patients and medical staff. A multivariate ranking analysis was applied to investigate the survey with the goal of comparing the wards from both patients' and staff's points of view. Such a global assessment can be easily used to rank the necessity or the priority of amendments or improvements in the buildings and also to determine the most important features both singularly and globally to be taken into consideration.

6.3.1 Introduction

Hospitals are primarily and traditionally a place for people to recover from illness or disease, but a new concept requires such buildings to offer more, particularly following several studies that have demonstrated that improved environmental quality can increase productivity and reduce recovery times. Even the aesthetics of a building can influence the health and well-being of patients and staff: a Norwegian study (Caspari et al., 2006) presented a survey on aesthetic planning and the importance of aesthetics in the strategic planning of hospitals in that country.

An interesting literature review investigates the concept of 'comfort' for building occupants, analysing whether all environmental conditions have an equal impact on human comfort and which nonenvironmental conditions affect indoor quality (Frontczak and Wargocki, 2011). The review analyses the impact of individual characteristics of building occupants (gender, age, country of origin), building-related factors (room interior, type of building, possibility of user control), outdoor climate and season on satisfaction with Indoor Environmental Quality (IEQ).

As part of the concept of patient-centred care, the humanization of health-care settings has to be taken into account. In an Italian study, a number of orthopaedics units with different levels of environmental humanisation were selected from three different hospitals (Fornara et al., 2006): the purpose of this study was to develop perceived hospital environmental quality indicators and rank hospital settings, to compare hospitals in terms of spatial-physical humanization. An extension of Fornara et al. (2006) in a different cultural context was performed in Portuguese hospitals (Andrade et al., 2012): physical and social environments were evaluated by hospital users and objectively evaluated by architects to establish the Environmental Quality Perception (EQP) of hospital users.

Participation of patients and medical staff in the design process, in what is known as a 'user-centred design', could improve the quality of stay and work in such buildings. A hospital hosts three different categories of people: medical staff, patients and visitors. All of them have different expectations and needs with regard to the building: for staff the hospital is their workplace and therefore a permanent environment; for patients it is usually a temporary place to stay while waiting to return home. Environmental conditions inside a hospital may be secondary for patients, as the skill of staff is the most important aspect. As such their judgement of the building might change depending on the performance of doctors. Staff, on the other hand, might be more objective with regard to indoor conditions and building-related factors (privacy, cleanness, orientation, appearance, etc.), although the relationship between colleagues can influence their judgement, as often happens in the workplace. Finally, visitors might be interested primarily in the comfortable appearance of rooms, in general, and outpatient rooms, in particular (in this work visitors were omitted from the investigation).

A comparison of patient and nurse perceptions of the quality of nursing services, satisfaction and intention to revisit a hospital was analysed in a piece of Korean research using a specific survey (Lee and Yom, 2007). Performance was rated relatively lower than both patient and nurse expectations linked to the overall quality of nursing services. Finnish university hospital registered nurses participated in a specific survey regarding their work environment and nursing outcomes (Hellgren et al., 2011): it showed that quality of care is one of their main complaints and that clinical work is influenced by nurses' complaints. Seasonal variations can influence perception of the thermal environment. A Swedish study found that the difference between staff and patients' perception of indoor air temperature differed more during winter than summer, despite recorded temperatures being similar in both seasons (Skoog et al., 2005).

Indoor environmental quality in hospital operating theatres was analysed in Greek hospitals (Dascalaki et al., 2009): some of the factors that appear to predispose people to report a higher number of discomfort symptoms are gender (women reported symptoms more frequently than men), room size, space availability, occupancy and use of indoor environmental controls. A patient-adjustable comfort warming system was evaluated to study comfort warming rather than therapeutic warming in a survey that investigated satisfaction with thermal comfort and level of anxiety. A group of patients equipped with a patient adjustable warming system was compared to another equipped with a blanket warming control: the first group perceived greater control and satisfaction (O'Brien et al., 2010).

Ventilation in healthcare facilities is extremely important for both patients and staff as it provides thermal comfort and prevents harmful emissions of airborne pathogenic materials. Ventilation requirements also depend on geographical location, as well as economic background. To improve ventilation studies of open ward facilities in the tropics, a Malaysian study (Yau et al., 2011) investigated the ventilation of multiple-bed hospital wards considering the design, indoor conditions and engineering controls.

The effect of the condition, performance and modernity of ventilation systems on the perceived indoor air quality as related to symptoms of hospital employees was analysed in a Finnish study (Tervo-Heikkinen et al., 2008). Well-ventilated environments resulted in fewer complaints and symptoms.

The study presented in De Giuli et al. (2013) reports an analysis, using both an objective and subjective approach, of the indoor environmental quality of a number of medical wards within Padua Hospital. Evaluating indoor conditions by means of surveys alongside measuring campaigns is a widely used approach that has been extensively tested, for example, in educational buildings (Corgnati et al., 2007; De Giuli et al., 2012).

6.3.2 Description of the Case Study

This study was conducted in three med wards of Padua Hospital, located in the city centre: Orthopaedics, Paediatrics, and Internal Medicine (Figure 6.9). Padua is a town in the north-east of Italy and it is characterized by a rather continental climate, with cold and sometimes even humid winter and hot-humid summer. The Orthopaedics building (Figure 6.9a) is five floors high, but only the fourth floor has been investigated. Patients' rooms are facing north and south and the building has been recently renovated.

Internal Medicine is located at the ninth floor of a building eleven floors high (Figure 6.9b). The beds are located all along a central aisle and they are all south-east oriented, except two rooms, which are north-west oriented (Figure 6.10b). Paediatrics is a five-floor building (Figure 6.9c), recently renovated, with all the inpatient rooms south oriented, while the offices face north (Figure 6.10c).

FIGURE 6.9
Plant map of the three med wards analyzed: Orthopaedics (a), Internal Medicine (b) and Paediatrics (c).

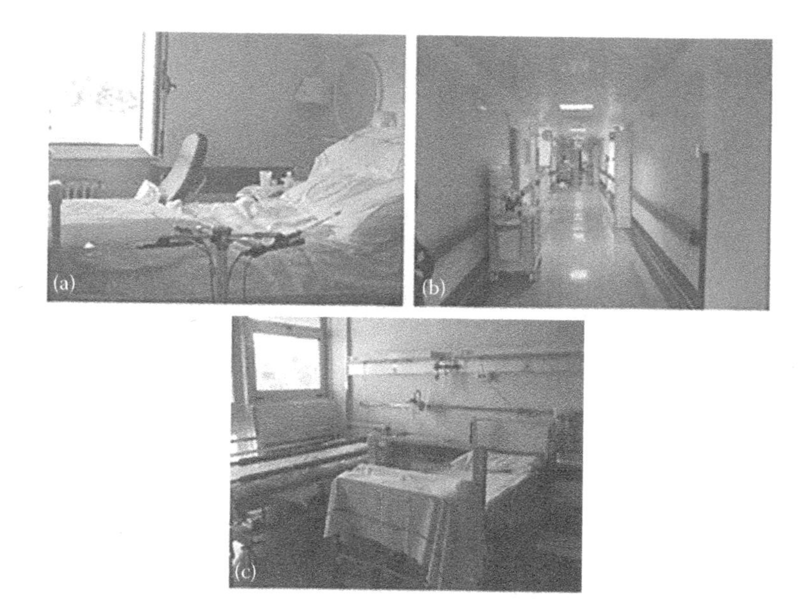

FIGURE 6.10
Internal appearance of the analyzed med wards: Orthopaedics (a), Internal Medicine (b) and Paediatrics (c).

Both medical staff and patients have been involved in the evaluation of the buildings and of indoor microclimatic conditions. The sample comprised 55 people belonging to the staff (26 for Orthopaedics, 16 for Internal Medicine and 13 for Paediatrics) and 35 patients (20 of Orthopaedics and 15 of Internal Medicine).

6.3.3 Materials and Methods

Two different surveys were developed, one for patients and one for doctors and nurses. The surveys focus on many aspects that can be summarized as follows: general information (gender, age, state of health and psychological state), satisfaction of the building (services, cleanness, etc.), and indoor environmental comfort (thermal, air quality, acoustical and visual). In the present work we focussed on ranking analysis from a set of 16 items belonging to two main domains (Table 6.7): satisfaction and discomfort.

Respondents were asked how satisfied and how often they were annoyed for each of these items and the response options range from 1 (completely dissatisfied/never) to 7 (completely satisfied/always). Note that for the set of the first 10 items on satisfaction the rule 'the higher the better' holds, while the rule 'the lower the better' holds for the last six frequency discomfort items. The judgement scale of 7 points refers to the one in the Standard UNI EN ISO 10551:2002 (CEN, 1995). The questions are chosen considering which aspects can influence human comfort beyond the ones strictly connected to indoor microclimatic conditions.

TABLE 6.7

Definitions of Items Investigated for Both Staff and Patients

Item	Domain	Definition
Orientation	Satisfaction	Ease of finding your way around
Cleanness	Satisfaction	Cleanness of the hospital
Common spaces	Satisfaction	Amount of outpatient areas
Homey	Satisfaction	Homely atmosphere in patients' rooms
Room size	Satisfaction	Patients' room sizes
Privacy	Satisfaction	Privacy inside patients' rooms
Building	Satisfaction	Hospital structure on the whole
Temperature	Satisfaction	Indoor air temperature
Daylight	Satisfaction	Daylight availability in the medical ward
Blinds	Satisfaction	Available shading devices
Draughts	Discomfort	Presence of draughts
Poor air	Discomfort	Frequency of poor air quality
Day noise	Discomfort	Frequency of acoustic discomfort during the day
Night noise	Discomfort	Frequency of acoustic discomfort during the night
Daylight	Discomfort	Visual discomfort due to daylight
Artificial light	Discomfort	Frequency visual discomfort due to artificial light

6.3.4 Ranking Analysis Results

A first look at satisfaction and discomfort data is provided by radar graphs in Figures 6.11 and 6.12, where score sample means and top-three-box percentages-T3B (i.e., the percentage of scores greater than 4) by medical ward and type of respondent are reported. As pointed out by De Giuli et al. (2013), all things being equal, medical ward patients appear generally as more satisfied and less frequently discomforted than the medical staff. Moreover, the most

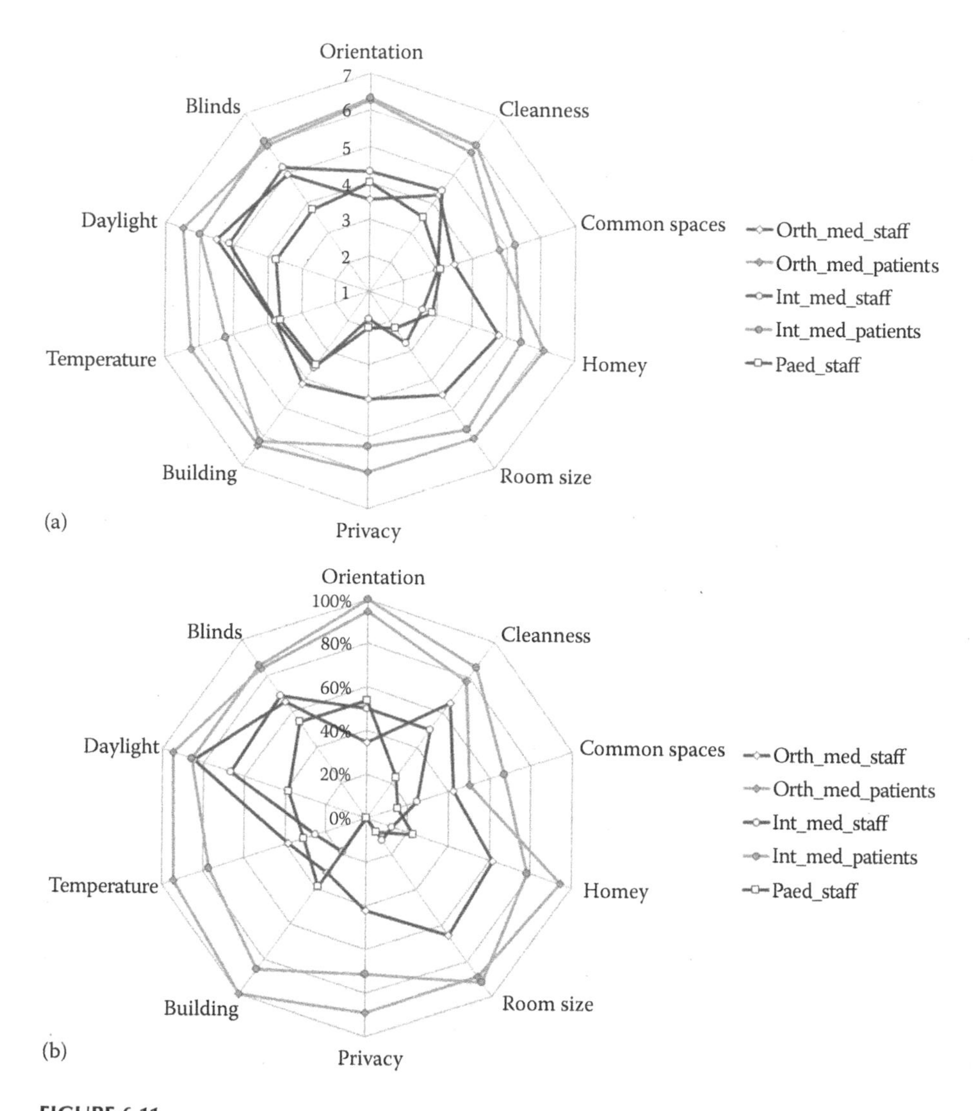

FIGURE 6.11
Radar graphs for satisfaction items by medical ward and type of respondent. (a) Item-by-item sample score means. (b) Item-by-item T3B percentages.

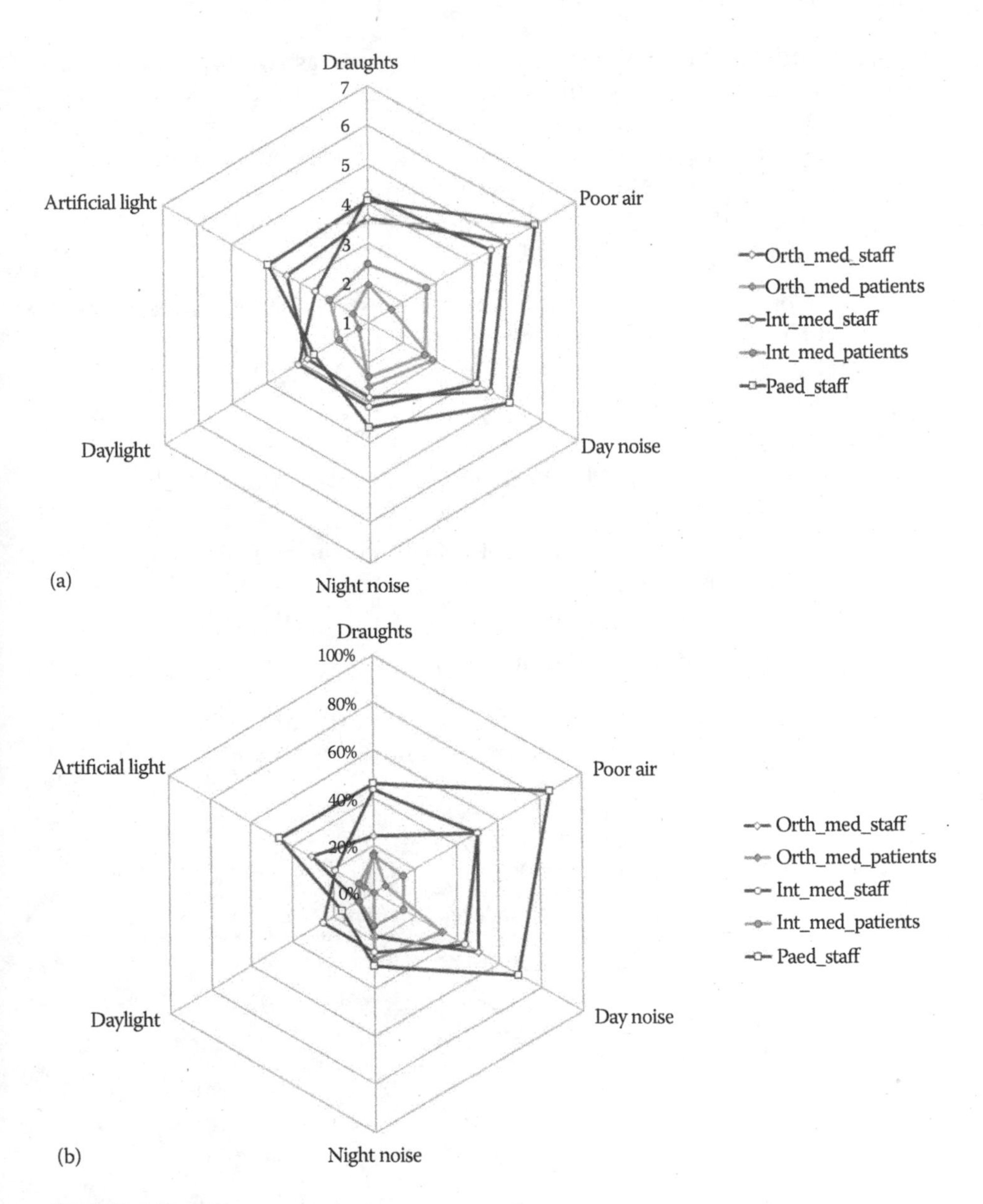

FIGURE 6.12
Radar graphs for discomfort items by medical ward and type of respondent. (a) Item-by-item sample score means. (b) Item-by-item T3B percentages.

and less appreciated and comfortable wards seem to be Orthopaedics and Paediatrics, which are seemingly denoted by the best and the worst assessments respectively.

The main objective of this work was to rank the three investigated medical wards from both patients' and staff's point and views using their assessments on satisfaction and discomfort items. For this goal, as two types of respondent were jointly enrolled in the survey, a stratified global ranking analysis is advised. We recall that by 'stratified analysis' we mean that several separated ranking analyses for each stratum (type of respondent) along with an overall ranking analysis have been performed.

The setting of the perceived indoor environmental quality ranking problem (see Chapter 3) can be summarized as follows.

- Type of design: one-way MANOVA with stratification (three independent samples, i.e., three multivariate populations to be ranked, and one confounding factor to take into account, i.e., the specific type of respondent)
- Domain analysis: no
- Type of response variables: ordered categorical (16 ordinal 1–7 scores; 10 comfort items and 7 discomfort items)
- Ranking rule: 'the higher the better' (comfort items) and 'the lower the better' (discomfort items)
- Combining function: Fisher
- B (number of permutations): 1000

Details on multivariate and univariate medical ward comparisons via directional permutation tests are presented in Tables 6.8 and 6.10, Tables 6.9

TABLE 6.8

Details of Stratified Global Ranking Results for Satisfaction Items ($\alpha = 5\%$)

Medical Ward	Orth_Med	Int_Med	Paed
Global Ranking	**1**	**2**	**3**
Medical Staff	1	1	3
Patients	1	2	–
All Respondents	*Orth_Med*	*Int_Med*	*Paed*
Orth_Med		**.020**	**.003**
Int_Med	1.00		**.003**
Paed	.999	.999	
Medical Staff	*Orth_Med*	*Int_Med*	*Paed*
Orth_Med		**.017**	**.009**
Int_Med	1.00		.081
Paed	.944	.809	
Patients	*Orth_Med*	*Int_Med*	*Paed*
Orth_Med		**.021**	–
Int_Med	.917		–
Paed	–	–	

TABLE 6.9
Satisfaction Item-by-Item Univariate Directional Permutation p-Values

	Orth_Med	Int_Med	Paed	Orth_Med	Int_Med	Paed	Orth_Med	Int_Med	Paed
Orientation					*Cleanness*			*Common Spaces*	
Orth_Med		.933	.108		.673	**.008**		.362	.057
Int_Med	.066		**.007**	.326		**.003**	.637		.110
Paed	.891	.992		.991	.996		.942	.889	
Homey					*Room Size*			*Privacy*	
Orth_Med		**.000**	**.000**		**.000**	**.000**		**.001**	**.000**
Int_Med	.999		.046	.999		**.004**	.998		**.020**
Paed	.999	.953		.999	.995		.999	.979	
Building					*Temperature*			*Daylight*	
Orth_Med		.167	**.003**		.197	**.020**		.208	**.000**
Int_Med	.832		.027	.802		.050	.791		**.001**
Paed	.996	.972		.979	.949		.999	.998	
Blinds									
Orth_Med		.749	**.002**						
Int_Med	.250		**.000**						
Paed	.997	.999							

TABLE 6.10

Details of Stratified Global Ranking Results for Discomfort Items ($\alpha = 5\%$)

Medical Ward	Orth_Med	Int_Med	Paed
Global Ranking	**1**	**1**	**3**
Medical Staff	1	2	2
Patients	1	2	–
All Respondents	*Orth_Med*	*Int_Med*	*Paed*
Orth_Med		1.00	.998
Int_Med	.378		.992
Paed	**.004**	**.006**	
Medical Staff	*Orth_Med*	*Int_Med*	*Paed*
Orth_Med		.905	.947
Int_Med	.573		.899
Paed	.119	.174	
Patients	*Orth_Med*	*Int_Med*	*Paed*
Orth_Med		**.021**	–
Int_Med	.857		–
Paed	–	–	

and 6.11, respectively for satisfaction and discomfort items. As the survey presented several missing values for both staff and patients, before performing ranking analysis we replaced them by using the related sample median.

Ranking analysis results allow us to easily rank the three considered medical wards from both patients' and staff's points of view using their assessments on satisfaction and discomfort items. The most satisfying ward is Orthopaedics, followed by Internal Medicine, while the worst perceived ward is Paediatrics. As far as the most comfortable ward, Orthopaedics and Internal Medicine are tied and both ranked at the first position while Paediatrics is evaluated as the worst one more time. It is also worth noting that the stratified ranking analysis provided several interesting insights into how staff and patients evaluate the medical wards differently. In fact, the within-strata estimated rankings are not exactly the same, although it must be emphasized that they are consistent with one another.

6.3.5 Conclusions

This study presents the results of an extended environmental analysis in three medical wards of Padua hospital. The work consisted of a survey administration to either patients or medical staff. Innovative multivariate nonparametric ranking methods have been applied to analyse the survey, comparing the wards from both patients' and staff's points of view. Ranking results allowed us to rank the medical wards, showing that Orthopaedics is

TABLE 6.11

Discomfort Item-by-Item Univariate Directional Permutation p-Values

	Orth_Med	**Int_Med**	**Paed**	**Orth_Med**	**Int_Med**	**Paed**	**Orth_Med**	**Int_Med**	**Paed**
Draughts					*Poor Air*			*Day Noise*	
Orth_Med		.918	.979		.719	.998		.253	.994
Int_Med	.081		.828	.280		.998	.736		.997
Paed	**.020**	.171		**.001**	**.001**		**.005**	**.002**	
Night Noise					*Daylight*			*Artificial Light*	
Orth_Med		.502	.948		.729	.894		.323	.992
Int_Med	.497		.917	.270		.762	.676		.995
Paed	.051	.082		.105	.237		**.007**	**.004**	

the most preferred and comfortable ward while the worst perceived ward is Paediatrics. Estimated ranking from medical staff and patient assessments have proved to be consistent but not exactly the same.

The ranking results are in agreement with results found by De Giuli et al. (2013), who noted that the staff mostly complained about lack of privacy, room size, amount of common spaces, poor air quality and acoustic discomfort. For the staff, Orthopaedics is the ward showing the highest level of satisfaction regarding building-related aspects, while Internal Medicine is the ward showing the lowest frequency of discomfort. On the other hand, patients expressed a higher level of satisfaction towards building-related aspects and perceived lower frequency of environmental discomfort: this might be related to the different role that the hospital has for the staff with respect to that for patients. Patients have a short stay in the hospital and they are probably more interested in their health than in indoor conditions, while the staff continuously works in the hospital; therefore the environmental conditions inside the building could have an influence on their working activity and well-being.

References

Andrade C., Lima M. L., Fornara F., Bonaiuto M. (2012). Users' views of hospital environmental quality: Validation of the perceived hospital environmental quality indicators (PHEQIs). *Journal of Environmental Psychology*, 32, 97–111.

Astolfi A., Pellerey F. (2008). Subjective and objective assessment of acoustical and overall environmental quality in secondary school classrooms. *Journal of the Acoustical Society of America*, 123(1), 163–173.

Astolfi A., Corgnati S. P., Lo Verso V. R. M. (2003). Environmental comfort in university classrooms and thermal, acoustic, visual and air quality aspects. In *Proceedings of the 2nd International Conference on Research in Building Physics*, Leuven, Belgium, September.

Bakó-Biró Z., Clements-Croome D. J., Kochhar N., Awbi H. B., Williams M. J. (2012). Ventilation rates in schools and pupils' performance. *Building and Environment*, 48, 215–223.

Caspari S., Eriksson K., Naden D. (2006). The aesthetic dimension in hospitals and an investigation into strategic plans. *International Journal of Nursing Studies*, 43, 851–859.

Chen L., Jennison B. L., Yang W., Omaye S. T. (2000). Elementary school absenteeism and air pollution. *Inhalation Toxicology*, 12(11), 997–1016.

Corgnati S. P., Filippi M., Viazzo S. (2007). Perception of the thermal environment in high school and university classrooms: Subjective preferences and thermal comfort. *Building and Environment*, 42, 951–999.

Corgnati S. P., Ansaldi R., Filippi M. (2009). Thermal comfort in Italian classrooms under free running conditions during mid seasons: Assessment through objective and subjective approaches. *Building and Environment*, 42, 785–792.

Dascalaki E. G., Gaglia A. G., Balaras C. A., Lagoudi A. (2009). Indoor environmental quality in Hellenic hospital operating rooms. *Energy and Buildings*, 41, 551–560.

De Carli M., De Giuli V., Zecchin R. (2008). Review on visual comfort in office buildings and influence of daylight in productivity. In *Indoor Air 2008*: The 11th international conference on indoor air quality and climate, Copenhagen, Denmark, August 16–21, 2008.

De Giuli V., Da Pos O., De Carli M. (2012). Indoor environmental quality and pupil perception in Italian primary schools. *Building and Environment*, 56, 335–345.

De Giuli V., Zecchin R., Salmaso L., Corain L. (2013). Measured and perceived indoor environmental quality: Padua Hospital case study. *Building and Environment*, 59, 211–226.

De Giuli V., Zecchin R., Corain L., Salmaso L. (2014a). Measurements of indoor environmental conditions in Italian classrooms and their impact on children. *Indoor and Built Environment*, April 24, 2014.

De Giuli V., Zecchin R., Salmaso L., Corain L. (2014b). Measured and perceived environmental comfort: Field monitoring in an Italian school. *Applied Ergonomics*, 45, 1035–1047.

Duran-Narucki V. (2008). School building condition, school attendance and academic achievement in New York City public schools: A mediation model. *Journal of Environmental Psychology*, 28, 278–286.

Evans G. W., English K. (2002). The environment of poverty: Multiple stressor exposure, psychophysicological stress, and emotional adjustment. *Child Development*, 73(4), 1238–1248.

Fornara F., Bonaiuto M., Bonnes M. (2006). Perceived hospital environment quality indicators: A study of orthopaedics units. *Journal of Environmental Psychology*, 26, 321–334.

Frontczak M., Wargocki P. (2011). Literature survey on how different factors influence human comfort in indoor environments. *Building and Environment*, 46, 922–937.

Hathaway W. E. (1983). Lights, windows, color: Elements of the school environment. In *59th Annual Meeting of the Council of Educational Facility Planners*, International (59th, Columbus, OH, September 26–29, 1982) (pp. 28), Alberta, Canada.

Hellgren U. M., Hyvarinen M., Holopainen R., Reijula K. (2011). Perceived indoor air quality, air-related symptoms and ventilation in Finnish hospitals. *International Journal of Occupational Medicine and Environmental Health*, 24(1), 48–56.

Houghton F., Gleeson M., Kelleher K. (2003). The use of primary/national school absenteeism as a proxy retrospective child health status measure in an environmental pollution investigation. *Public Health*, 117(6), 417–423.

Humphreys M. A. (2005). Quantifying occupant comfort: Are combined indices of the indoor environment practicable? *Building Research & Information*, 33(4), 317–325.

Lang D. C. (2002). Teacher interactions within the physical environment: How teachers alter their space and/or routines because of classroom character. PhD thesis. Seattle: University of Washington.

Leaman A., Bordass W. (1999). Productivity in buildings: The 'killer' variables. *Building Research & Information*, 27(1), 4–29.

Lee M. A., Yom Y. H. (2007). A comparative study of patients' and nurses' perceptions of the quality of nursing services, satisfaction and intent to revisit the hospital: A questionnaire survey. *International Journal of Nursing Studies*, 44, 545–555.

Lundquist P., Kjelleberg A., Holmberg K. (2002). Evaluating effects of the classroom environment: Development of an instrument for the measurement of self-reported mood among school children. *Journal of Environmental Psychology*, 22, 289–293.

Mendell M. J., Heath G. A. (2005). Do indoor pollutants and thermal conditions in schools influence student performance? A critical review of the literature. *Indoor Air*, 15(1), 27–52.

Mumovic D. (2009). Winter indoor air quality, thermal comfort and acoustic performance of newly built secondary schools in England. *Building and Environment*, 44, 1466–1477.

Nichols J. D. (2003). Prediction indicators for students failing the state of Indiana high school graduation exam. *Preventing School Failure: Alternative Education for Children and Youth*, 47(3), 112–120.

O'Brien D., Greenfield M. L., Anderson J. E., Smith B. A., Morris M. (2010). Comfort, satisfaction and anxiolysis in surgical patients using a patient-adjustable comfort warming system: A prospective randomized clinical trial. *Journal of PeriAnesthesia Nursing*, 25(2), 88–93.

Schneider M. (2003). *Linking School Facility Conditions to Teacher Satisfaction and Success*. Washington, DC: National Clearinghouse for Educational Facilities.

Skoog J., Fransson N., Jagemar L. (2005). Thermal environment in Swedish hospitals. Summer and winter measurements. *Energy and Buildings*, 37, 872–877.

Tervo-Heikkinen T., Partanen P., Aalto P., Vehvilainen-Julkunen K. (2008). Nurses' work environment and nursing outcomes: A survey study among Finnish university hospital registered nurses. *International Journal of Nursing Practice*, 14(5), 357–365.

Toutkoushian R. K., Curtis T. (2005). Effects of socioeconomic factors on public high school outcomes and rankings. *Journal of Educational Research*, 98, 5.

Winterbottom M., Wilkins A. (2009). Lighting and discomfort in the classroom. *Journal of Environmental Psychology*, 29, 63–75.

Yau Y. H., Chandrasegaran D., Badarudin A. (2011). The ventilation of multiple-bed hospital wards in the tropics: A review. *Building and Environment*, 46, 1125–1132.

7

Evaluating Customer Satisfaction in Public and Private Services

The permutation approach for ranking multivariate populations can be quite useful in the framework of customer satisfaction evaluation because many times these kinds of studies are denoted by a multivariate ordered categorical response for which the traditional inferential parametric approach lacks effective tools. Moreover, very often the items on which subjects/customers/users are asked to express evaluations must be ordered with respect to one another, from most to least preferred/satisfactory.

In the following sections the multivariate ranking approach is applied to several real customer satisfaction case studies in which several public or private services must be evaluated and ranked using the related evaluations expressed by a sample of users via dedicated surveys. In two cases the extreme profile approach based on the nonparametric combination of dependent rankings is applied as well, highlighting how it may be useful to construct composite multivariate indicators aimed to synthesize either quality or satisfaction evaluations of products and services.

In the next section the ranking methods are applied first to a postdoc survey of the University of Ferrara (Italy) planned and conducted in 2004 with the aim of acquiring knowledge on the professional placing of postdocs, on the relationship between education received during the PhD programme and employment and on their satisfaction with various aspects of the education and research programme conducted during the doctorate programme. The second case study is concerned with comparing and ranking the perceived performance of the services supplied to Italian citizens in different areas by the so-called Territorial Services Centers (TSC). Next, within the winter sport tourism framework, we propose to measure customer satisfaction and service quality by means of multivariate ranking methods applied to the S.E.S.T.O. survey, a study focussed on customer satisfaction with a ski teaching service.

The fourth application is related to an observational survey on tourists' opinions about the Natural Park 'Tre Cime' (District of Sesto Dolomites/Alta Pusteria, Italy), denominated Sesto Nature survey. This study was undertaken in 2010 to assess the quality of structures and services offered inside the natural area. Afterwards, we focus on a survey in which the main goal is to identify homogeneous segments of consumers based on preferences and consumption habits toward demand for wine and specifically for Veneto's

Passito produced in north-east Italy. Finally, in the last case study we deal with a sensory survey on odour perceptions to evaluate the impact on the local population of various malodourous sources located in the territory of two towns.

7.1 Survey on Job Satisfaction of PhD Graduates from the University of Ferrara

In the context of research doctorate evaluation, a sample survey of postdocs from the University of Ferrara was planned and conducted in 2004 with the objective of gathering information on the professional placing of postdocs, on the relationship between education received during the PhD programme and employment, and on their satisfaction with a number of aspects of the education and research programme carried out during the doctorate programme (Arboretti Giancristofaro et al., 2007).

The application of several nonparametric statistical methods, such as Nonparametric Combination (NPC) tests, extreme profile ranking and multivariate ranking methods allows us to gain insights into which kinds of possible differences in job satisfaction occur among different doctorate areas.

7.1.1 Italian Context of Research Doctorate Evaluation

Set up in Italy in 1980, research doctorates (PhDs) represent the third level of university education and constitute a strategic resource in the university for the provision of high standards of education and scientific research methods. In Italy the PhD has established itself predominantly as an academic qualification, with the PhD researcher's career naturally continuing in the academic field or in research institutions, unlike in other countries (e.g. the United States, Germany, United Kingdom) in which the PhD qualification also holds recognized professional value outside of the academic context, in industry or in administration. The Italian university and research system's growing difficulty of absorbing growing numbers of postdocs, together with growing awareness that PhDs can make an important contribution to the cultural and technological innovation, for which the entire economic-production system perceives the need, have represented recent incentives for universities and central authorities to support initiatives for the appreciation, recognition and improvement of the education programme represented by PhDs.

The recent coming into force of university autonomy, together with the gradual changeover to a system of public financing based on the evaluation of achieved results, are further reasons for the Italian university-research system to promote activities for the evaluation of the education programmes provided at various levels, and therefore also PhDs.

Evaluating the quality of a PhD is somewhat complex and includes several dimensions of analysis (adequacy of education content, success of entry into employment, etc.) and several subjects involved in the education process at various levels: PhD researchers and postdocs; teaching staff and doctorate tutors; PhD structures, both internal (departments, institutes) and external in which study and research activities take place and the university-research system and labour market (public and private sector subjects operating in the academic and research field and in the labour market) in which postdocs are placed. By considering the PhD as a process characterized by elements of initial inputs and final outcomes, evaluation of the doctorate is based on process quality indicators and outcome indicators, taking elements of initial inputs (input characteristics of PhD researchers, resources available for the PhDs, characteristics of the territorial context, etc.) into account. In some cases, to define and calculate quality indicators reference is made to objective data (structural data, etc.) that can be obtained from existing sources (e.g. administrative sources); in other cases reference is made to subjective data (opinions, suggestions, judgements of satisfaction) perceived and declared by the various subjects involved (PhD researchers, teachers, etc.), obtained by means of ad hoc surveys and instruments (e.g. sample surveys with a questionnaire).

7.1.2 Survey Description

The postdoc survey at the University of Ferrara involved a representative sample of 120 PhD holders, selected from the four cohorts of 288 postdocs who earned the degree from 2001 to 2004. The various PhDs were grouped into four macro areas: economic–legal (EL), medical–biological (MB), scientific-technological (ST) and the humanities (HU). A random sample of 30 postdocs was extracted for each year, stratified by doctorate area with proportional allocation.

The survey was carried out by means of Computer-Assisted Telephone Interviews (CATI). The electronic questionnaire was divided into six sections: personal details and education, post-doctorate training, employment condition and characteristics, education used at work, job searching and opinion of education received on the doctorate course. To observe postdoc satisfaction with regard to various aspects of their work, the education received and the PhD organization, a scale of scores from 1 to 4 was used (not at all, not very, quite, very satisfied).

7.1.3 Extreme Profile Ranking Results

Extreme profile ranking analysis of the gathered data was structured into two different steps. The first step aims to highlight differences among 'scientific–technological', 'medical–biological' and 'economic–legal' doctorates applying NPC tests. The 'humanities' area, as it is present only in the

2004 cohort, was not included in the analysis by doctorate area. In the second step, global satisfaction indicators based on the extreme profile ranking method were constructed to synthesize PhD students' opinions on several aspects related to the education–employment relationship and employment expectations and opportunities.

With respect to the first phase of the analysis, the results by doctorate area are related to four aspects of postdoc satisfaction: education–employment relationship, teaching and the research work carried out during doctorate, doctorate structures and services, employment expectations and opportunities. The objective of the analysis was to verify the presence of differences in the satisfaction of postdocs belonging to different areas and to understand if any observed differences are linked to the particular nature of the disciplines or if they reflect differences in quality of the courses' internal and external effectiveness. Tables 7.1 through 7.4 show the percentages of very satisfied and quite satisfied postdocs in relation to the variables regarding the four aspects under examination.

Figure 7.1 and Tables 7.5 through 7.7 illustrate the results of tests of hypotheses for the comparison among the three macro areas regarding the four aspects of satisfaction indicated previously, following the nonparametric procedure. Moreover, closed testing based on the permutation minP method (Tippett's test) was applied to control for multiplicity.

With reference to the relationship between PhD education and employment, postdocs (excluding subjects in training, with study grants, research allowances and/or unpaid collaborations) expressed an evaluation of the *coherence* between education and employment, of *use* in employment of the abilities acquired during the PhD programme, and of the *adequacy* of

TABLE 7.1

Education–Employment Relationship

Education–Employment Relationship[a] **% Very Satisfied (% *quite satisfied*)**	**Medical–Biological Area** $N = 20$	**Scientific–Technological Area** $N = 23$	**Economic–Legal Area** $N = 23$	**Total for All Areas** $N = 66$
Education/employment coherence	30.0 (35.0)	60.9 (21.7)	91.3 (8.7)	62.1 (21.2)
Use of education at work	10.0 (45.0)	39.1 (17.4)	65.2 (30.4)	39.4 (30.3)
Adequacy of education with regard to employment	25.0 (30.0)	26.1 (30.4)	56.5 (34.8)	36.4 (31.8)

[a] Subjects in training, with study grants, research allowances or unpaid collaborations are excluded.

TABLE 7.2

Teaching and Research Work Conducted during Doctorate Programme

Teaching and Research Work Carried Out during Doctorate Programme % Very Satisfied (% *quite satisfied*)	**Medical–Biological Area**	**Scientific–Technological Area**	**Economic–Legal Area**	**Total for All Areas**
Courses and seminars (*N* MB = 26; *N* ST = 27; *N* EL = 27)	11.5 (57.7)	14.8 (48.1)	37.0 (55.6)	21.2 (53.8)
Individual research work (*N* MB = 41; *N* ST = 41; *N* EL = 32)	48.8 (34.1)	53.7 (41.5)	59.4 (37.5)	53.5 (37.7)
Research work as part of groups of lecturers and/or students (*N* MB = 28; *N* ST = 18; *N* EL = 21)	39.3 (50.0)	50.0 (27.8)	33.3 (57.1)	40.3 (46.3)
Stays abroad or in structures outside the university (*N* MB = 21; *N* ST = 29; *N* EL = 14)	95.2 (0)	72.4 (20.7)	42.9 (57.1)	73.4 (21.9)
Doctorate thesis (*N* MB = 41; *N* ST = 41; *N* EL = 32)	46.3 (46.3)	53.7 (39.0)	62.5 (37.5)	53.5 (41.2)

TABLE 7.3

Doctorate Structures and Services

Doctorate Structures and Services % Very Satisfied (% *quite satisfied*)	**Medical–Biological Area**	**Scientific–Technological Area**	**Economic–Legal Area**	**Total for All Areas**
Libraries (*N* MB = 34; *N* ST = 36; *N* EL = 25)	38.2 (52.9)	25.0 (55.6)	36.0 (48.0)	32.6 (52.6)
Lecture theatres (*N* MB = 30; *N* ST = 33; *N* EL = 24)	6.7 (73.3)	6.1 (57.6)	16.7 (66.7)	9.2 (65.5)
Study areas (*N* MB = 33; *N* ST = 32; *N* EL = 24)	9.1 (54.5)	34.4 (34.4)	16.7 (54.2)	20.2 (47.2)
Laboratories (*N* MB = 33; *N* ST = 32; *N* EL = 6)	30.3 (45.4)	40.6 (40.6)	0 (16.7)	32.4 (40.8)
Information received on courses and teaching (*N* MB = 33; *N* ST = 39; *N* EL = 29)	21.2 (36.4)	7.7 (48.7)	31.0 (44.8)	18.8 (43.6)
Help with bureaucratic matters (*N* MB = 41; *N* ST = 41; *N* EL = 32)	17.5 (60.0)	19.5 (48.8)	28.1 (43.7)	21.2 (51.3)
Support of lecturers, tutors, coordinator (*N* MB = 40; *N* ST = 41; *N* EL = 32)	43.9 (39.0)	53.7 (29.3)	65.6 (28.1)	53.5 (32.5)

TABLE 7.4
Employment Expectations and Opportunities

Employment Expectations and Opportunities % Very Satisfied (% *quite satisfied*)	Medical–Biological Area $N = 41$	Scientific–Technological Area $N = 41$	Economic–Legal Area $N = 32$	Total for All Areas $N = 114$
Opportunities in the academic world	4.9 (22.0)	17.1 (22.0)	31.3 (43.8)	16.7 (28.1)
Opportunities in the labour market	4.9 (22.0)	4.9 (41.5)	3.1 (50.0)	4.4 (36.8)
Opening towards the scientific community	31.7 (41.5)	31.7 (51.2)	37.5 (53.1)	33.3 (48.3)

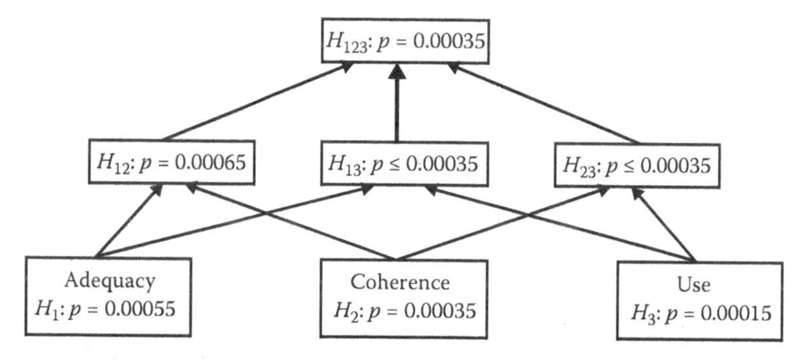

FIGURE 7.1
Closed testing based on Permutation MinP (Tippett's test) for composite hypotheses: Satisfaction with education–employment relationship.

TABLE 7.5
Raw and Adjusted p-Values: Satisfaction with Education–Work Relationship

Partial Tests	Raw p-Values	Adjusted p-Values
Adequacy	$p = .00055$	$p = .00065$
Coherence	$p = .00035$	$p = .00035$
Use	$p = .00015$	$p = .00035$

the PhD training for the work carried out. Both the global test regarding the hypothesis of equality in the multivariate distribution of the three categorical variables (with the following responses: not at all, not very, quite, very satisfied) and the partial tests regarding the contribution of each variable are statistically significant results at level $\alpha = 0.05$, after controlling for multiplicity (Figure 7.1), thus highlighting a difference in satisfaction among the three analysed macro areas. Note that those nodes in Figure 7.1 shown as

TABLE 7.6

Raw and Adjusted p-Values: Satisfaction with Teaching and Research Work

	Raw p-Values	Adjusted p-Values
Overall test	$p = .01435$	
Partial tests		
Courses and seminars	$p = .04495$	n.s.
Individual research	n.s.	n.s.
Group research	n.s.	n.s.
Outside stay	$p = .00295$	$p = .01435$
Doctorate thesis	n.s.	n.s.

Note: n.s. = not significant.

TABLE 7.7

Iterated Combination Procedure: Satisfaction with Structures and Services

	Combining Function		
p-Value of Overall Test	**Tippett**	**Fisher**	**Liptak**
First iteration	.20873	.06864	**.03595**
Second iteration	.06764	.06974	.06994

'$p \leq$' rather than '$p =$' are calculated by implication, not directly, according to remarks in Chapter 3, Section 3.1.6.2. Table 7.7 shows the adjusted p-values for partial tests.

In relation to opinions expressed about teaching and research work carried out during the PhD, three variables recoded as binary with missing values were analysed (courses and seminars, group research and activities outside the university; with values: 0 = not at all, not very, quite satisfied, 1 = very satisfied), and two ordered categorical variables without missing data (individual research and PhD thesis; with the following responses: not at all, not very, quite, very satisfied). Table 7.6 shows the presence of a statistically significant difference at a global level among the macro areas, with a particular contribution from the satisfaction expressed in relation to study periods abroad or in institutions other than the University of Ferrara.

No statistically different evaluation among the EL, ST and MB PhDs emerges in relation to the courses' support services and structures. For this analysis six variables recodified as binary with missing values were considered (libraries, lecture theatres, study areas, laboratories, information about teaching and bureaucratic support; with values: 0 = not at all, not very, quite satisfied, 1 = very satisfied), and a categorical variable without missing data (support from lecturers and tutors; with the following responses: not at all, not very, quite, very satisfied). The combination algorithm with different combining functions was iterated until the final overall p-values became reasonably invariant (Table 7.7). In this way a preliminary, significant result from Liptak's combining function was no longer significant with the second iteration.

The final aspect regarding PhD researchers' expectations of employment after earning the degree (three categorical variables: opportunities in the academic field, opportunities in the labour market and opening towards the scientific community, with the following responses: not at all, not very, quite, very satisfied) again shows a statistically significant difference at a global level among the three PhD areas, to which opinions of opportunities in the academic field contribute very significantly (Table 7.8).

To construct a global satisfaction indicator, the NPC ranking methodology was also applied considering the strong satisfaction profile. For the three doctorate areas, Table 7.9 shows the median values and the interquartile range of the global satisfaction score expressed on a scale of 0–1 in relation to the three aspects regarding the education–employment relationship and the three aspects regarding prospects offered by the doctorate. The box plots shown in Figures 7.2 and 7.3 illustrate the distribution of the two global satisfaction indexes in the three considered groups.

TABLE 7.8

Raw and Adjusted p-Values: Satisfaction with Employment Expectations and Opportunities

	Raw p-Values	Adjusted p-Values
Overall test	$p = .00035$	
Partial tests		
Academic opportunity	$p = .00015$	$p = .00035$
Labour market opportunity	n.s.	n.s.
Scientific community opening	n.s.	n.s.

Note: n.s. = not significant.

TABLE 7.9

Extreme Profile Ranking Analysis

Global Index of Satisfaction *Median (first quartile–third quartile)*	Medical–Biological Area	Scientific–Technological Area	Economic–Legal Area	Total for All Areas
Education–employment relationship (N MB = 20; N ST = 23; N EL = 23)	0.34 (0.12–0.53)	0.47 (0.37–0.63)	0.66 (0.53–0.79)	0.53 (0.36–0.66)
Employment expectations and opportunities (N MB = 41; N ST = 41; N EL = 32)	0.26 (0.20–0.41)	0.32 (0.26–0.47)	0.41 (0.30–0.50)	0.32 (0.26–0.46)

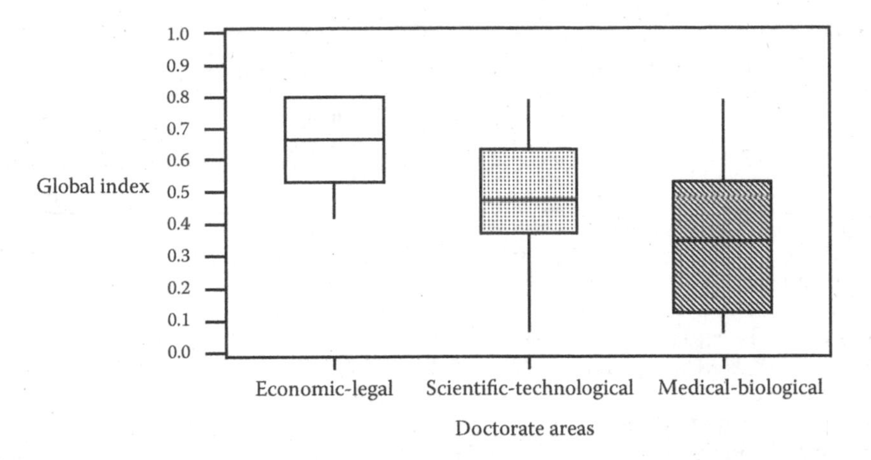

FIGURE 7.2
Box plot of global index of satisfaction with education–employment relationship.

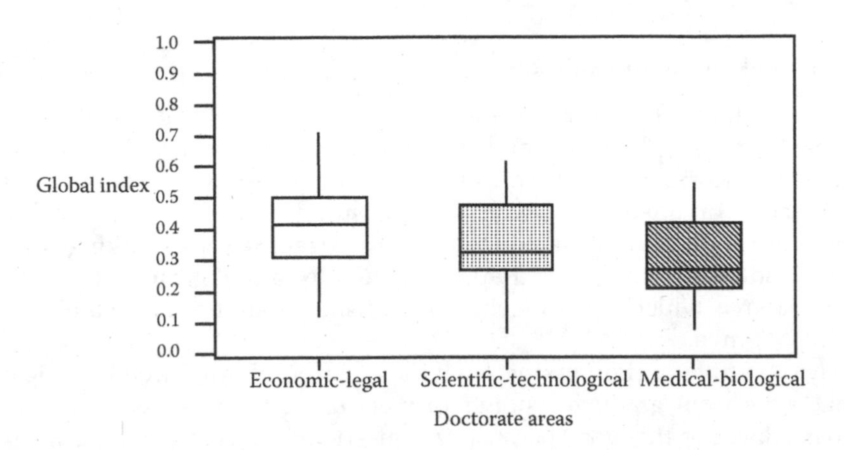

FIGURE 7.3
Box plot of global index of satisfaction with employment expectations and opportunities.

With reference to subjects who stated they had a stable job placement at the time of the survey (excluding therefore postdocs in training, with study grants, research allowances and/or unpaid collaborations) and for whom information regarding the education–employment relationship was available, the nonparametric correlation between the global indexes obtained using the NPC ranking methodology was calculated. The first satisfaction index concerned satisfaction with coherence, use and adequacy of education in relation to employment; the second concerned prospects offered by the doctorate. This analysis revealed a moderate, positive correlation between the two indexes and by separating data for the three doctorate areas a more substantial and statistically significant correlation is found for postdocs in the MB area (Table 7.10).

TABLE 7.10

Nonparametric Correlation between Global Indexes of Satisfaction

	Medical–Biological Area $N = 20$	Scientific–Technological Area $N = 23$	Economic–Legal Area $N = 23$	Total for All Areas $N = 66$
Global index of satisfaction with education – employment relationship	.34	.47	.66	.53
Median (first quartile–third quartile)	(.12–.53)	(.37–.63)	(.53–.79)	(.36–.66)
Global index of satisfaction with employment expectations and opportunities	.21	.27	.43	.30
Median (first quartile–third quartile)	(.15–.32)	(.21–.39)	(.26–.49)	(.21–.43)
Spearman's Correlation	.66	.11	.28	.43
(Monte Carlo estimate of p-value)	(.0016)	(n.s.)	(n.s.)	(.0002)

Note: n.s. = not significant.

7.1.4 Ranking Analysis Results

A first look at satisfaction data is provided by radar graphs in Figure 7.4, where score sample means and top-two-box-T2B percentages (i.e. the percentage of scores greater than 2 and 6, the latter case with reference to the satisfaction on adequacy, which was measured on a 1–10 rating scale) by PhD area are reported. As pointed out by extreme profile ranking results, PhD graduates from EL area appear generally more satisfied than those from ST area, which in turn seem more satisfied than those who took a PhD in the MB area.

The objective of the ranking analysis was to rank the three investigated PhD areas from graduates' points of view using their assessments on job satisfaction. For this goal, because the questionnaire contained a list of satisfaction items to be classified into two sets, that is, education–employment relationship and satisfaction with employment expectations and opportunities, a domain global ranking analysis is here advised. We recall that by 'domain analysis' we mean that several intermediate ranking analyses for each domain along with an overall ranking analysis should be performed.

The setting of the postdoc satisfaction ranking problem (see Chapter 3) can be summarized as follows.

- Type of design: one-way ANOVA (three independent samples, i.e. three multivariate populations to be ranked)
- Domain analysis: yes (two domains, education–employment relationship and employment expectations and opportunities)
- Type of response variables: ordered categorical (five ordinal 1–4 scores and one ordinal 1–10 score)

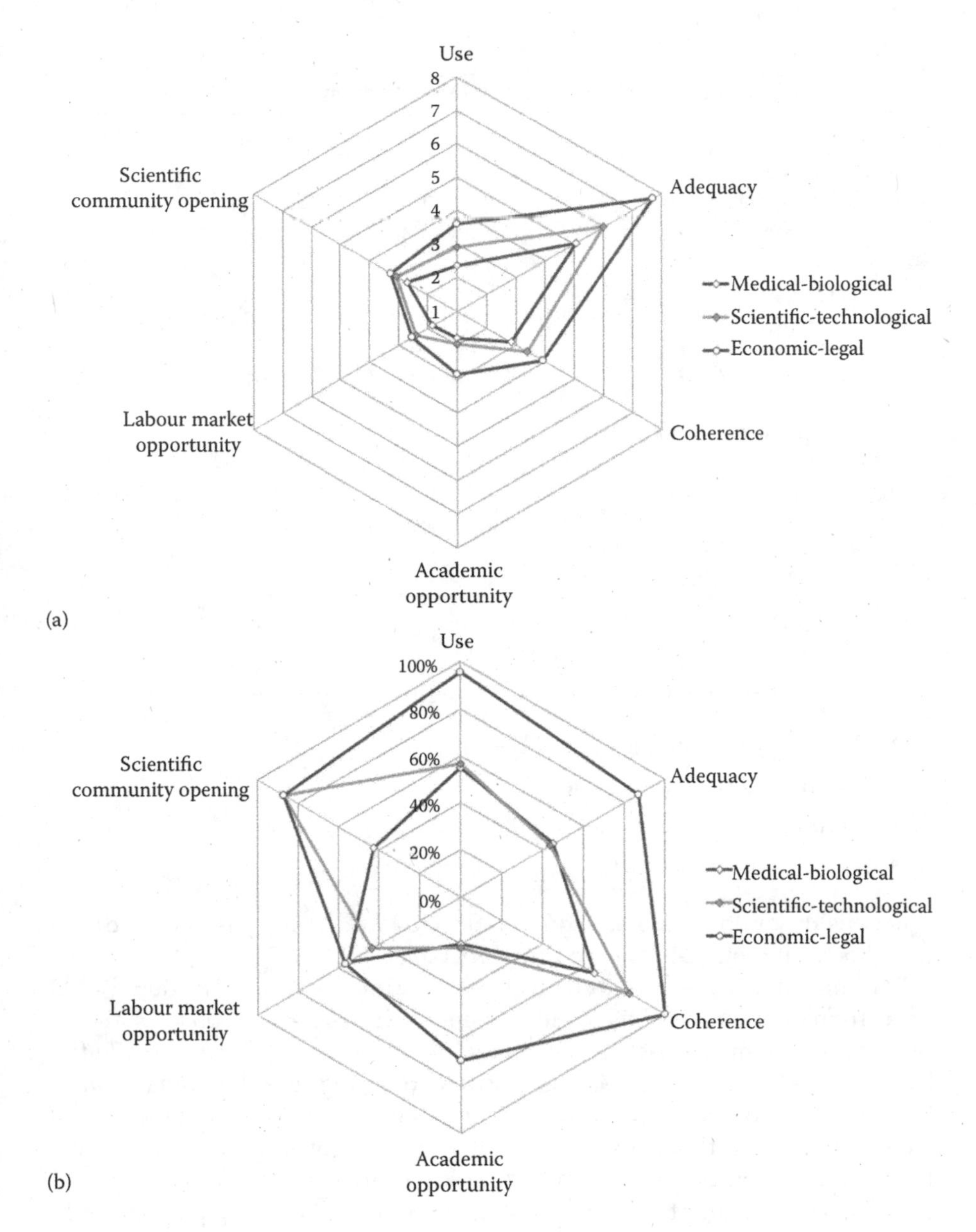

FIGURE 7.4
Radar graphs for satisfaction aspect by PhD area. (a) Item-by-item sample score means. (b) Item-by-item T2B percentages.

TABLE 7.11
Details of Global Ranking Results on PhD Satisfaction Data ($\alpha = 5\%$)

PhD Area	**Med-Bio**	**Sci-Tech**	**Eco-Leg**
Global Ranking	**3**	**2**	**1**
EDU-EMPL Relationship	3	2	1
Expect & Opportunities	2	2	1
All Aspects			
Med-Bio		1.00	1.00
Sci-Tech	**.016**		1.00
Eco-Leg	**.002**	**.001**	
EDU–EMPL Relationship			
Med-Bio		1.00	1.00
Sci-Tech	.032		1.00
Eco-Leg	**.002**	**.001**	
Expect & Opportunities			
Med-Bio		.978	1.00
Sci-Tech	.037		1.00
Eco-Leg	**.002**	.023	

- Ranking rule: 'the higher the better'
- Combining function: Fisher
- B (number of permutations): 2000
- Significance α-level: 0.05

Details on multivariate and univariate PhD area comparisons via directional permutation tests are presented in Tables 7.11 and 7.12.

Ranking analysis results allow us to easily rank the three considered PhD areas from graduates' points of views using their assessments on job satisfaction items. The most satisfied area is the EL, followed by ST area, while the least satisfied is the MB area. It is worth also noting that the domain ranking analysis provides several interesting insights on how graduates evaluate satisfaction on their jobs with respect to different satisfaction domains. In fact, the within-domain estimated rankings are different from the global ranking; in particular the ST area is tied at the second ranking position with the MB area.

7.1.5 Discussion and Conclusions

Comparison of the three PhD areas considered in the survey (EL, ST, MB), carried out using the nonparametric procedure based on the combination of dependent permutation tests and the permutation minP method (Tippett's test), represented a significant aspect of the analysis of the gathered data,

TABLE 7.12

Satisfaction Aspect-by-Aspect Univariate Directional Permutation *p*-Values

	Med-Bio	**Sci-Tech**	**Eco-Leg**	**Med-Bio**	**Sci-Tech**	**Eco-Leg**	**Med-Bio**	**Sci-Tech**	**Eco-Leg**
		Use			**Adequacy**			**Coherence**	
Med-Bio		.976	.999		.873	.999		.955	.999
Sci-Tech	.023		.998	.126		.999	.045		.994
Eco-Leg	**.000**	**.002**		**.001**	**.001**		**.000**	**.006**	
		Academic Opportunity			**Labour Market Opportunity**			**Scientific Community Opening**	
Med-Bio		.756	.999		.967	.981		.934	.986
Sci-Tech	.244		.997	.033		.621	.065		.777
Eco-Leg	**.000**	**.003**		**.018**	.379		**.014**	.222	

in that it aimed to highlight any differences in the interviewed postdocs' multivariate and univariate satisfaction profiles. It should be pointed out that the subsequent interpretation of the differences that were statistically significant do in any case require adequate caution, particularly given that the analysed variables represent quality indicators for the PhD courses. The variables' subjective nature and the survey's observational context do in fact lead to particular care being required when interpreting the results of the statistical analysis. Postdocs from the humanities and EL areas are generally expected to be less critical than colleagues from the scientific areas, whatever the objective quality of the PhD courses. Results from the comparison therefore require careful consideration by those running the PhD courses to identify the appropriate means of interpretation.

Opinions expressed on teaching and research work carried out during PhD courses can provide information about student perception of the *internal effectiveness* of PhDs, intended as the suitability of course methods to the stated education objectives. In general, postdocs stated they were quite satisfied with the activities carried out (Table 7.2). The greatest appreciation is observed for the PhD thesis and for work carried out abroad or in institutions outside the university. Differences in satisfaction profiles for the three areas regarding the aspects of internal effectiveness considered in the survey are substantially due to the opinion of activities carried out externally (Table 7.6), with almost total satisfaction in the MB area (95.2% very satisfied), and slightly more contained satisfaction in the EL area (42.9% very satisfied) (Table 7.2). Even the quality of research conducted individually, with or without the supervision of lecturers, is perceived positively (Table 7.2). Though all respondents stated that they had carried out research individually, just over half (57.3%) reported experience of research within groups of lecturers and/or students: of these 86.6% were very/quite satisfied with this type of initiative. Less positive is the opinion of the education received: a quarter of interviewees held this aspect to be not very or not at all satisfactory and slightly more than a fifth were highly satisfied.

Indications as to the *external effectiveness* of PhDs, understood to mean the usefulness and usability of a PhD qualification for insertion into the academic or employment field, can be derived from a combination of questions regarding coherence between employment and study, use of competencies acquired during studies, and the adequacy of education in relation to employment (Tables 7.1 and 7.5, Figure 7.2). On the whole the satisfaction area (very/quite satisfied) was high, though characterized by significant differences among the areas, with a growing trend as we move from the MB area to the ST and EL areas. More positive opinions are found in relation to coherence (62.1% very satisfied and 21.2% quite satisfied), less positive for the level of use of acquired techniques (69.7% very/quite satisfied). Room for improvement can also be identified with regard to adequacy of education (31.8% not satisfied). Differences among the areas are also highlighted by the distribution of the global satisfaction index expressed on a scale of 0–1 (Table 7.9,

Figure 7.2). In the EL area, as well as a shifting of the distribution towards high values of satisfaction, less variability in scores is found, particularly compared with the MB area.

More information on PhD qualifications' external effectiveness comes from the evaluation of the satisfaction levels with prospects offered by the PhD in terms of opportunities for insertion into the academic field or labour market or as an opening towards the scientific community (Tables 7.4 and 7.8). The low percentages of those who consider themselves very satisfied with these aspects denote a certain lack of confidence in the absorption ability of both the university (55.2% not very/not at all satisfied) and even more so the labour market in general (58.8% not very/not at all satisfied). Particularly contributing to the differentiation of satisfaction profiles in the three PhD areas regarding opportunities offered by PhDs are opinions on the possibility of continuing an academic career, held to be more plausible by postdocs in the EL area than by colleagues in the scientific area. Distribution of the global satisfaction index expressed on a scale of 0–1 (Table 7.9, Figure 7.3) stands level with relatively low values for the three areas, which also highlights substantial homogeneity in postdoc perceptions. A percentage of 71.9 of interviewees in the current survey remain in the academic field after earning their degree (this percentage varies from 81.2% in the EL area to 70.7% in the ST area and 65.9% in the MB area), often only through unpaid collaborations or postdoc study grants and research allowances. Furthermore, 47.7% of those who do not yet have a structured working position state they are only interested in jobs in the academic field, and this percentage varies from 31.2% in the ST area to 47.4% in the MB area, up to 77.8% in the EL area. Faced with this information, the organization of better structured education programmes would seem to be desirable in the PhD field, at the same time encouraging awareness of the qualification outside academic circles. The organization of PhDs into PhD Schools as well as improvement of the teaching programme is a move in the right direction, with one of its objectives being a more organic and widespread relationship between university PhDs and outside environments, both professional and those involving research. Publicizing of the PhD qualification should therefore be accompanied by specific partnership initiatives between academic and nonacademic fields. In a recent paper, the National Committee for the Evaluation of the University System (NCEUS) actually highlighted the fact that finding employment is still an uncertain matter for PhD researchers. The percentage of Italy's workforce constituted by scientists and engineers, a datum that implicitly represents the technological level of the production system, is one of the lowest among major European countries (MIUR, 2003), highlighting the fact that companies lack preparation when it comes to employing highly qualified personnel.

Regarding the solutions based on the nonparametric combination of dependent partial rankings for ordered variables, we can outline the flexibility of the method to construct a combined indicator starting from k dependent indicators based on ordered variables. This procedure can be used both to derive a global index of satisfaction used in comparisons among different groups or compared with an optimal desired value of satisfaction, and to construct a global ranking of units that can be useful when exploring clusters of units with a lower or upper global level of satisfaction.

7.2 Performance and Perceived Quality of Services Delivered by Italian Public Administrations

The aim of this work was to compare and rank the perceived performance of the services supplied to Italian citizens in different areas by the so-called Territorial Services Centers (TSC). Actually, this is a study connected with an Italian national project known as IQuEL-2010, aimed to solve some problems associated to the digital divide by TSC (Bolzan et al., 2014). IQuEL, which stands for Innovation and Quality delivered and perceived by local authorities, is one of the six winning projects under the Italian Programme Elisa (*Enti Locali Innovazione di Sistema*, i.e. Local Authorities for System Innovation), which was funded by the Italian Presidency of the Council of Ministers. The purpose of IQuEL is to develop a system for monitoring and evaluating the quality of services provided by the offices of the Public Administration through dedicated tools. The data collected are used for the definition and implementation of concrete actions to improve its organization and personalization of services for citizens.

A specific survey was carried out by sample of local operators operating in the three Italian provinces of Padova, Parma and Pesaro–Urbino. We applied the multivariate ranking method on a set of nine quality dimensions related the public services supplied. The results show important differences among the provinces, at least for six out of nine TSC abilities or performances, and allow us to rank the provinces according to the more or less perceived quality of TSC services.

7.2.1 Introduction

During 2010, the most frequent type of interaction of enterprises with public administration (PA) in the EU27 using the Internet was downloading electronic forms (76%), followed by obtaining information (74%) and submitting completed forms (69%). More than 90% of all enterprises in Slovakia, Lithuania, Finland and Sweden reported that they used the Internet to obtain information from public authorities' websites in 2010, while it was fewer than half of enterprises in Romania and the Netherlands. On the other hand, more than 90% of enterprises in the Netherlands, Lithuania, Greece, Poland and Finland reported that they used the Internet in 2010 to submit completed forms electronically to public authorities. In Italy and Romania it was less common (both 39%). In Italy, after 10 years since the start of the eGovernment Italian project, only 28.9% of population contact the PA services electronically to find information and not more than 10% to fill and send documents and submit complete forms online (Rapporto eGov Italia, 2010). IQuEL Project in 2009 involved large local units (commons, provinces and regions) of around 15 million people. This is a trial to be expanded to all the local units in the nation. The goal was to monitor and evaluate the quality of the services supplied from PA by Territorial Services Centers (TSC). The TSCs are the government's executive branch of the IQuEL project aimed to solve some problems associated with the digital divide existing in fringe areas. IQuEL uses a quali-quantitative approach on services supplied, based on a series of statistical indicators shared with the stakeholders. The indicators are about the level of access, of specific performance for each channel of distribution and of Customer Satisfaction about Citizen Relationship Management services.

7.2.2 Survey Description and Ranking Results

A particular experiment conducted in three provinces in the north and centre of Italy – Padova (PD), Parma (PR) and Pesaro-Urbino (PU) – involved in the project was evaluated by a sample of local operators (nPd = 25, nPr = 42; nPU = 13) on each of nine abilities to communicate and to transfer information from TSC: (1) Precision of communication at start point of service; (2) Politeness of TSC's office worker; (3) Clarity in the application forms; (4) Performance of TSC's office worker; (5) Friendly use and quality of the service by the TSC; (6) Timeliness to assistance; (7) Quality of assistance for a clarification; (8) Level of competitiveness with respect to market needs and (9) Level of global satisfaction towards the service provided. The answer for each dimension ranged in the interval 0–5. Table 7.13 displays the sample medians of all investigated abilities by each province.

The setting of the perceived quality on public services ranking problem (see Chapter 3) can be summarized as follows.

- Type of design: one-way ANOVA (three independent samples, i.e. three multivariate populations to be ranked)
- Domain analysis: no
- Type of response variables: ordered categorical (nine ordinal 0–5 scores)
- Ranking rule: 'the higher the better'
- Combining function: Fisher
- B (number of permutations): 2000
- Significance α-level: 0.1

TABLE 7.13

Medians of the Nine Investigated Abilities from TSC by Province

	Province		
Abilities from TSC	PD	PR	PU
Precision of communication	4	3	4
Politeness of TSC's office worker	4	4	4
Clarity in the application forms	4	3	4
Performance of TSC's office worker	4	4	4
Friendly use and quality of the service	4	3	4
Timeliness to assistance	4	3	4
Quality of assistance	4	4	4
Level of competitiveness	4	3	4
Global satisfaction	4	3	4

TABLE 7.14

Details of Global Ranking Results on Public Services Perceived Quality Data (α = 10%)

Province	PD	PR	PU
Global ranking	**1**	**3**	**1**
PD		**.028**	.790
PR	.973		.989
PU	.259	**.021**	

Details on multivariate and univariate PhD area comparisons via directional permutation tests are presented in Tables 7.14 and 7.15.

The analysis by means of the NPC ranking method gives both PD and PU on the first position in the ranking.

7.2.3 Conclusions

The results show important heterogeneity among the three Italian provinces, at least for six of nine TSC abilities or performances (see Table 7.15) producing a global satisfaction ranking displayed in Table 7.14. The Performance of TSC office worker, the Level of competitiveness with respect to market needs and the Global satisfaction seem to be important because they give a good 'image' of TSC to the local community, in particular to attract new enterprises to the area. The main differences among provinces are due to the specificity expressed by areas but also to the different policies to training personnel. Before the start of IQuEL project, PD and PU had to work considerably to promote the competences and skills of the municipal managers as those are the keys to organize all the services by means of

TABLE 7.15

Satisfaction Aspect-by-Aspect Univariate Directional Permutation p-Values

	Precision of Communication			Politeness to CST's Office			Clarity in the Application Forms		
Province	PD	PR	PU	PD	PR	PU	PD	PR	PU
PD		.324	.739		.116	.618		**.004**	.461
PR	.676		.836	.883		.863	.996		.978
PU	.260	.164		.382	.137		.539	**.021**	
	Performance to CST's Office			**Friendly Use and Quality of the Service**			**Timeliness to Assistance**		
Province	PD	PR	PU	PD	PR	PU	PD	PR	PU
PD		**.005**	.788		**.017**	.714		**.010**	.405
PR	.994		.988	.982		.971	.989		.937
PU	.212	**.012**		.286	**.028**		.595	.062	
	Quality of Assistance			**Level of Competitiveness**			**Global Satisfaction**		
Province	PD	PR	PU	PD	PR	PU	PD	PR	PU
PD		.140	.744		**.006**	.855		**.022**	.925
PR	.860		.910	.994		.995	.977		1.00
PU	.255	.090		.144	**.005**		.074	**.000**	

an envisioned New Public Management aimed at changing perspectives: from an administration based on authority to one that provides services to citizens or enterprises (Bolzan, 2010). In general the analysis showed a positive evaluation on TSC, emphasizing some critical situation (insufficient information about the procedures to use it, the operative and communicative interface device which is very formal and not of friendly use and again feedback not expected). Information about the workers showed a low profile of competence about the most innovative aspects of the network. Such problems might be addressed by updating and training (see also Bolzan, 2010; Bolzan and Boccuzzo, 2011).

7.3 Customer Satisfaction on Ski Teaching Service: S.E.S.T.O. Survey

In the sports tourism field, the customer satisfaction and service quality dimensions are crucial points to deliver a high-quality service and to be competitive. Measuring such dimensions with appropriate tools is therefore of fundamental importance. In a winter sports tourism framework, we propose to measure customer satisfaction and service quality by means of multivariate ranking methods applied to the S.E.S.T.O. survey, a study focussed on customer satisfaction on a ski teaching service. The S.E.S.T.O. (Statistical Evaluation of a Skischool from Tourists' Opinions) survey is the first Italian survey on the evaluation of ski instructors. This study is innovative at a national level: it is the first systematic study conducted in different schools, with qualitative evaluation, using a questionnaire scientifically designed to measure satisfaction and quality perceived by the users. A multivariate ranking analysis has been applied to investigate the survey with the goal of comparing the different time periods, that is, weeks, during which the ski teaching service is provided. Such types of global comparisons can be easily used to determine whether during the high season weeks versus the low-season weeks the global satisfaction is perceived in the same way or if there are some criticisms on the service as a whole or on some partial aspect of it.

7.3.1 Literature Review

The past decades have seen an increased trend towards more active holidays, with a growing interest within academic circles (Smith and Jenner, 1990), particularly for the sports tourism field. One definition of sports tourism is 'travel for non-commercial reasons, to participate or observe sporting activities away from the home range' (Hall, 1992). Weed and Bull (2004) suggest five types of sports tourism: tourism with sports content, sports participation tourism, sports training, sports events and luxury sport tourism. The research is usually guided by the definitions adopted. The definition of 'sports tourism' is the starting point in Weed (2009); his paper, a meta-review of 18 reviews from 1990 to 2008, aimed at tracing the different paths taken by researchers in sports tourism. Delineating the contributions in sports tourism, Weed (2006, 2009) describes the 'event sports tourism' as the main researched area followed by 'active sport tourism', particularly golf and ski tourism. Golf and ski tourism have been categorized as 'active sport tourism' (Gibson, 2002) or 'sports participation tourism' (Weed and Bull, 2004) by several authors (Chalip, 2001; Gibson, 2003; Jackson and Weed, 2003). Moreover, it is worth citing some studies that have dealt with the behaviour of sports tourists, including Petrick and Backman's (2002a,b,c) work on the satisfaction and value perceived by golf tourists and Williams and Fidgeon's (2000) research on the barriers that keep many potential skiers off the slopes and trials.

As stated by Chalip (2001), sports tourism is 'multifaceted', with authors performing sports tourism research from different disciplinary perspectives. Weed (2009) invited scholars from other disciplines to contribute to sports tourism research. Finally, Weed (2009) also intended to highlight the scarcity of studies related to customer satisfaction, particularly for winter sports. In a study addressing participation constraint in potential skiers, Williams and Fidgeon (2000) stated that 'so much of the breaking down of the barriers to skiing evolves around treating new skiers in friendly and hospitable ways' (p. 390). Accomplishing this goal seems important not only in terms of customer service marketing that makes skiers feel comfortable, but also in evaluating and monitoring customer satisfaction and service quality. Within the sports tourism industry, the quality of services provided is a relevant issue in maintaining competitiveness (Kouthouris and Alexandris, 2005; Shonk and Chelladurai, 2008). Studies on perception of service quality by participants include the study of service quality within health and fitness centers (Alexandris et al., 2004), golf courses (Crilley et al., 2002), recreational and leisure facilities (Ko and Pastore, 2004) and spectator sports (McDonald et al., 1995; Wakefield et al., 1996; Kelley and Turley, 2001; Greenwell et al., 2002a,b). Considering service quality, the ways to satisfy customers with quality service can, of course, determine the success of a sports organization (Ko and Pastore, 2004). The same concept seems to be relevant to Matzler et al. (2008) in a study on customer satisfaction with alpine areas; they claimed that winter tourism is crucial for the eastern alpine region economy, in particular alpine skiing activities that attract many tourists (Richards, 1996; Weiermair and Fuchs, 1999; Williams and Fidgeon, 2000; Dolnicar and Leisch, 2003; Franch et al., 2003; Matzler and Siller, 2004; Matzler et al., 2004). Matzler et al. (2008) also reported how 'more and more winters with little snow and the rapid growth of long-distance travel (Pechlaner and Tschurtschenthaler, 2003) increase competition between alpine ski areas. In this competitive market environment, destination success depends strongly on a thorough analysis of tourist motivation, customer satisfaction, and loyalty (Yoon and Uysal 2005)' (p. 403). In the field of sports participation tourism, in particular a winter sport like skiing, the paper aims to analyse customer satisfaction data by means of innovative statistical tools on several quality aspects of the ski school services in the Alto Adige, a northern Italy alpine region. Requirements for high-quality service are also specified by ISO 9001 document (2008). The European regulation ISO 9001 states that an organization needs to show its ability to regularly provide a product that satisfies customers' requirements and wishes to increase customers' satisfaction. The former seems to be related to monitoring of quality and the latter to improvement of quality. It is generally accepted and promoted the application of methods such as statistical techniques in order to monitor, analyse and improve service and customer satisfaction. The paragraphs in Data Analysis subsection of ISO 9001 speak for themselves: the priority is to provide information regarding customers' satisfaction. The organization must monitor the information relating to customers' perception of whether the organization has satisfied their requirement. To collect information and monitor customers' perception we can use many sources such as customer satisfaction surveys. In such a framework, with sport tourism literature lacking in customer satisfaction studies, especially for winter sports and the requirements of ISO 9001, our aim was to monitor and measure customers' perception of the Alto Adige ski schools by means of innovative statistical analysis such as the multivariate ranking methods.

7.3.2 Survey Description

A survey was conducted in the ski schools of Alto Adige area in northern Italy. The parents of children younger than the age of 13, who participated in week-long ski courses organized in the ski schools, were asked to answer a questionnaire to express their level of satisfaction about some aspects of the service. This study was conducted in six runs covering nine weeks (Table 7.16)

TABLE 7.16

Details on the Six Investigated Weeks in Ski Teaching Service Survey

Period No.	Date	Season Level
1	27/12/2010–09/01/2011	Peak
2	14/02/2011–20/02/2011	Medium
3	19/12/2011–01/01/2012	Medium-Peak
4	09/01/2012–22/01/2012	Medium
5	06/02/2012–12/02/2012	Low
6	20/02/2012–26/02/2012	Peak

of ski teaching service during the winter season 2010–11 and 2011–12, and involved a sample of 416 children/parents.

The first part of the questionnaire is about demographics and other questions useful for segmentation and to set the covariates. The second part asks questions about the service. We identified three phases of the service, each with specific quality dimensions:

1. *Booking service,* with quality measured by adequate opening times; clarity and completeness of informative brochures; staff (clarity and completeness of information provided, courtesy and helpfulness)
2. *Course organization,* with the following quality dimensions: ways to organize courses; skill homogeneity of groups after selection; events planned with courses (torchlit descents, competitions,...); slope enrichment (inflatables, snow sculptures,...)
3. *Carrying out classes,* with quality measures based on: effective teaching (clarity of notions, courtesy and helpfulness of teachers); safety (adequate slopes and lifts, subjective perception of safety); users' general satisfaction (enjoyment and fun, increased passion for skiing, children's comfort,...)

Each dimension was investigated with specific questions and adequate scales.

To evaluate and compare the perceived satisfaction under different time periods in which the ski teaching service was provided, a multivariate ranking analysis was performed in which we focussed on the classes' items. Six different aspects of the ski teaching were investigated as reported in Table 7.17.

To measure customer satisfaction with regard to various aspects of the ski teaching service, a scale of scores from 0 (not at all) to 10 (very much) was used.

TABLE 7.17

Definitions of Items Investigated in Ranking Analysis of Ski Teaching Service Survey

Item	Definition
Learning	Instruction was clear?
Teacher	The instructor was polite and helpful?
Slopes	Slopes/structures used during lessons were suitable?
Safety	You felt safe doing the various activities during lessons?
Fun	You enjoyed the lessons?
Involvement	The course increased your interest in the sport?

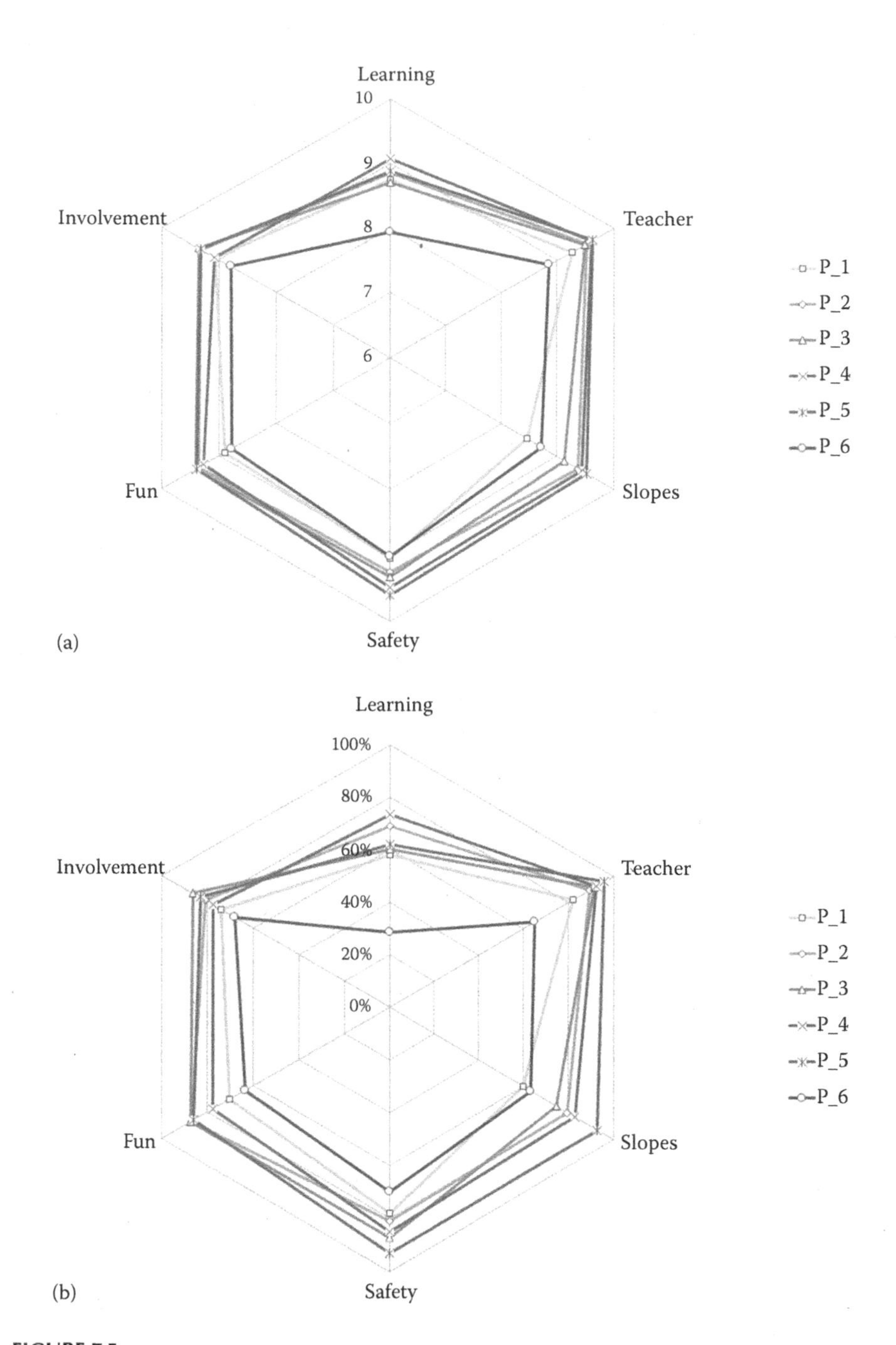

FIGURE 7.5
Radar graphs of score means and top-two-box percentages by teaching period. (a) Item-by-item sample score means. (b) Item-by-item T2B percentages.

7.3.3 Ranking Analysis

A first look at customer satisfaction data is provided by radar graphs in Figure 7.5, where score sample means and top-to-box-T2B percentages (the percentage of scores greater than 9) by teaching periods are reported. From a descriptive point of view it appears that the most critical periods seem to be 1 and 6 because the related score means and top-to-box percentages tend to take generally lower values than all the other times, which in turn seem to show a more or less similar performance. The radar graphs seem to suggest the satisfaction dimensions that most discriminate the perceived quality toward ski teaching service are Learning and Teacher.

The main objective of this work was to rank six teaching periods from the standpoint of customer satisfaction using the customers' assessments on several aspects of satisfaction with the ski teaching service. The setting of the ski teaching satisfaction ranking problem (see Chapter 3) can be summarized as follows.

- Type of design: one-way ANOVA (six independent samples, i.e. six multivariate populations to be ranked)
- Domain analysis: no
- Type of response variables: ordered categorical (six ordinal 1–10 scores)
- Ranking rule: 'the higher the better'
- Combining function: Fisher
- B (number of permutations): 2000
- Significance α-level: 0.05

Details on multivariate and univariate PhD area comparisons via directional permutation tests are presented in Tables 7.18 and 7.19.

Ranking analysis results allow us to easily rank the teaching periods under investigation using customers' judgements on satisfaction items.

TABLE 7.18

Details of Global Ranking Results on Six Periods of Ski Teaching Service ($\alpha = 5\%$)

Teaching Period	**1**	**2**	**3**	**4**	**5**	**6**
Global ranking	**5**	**1**	**1**	**1**	**1**	**6**
1		1.00	1.00	1.00	.985	1.00
2	.130		1.00	1.00	1.00	**.020**
3	.155	1.00		1.00	1.00	**.008**
4	.145	1.00	1.00		1.00	**.020**
5	.097	1.00	1.00	1.00		**.005**
6	1.00	1.00	1.00	1.00	1.00	

TABLE 7.19

Item-by-Item Univariate Directional Permutation p-Values

Teaching Period	1	2	3	4	5	6
Global ranking	**5**	**1**	**1**	**1**	**1**	**6**
1		1.00	1.00	1.00	.985	1.00
2	.130		1.00	1.00	1.00	**.020**
3	.155	1.00		1.00	1.00	**.008**
4	.145	1.00	1.00		1.00	**.020**
5	.097	1.00	1.00	1.00		**.005**
6	1.00	1.00	1.00	1.00	1.00	

The periods where we observe the highest satisfaction are from 2 to 5, while the less satisfactory periods are 1 and 6. It is worth noting that the worst periods are actually those occurring during the peak season; conversely, the best periods match the medium- or low-season weeks. Moreover, according to Carrozzo et al. (2012), the number six teaching period can be considered as an 'out-of-control' occurrence, meaning that the global multivariate satisfaction level can be considered significantly lower than the satisfaction observed during the low-season periods where we can assume that the ski teaching service is provided with the utmost care and efficiency.

7.3.4 Conclusions

The main goal of this work was to investigate whether during the high-season weeks versus the low-season weeks the global satisfaction is perceived in the same way or there are some criticisms on the service as a whole or on partial aspects of it. Ranking analysis results highlight that the perceived quality of ski teaching appears to be different during peak periods and low seasons, which is an indication of some criticism of the teaching service.

To exploit its potential fully, a survey like this and the related ranking analysis should be repeated in different years: only by comparing the rankings in different moments in time can we understand trends of improvement/constant level/decline, thus monitoring and improving quality continuously. This is particularly important to assess quantitatively whether actions taken by the school managers have been appreciated by the clients.

Still another possible development would be to define rankings for the whole tourist package offered to skiers, not just to evaluate ski schools, however important. Ultimately, one could investigate the quality perceived by those who do not ski and yet visit these mountain resorts. This perspective suggests a much broader vision where, along with the activities of a ski school, other services are monitored: facilities such as hotels and accommodation, entertainment for nonskiers, tours. Finally, quality could be analysed in terms of perception by tourists, but also listening to the opinion of those who provide the service: teachers on courses, technical staff on slopes and ski facilities, and so forth. This leads to the fundamental concept of total quality, once again evaluated in terms of progress over time.

7.4 Customer Satisfaction Evaluation of Tourist Facilities: Sesto Nature Survey Case Study

For a natural protected area, as a system that produces goods and services, it is necessary to monitor its quality, to set goals for improvement needed and to implement monitoring tools also with reference to the quality perceived by users. Of course, a natural park has some specific characteristics that must be taken into account when implementing a project to monitor the quality of its services. The ever increasing flow of visitors in natural reserves and alpine resorts, along with the growing number of excursionists, requires special care because these phenomena can modify the relationship with the geographical area examined, requiring ways to monitor touristic development.

The goal of this work is to present an extensive survey on customer satisfaction evaluation of tourist facilities in the context of the alpine natural park known as 'Tre Cime' (District of Sesto Dolomites/Alta Pusteria, Italy). An extreme profile ranking method and multivariate analysis were performed to gain insights into how tourist needs and expectations can be met, that is how to improve the satisfaction of natural park users.

7.4.1 Context of Statistical Evaluation of Natural Parks

Bearing in mind the natural tension between using the resources on one hand and preserving them on the other (how much can we exploit the environment without reducing its ecological and social value unacceptably?), over the last few years, new conceptual and managerial-operative orientations have emerged, such as Sustainable Tourism (optimal use and protection of natural resources, respect for the sociocultural identity of local communities, fair allocation of socioeconomic benefits) and Carrying Capacity of an area (Manning, 2007; with regard to its natural and sociocultural resources, its managerial resources and its recreational opportunities).

In recent years the management of a natural park has become increasingly complex because in addition to the purely naturalistic aspects, there are also social issues to keep in mind, with reference to the local population and tourists' experience. Administrations and institutions involved in managing natural contexts that are rich and attractive from the point of view of tourism are called to integrate the purely naturalistic aspects with social ones, in the sense that it is necessary to protect resources and ecological processes, but these resources and processes must also be directed to the benefit of society. Hence there is the need for institutions and administrations, to gather information to implement policies that not only protect the environment, but also promote sustainable tourism development and enhancement of the cultural integrity of local people.

In this context, the role of projects and statistical surveys that monitor, analyse and ultimately evaluate environmental and social processes involved with touristic activities is paramount to gather information helpful for any local government, public body or private citizen interested in managing or simply living in the area. In addition, the results of such studies help to define a series of statistical indicators, that is, objective ways to measure desirable conditions in several aspects of the management of natural assets (e.g. with reference to the experience the visitors have had, what percentage of tourists finds the signposting of mountain paths more than satisfactory?). Goals or improvement actions can be contemplated to reach specific target values for such indicators.

7.4.2 Sesto Nature Survey

In 2010, a survey was planned on tourists' opinions about the park 'Tre Cime' (District of Sesto Dolomites/Alta Pusteria, Italy), termed the Sesto Nature Survey. Two weeks in each month of July, August and September were randomly selected during which visitors were interviewed in some strategic points of the district. The questionnaire sections included (1) General information, (2) Information on the daily trip inside the district, (3) Opinion on walks or excursions along paths of the district, (4) Opinions on iron ways or rock climbs in the district and (5) Suggestions on improvement of services and protection of the district.

In the Sesto Nature Survey information was collected for a total of 262 respondents. Tables 7.20 and 7.21 show some descriptive characteristics of the sample.

The respondents had a mean age of 52 years, with a range between 12 and 85 years. The sample was substantially balanced between females and males. With reference to nationality, 61% of the respondents were Italian and 27% German. The sample showed a wide range of nationalities (11 countries), confirming the international knowledge of the natural area (Table 7.20). For about one third of the respondents, it was the first visit to the park (Table 7.21). Slightly more than half of the sample visited the natural park during August and about 36% of them visited the area with children or teenagers.

TABLE 7.20

Demographics Characteristic of the Respondents

	Learning						Teacher						Slopes					
	1	2	3	4	5	6	1	2	3	4	5	6	1	2	3	4	5	6
1		.544	.570	.897	.711	**.008**		.917	.898	.933	.928	.151		.999	.995	.998	.999	.793
2	.456		.549	.797	.568	**.017**	.082		.464	.588	.701	**.002**	**.000**		.026	.391	.675	**.001**
3	.429	.451		.860	.757	**.008**	.101	.536		.727	.630	**.005**	**.004**	.973		.916	.957	**.008**
4	.102	.202	.140		.198	**.000**	.067	.412	.273		.462	**.002**	**.001**	.609	.083		.632	**.002**
5	.289	.432	.242	.802		**.001**	.071	.299	.370	.537		**.002**	**.001**	.324	.043	.367		**.001**
6	.991	.983	.991	.999	.999		.848	.997	.995	.998	.997		.206	.999	.992	.997	.999	
	Safety						Fun						Involvement					
1		.765	.921	.978	.998	.504		.935	.980	.917	.925	.417		.851	.936	.525	.851	.217
2	.235		.670	.867	.978	.173	.065		.737	.592	.525	.064	.149		.594	.220	.443	.079
3	.079	.330		.867	.984	.044	.020	.263		.313	.425	**.004**	.063	.406		.059	.265	**.011**
4	**.021**	.132	.132		.802	**.008**	.082	.408	.686		.699	.027	.475	.779	.940		.822	.241
5	**.002**	**.021**	**.016**	.197		**.001**	.075	.475	.574	.301		.065	.148	.557	.734	.177		.043
6	.495	.827	.956	.991	.999		.583	.935	.996	.973	.935		.782	.920	.988	.758	.956	

TABLE 7.21

General Characteristics of the Visit in the Natural Park

	N	%
First Visit to the Park		
Yes	84	32.7
No	173	67.3
Total respondents	257	100
Missing: 5		
Month of Visit		
July	65	27.0
August	126	52.3
September	50	20.7
Total respondents	241	100
Missing: 21		
Presence of Children or Teenagers (≤17 years)		
Yes	89	35.7
No	160	64.3
Total respondents	249	100
Missing: 13		

Table 7.22 shows the distribution of the degree of seriousness of some potential problems reported by respondents.

From the fifth problem listed on, we can find questions that are not considered a problem by more than two thirds of the respondents. The second to fourth issues represent a problem with percentages ranging from 31% (almost one third of the respondents) and 43% (fewer than half the respondents). The overcrowded areas represent a problem for more than half the respondents.

7.4.3 Extreme Profile Ranking Results

To apply the extreme profile ranking method and to construct global satisfaction indicators, the 16 potential problems were divided into three sets: the first set related to problems of footpaths, routes and resting areas (emphasized in light grey); the second set related to congestion problems in the areas (emphasized with oblique lines) and the last set related to further remaining technical problems (emphasized in horizontal lines). For each set of the potential problems, a global satisfaction indicator was constructed. Descriptive statistics in Table 7.22 and box plots in Figures 7.6 and 7.7 illustrate the distribution of the three global satisfaction indexes for the total sample and for groups of respondents identified according to three stratification variables (nationality including Italian or German people, month of visit and presence or absence of children/teenagers).

TABLE 7.22

Degree of Gravity and Global Satisfaction Scores for Potential Problems

% Degree of Gravity for Potential Problems	Not a Problem	Small Problem	Serious Problem		Global Satisfaction Index
Overcrowded refuges ($N = 225$)	44.0	42.2	13.8	Mean	0.60
Presence of noisy visitors ($N = 228$)	57.0	33.8	9.2	Median	0.63
Visitors not respecting the environment ($N = 225$)	58.7	33.8	7.5	Min–max	0.09–0.81
Overcrowded footpaths and routes ($N = 227$)	56.0	37.4	6.6	Q_1–Q_3	0.48–0.73
Lack of supervision for schoolchildren ($N = 171$)	83.6	14.0	2.4		
Presence of litter along footpaths/routes ($N = 231$)	74.9	21.2	3.9	Mean	0.78
Presence of litter in resting areas ($N = 239$)	72.4	23.8	3.8	Median	0.79
Poor maintenance of footpaths/routes ($N = 226$)	85.8	11.5	2.7	Min–max	0.26–0.90
Poor maintenance of resting areas ($N = 233$)	86.3	12.9	0.8	Q_1–Q_3	0.69–0.90
Shortage of resting areas along footpaths ($N = 238$)	80.7	18.5	0.8		
Difficulty to find a parking space ($N = 239$)	69.0	26.0	5.0	Mean	0.83
Poor reception in refuges ($N = 235$)	80.4	15.3	4.3	Median	0.84
Too many rules and regulations to abide by ($N = 226$)	92.0	5.3	2.7	Min–max	0.36–0.92
Refuges non adequately equipped ($N = 215$)	87.9	9.8	2.3	Q_1–Q_3	0.77–0.92
Areas not equipped for children ($N = 189$)	85.7	12.2	2.1		
Poor maintenance of mountain lifts ($N = 194$)	95.9	3.6	0.5		

Distributions of the global satisfaction indexes showed relatively high values for the conditions of footpaths, routes and resting areas and for the set of more technical aspects, which also highlight substantial homogeneity in stratified respondents' perceptions. Less positive perceptions are recorded for the set of congestion problems in areas; in the stratified analysis Italians and visitors staying during the months of July and August reported more negative evaluations.

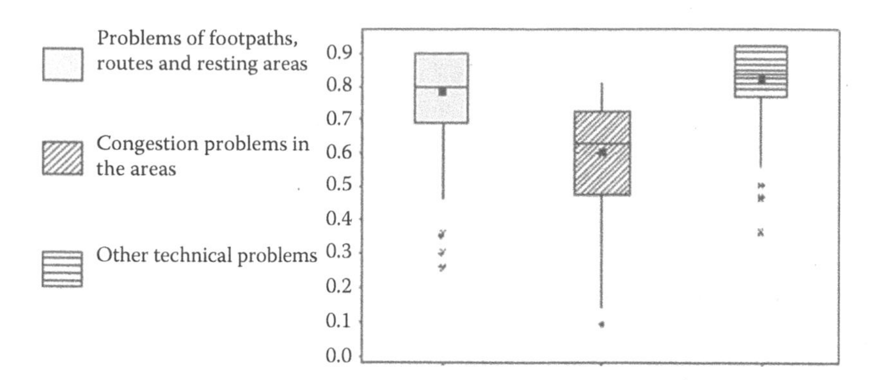

FIGURE 7.6
Box plot of the global satisfaction indexes for the three sets of potential problems (total sample).

7.4.4 Ranking Analysis Results

A first look at data for potential problems is provided by radar graphs in Figure 7.8, where problem-by-problem score sample means by tourist segment are reported. According to an extreme profile ranking analysis, from a descriptive point of view it appears that the most satisfied segment, that is, the group less worried about potential problems, seems to be the Italians travelling with no one younger than 18 years old, because the related score means tend to take generally small values. Conversely, foreign tourists accompanied with children/teens appear to be the least satisfied group because their score means are generally higher than those of other segments. In terms of the specific problems that most discriminate the perceived potential problems, the radar graphs seem to suggest that problem numbers 11, 12 and 13, that is, 'Difficulty to find a parking space', 'Poor reception in refuges' and 'Too many rules and regulations to abide by', are the most critical items.

The main goal of ranking analysis is to rank four tourist segments from the customer satisfaction standpoint using their assessments on the degree of seriousness of some potential problems. As the questionnaire contained a list of 16 problems classified into three sets, a domain global ranking analysis is here advised. We recall that by 'domain analysis' we mean that several intermediate ranking analyses for each domain (class of potential problems) along with an overall ranking analysis should be performed.

The setting of the tourist satisfaction ranking problem (see Chapter 3) can be summarized as follows.

- Type of design: one-way ANOVA (four independent samples, i.e. four multivariate populations to be ranked)
- Domain analysis: yes (three domains: footpaths, routes and resting areas, congestion problems in the areas and further technical problems)

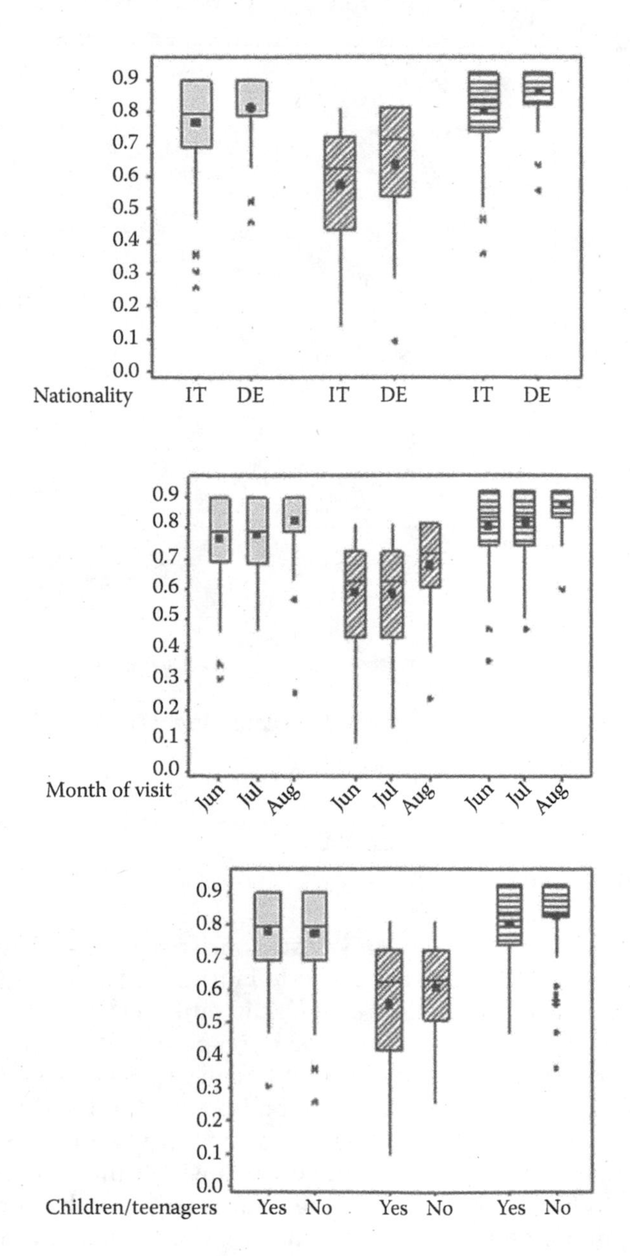

FIGURE 7.7
Box plots of the global satisfaction indexes for the three sets of potential problems (stratified samples).

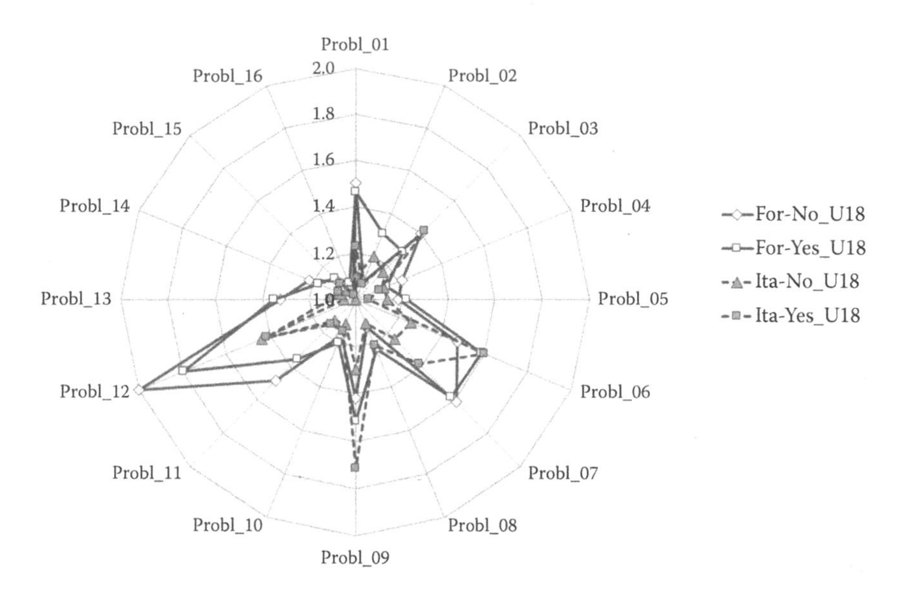

FIGURE 7.8
Radar graphs of problem-by-problem score means by tourist segment.

- Type of response variables: ordered categorical (16 ordinal 1–4 scores)
- Ranking rule: 'the lower the better'
- Combining function: Fisher
- B (number of permutations): 2000
- Significance α-level: 0.05

Details on multivariate and univariate tourist segment comparisons via directional permutation tests are presented in Tables 7.23 and 7.24.

Ranking analysis results allow us to easily rank the four tourist segments under investigation using their judgements on perceived potential problems. The most satisfied segments are the two Italian groups not accompanied by persons younger than 18 years, while the least satisfied and most worried appear to be the two foreign groups. It is also worth noting that the domain ranking analysis provides several interesting insights into how tourists evaluate the Sesto Nature Park with respect to different problem domains. In fact, the within-domain estimated rankings are not exactly the same; in particular, the 'further technical problems' domain appears as the most critical set of items with respect to the perceived potential problems.

7.4.5 Discussion and Conclusions

In assessments of quality of structures and services offered by natural parks, synthetic indicators of different problem areas can represent objective ways

TABLE 7.23

Details of Domain Global Ranking Results on Four Tourist Segments (α = 5%)

Tourist Segment	For-No_U18	For-Yes_U18	Ita-No_U18	Ita-Yes_U18
All problems	**3**	**3**	**1**	**1**
Footpaths, routes and resting areas	2	4	1	1
Congestion problems in the areas	2	4	1	2
Further technical problems	3	3	1	1
All Problems				
For-No_U18		1.00	**.003**	.022
Foc-Yes_U18	.240		**.003**	**.002**
Ita-No_U18	1.00	1.00		.879
Ita-Yes_U18	1.00	1.00	.234	
Footpaths, Routes and Resting Areas				
For-No_U18		.758	.043	.255
For-Yes_U18	.420		**.005**	**.003**
Ita-No_U18	1.00	.991		.627
Ita-Yes_U18	1.00	1.00	.493	
Congestion Problems in the Areas				
For-No_U18		1.00	.129	1.00
For-Yes_U18	.105		**.003**	.100
Ita-No_U18	1.00	.998		.974
Ita-Yes_U18	.298	1.00	.054	
Further Technical Problems				
For-No_U18		.575	**.003**	**.003**
For-Yes_U18	1.00		**.002**	**.002**
Ita-No_U18	1.00	1.00		.960
Ita-Yes_U18	1.00	1.00	.966	

to measure desirable conditions in several aspects of the management of natural assets.

We can outline the flexibility of the method of NPC of dependent partial rankings to construct combined indicators starting from several dependent indicators based on ordered variables. This procedure can be used both to derive a global index of satisfaction and to construct a global ranking of units that can be useful when exploring clusters of respondents with a lower or upper global level of satisfaction.

In the Sesto Nature Survey the methodological approach allowed the evaluation of how tourists perceive the presence in the natural park of three macro potential critical aspects: the first related to problems along footpaths, routes and resting areas; the second (associated with the lower evaluations) concerned with congestion problems in these areas and the

TABLE 7.24

Problem-by-Problem Univariate Directional Permutation p-Values

	For-No_ U18	For-Yes_ U18	Ita-No_ U18	Ita-Yes_ U18	For-No_ U18	For-Yes_ U18	Ita-No_ U18	Ita-Yes_ U18	For-No_ U18	For-Yes_ U18	Ita-No_ U18	Ita-Yes_ U18
	Probl_01				Probl_02				Probl_03			
For-No_U18		.521	**.001**	.033		.998	.947	.523		.392	.127	.811
For-Yes_U18	.478		**.000**	**.008**	**.002**		.083	**.001**	.608		.138	.895
Ita-No_U18	.999	.999		.942	.052	.916		.052	.872	.861		.971
Ita-Yes_U18	.966	.992	.058		.476	.999	.948		.188	.105	.028	
	Probl_04				Probl_05				Probl_06			
For-No_U18		.431	.196	.144		.618	.236	.063		.808	.157	.782
For-Yes_U18	.569		.369	.272	.381		.125	**.010**	.192		**.014**	.540
Ita-No_U18	.803	.630		.445	.764	.875		.143	.843	.986		.962
Ita-Yes_U18	.855	.728	.554		.936	.990	.857		.217	.460	.038	
	Probl_07				Probl_08				Probl_09			
For-No_U18		.707	**.002**	.082		.999	.534	.789		.887	.522	.995
For-Yes_U18	.293		**.000**	**.006**	**.001**		**.004**	**.015**	.113		.130	.953
Ita-No_U18	.998	.999		.934	.466	.996		.693	.478	.870		.994
Ita-Yes_U18	.917	.994	.065		.210	.984	.306		**.005**	.047	**.006**	

(Continued)

TABLE 7.24 (CONTINUED)

Problem-by-Problem Univariate Directional Permutation *p*-Values

	For-No_ U18	For-Yes_ U18	Ita-No_ U18	Ita-Yes_ U18	For-No_ U18	For-Yes_ U18	Ita-No_ U18	Ita-Yes_ U18	For-No_ U18	For-Yes_ U18	Ita-No_ U18	Ita-Yes_ U18
	Probl_10				**Prob1_11**				**Probl_12**			
For-No_U18		.483	.310	.391		.166	**.005**	**.002**		.042	**.000**	**.000**
For-Yes_U18	.517		.251	.285	.833		.025	**.014**	.958		**.000**	**.002**
Ita-No_U18	.689	.749		.577	.995	.975		.536	.999	.999		.827
Ita-Yes_U18	.608	.714	.998	.986	.464		.999	.998	.173		.974	.996
	Probl_13				**Probl_14**				**Probl_15**			
For-No_U18		.613	**.011**	.026		.357	.040	.067		.561	.239	.513
For-Yes_U18	.386		**.000**	**.004**	.642		.057	.108	.438		.152	.440
Ita-No_U18	.988	.999		.850	.960	.943		.687	.761	.848		.762
Ita-Yes_U18	.149		.932	.892	.313		.486			.559	.237	
	Probl_16											
For-No_U18		.770	.245	.434								
For-Yes_U18	.230		.075	.152								
Ita-No_U18	.755	.925		.722								
Ita-Yes_U18	.565	.847	.278									

third related with other technical problems. Another important aspect of the methodological approach was the possibility to stratify the respondents' evaluations with respect to variables potentially influencing the tourists' experience such as nationality, month of the visit and presence of children. The results outlined how the Italian people seemed to be more critical than the German tourists, how tourists' experience seems to improve in September and how the presence of children may increase the expectations.

A final additional advantage of this methodological approach is the possibility to identify specific target values for such global indicators as objective goals to reach with improvement actions.

It would be advisable to carry out surveys like this periodically: they provide valuable information at first, to support decisions regarding environmental and social aspects. Surely, the longitudinal approach (that is, repeating the survey at different times) has more than one advantage: it keeps the situation under control by taking several pictures of its status; it provides more robust information from a statistical standpoint (each sample is always subject to fluctuations) and it can show the effectiveness of the actions taken to improve the naturalistic experience, all this by using a tailored system of quality indicators. If this survey were to be repeated in the future, there would be room for several improvements: for example, it could be extended to include other aspects, such as the opinion of the workforce employed in the area.

7.5 Survey of the Potential of the Passito Wine Market

A survey was performed to study the demand for wine and specifically for Passito in Veneto, north-east Italy. The survey's main goal was to identify homogeneous segments of consumers based on preferences and consumption habits (Arboretti Giancristofaro et al., 2014).

Passito is an Italian word for wines made by the *appassimento* process whereby grapes are partially dried on straw mats or pallets in airy rooms or barns to concentrate the grapes' flavours and sweetness prior to vinification. As the grapes shrivel and lose water they become full of concentrated sugars and flavours. After anywhere from three to six months the semidried grapes are gently pressed and the juice fermented until it reaches the desired level of sweetness and alcohol. Most Passito wines will be kept in oak barrels for some time to develop additional flavours and complexity in addition to time resting in the bottle prior to release for sale. Italian wines made in the Passito style include both red and white wines.

7.5.1 Survey Description

The study on the potential of the Passito wine market in the Veneto region was conducted from April to August 2009. The data were collected through a Computer-Assisted Telephone Interview (CATI) survey, where a stratified by province of residence random sample of 349 potential Passito consumers was extracted. The consumers selected for the interview were chosen from all seven provinces of Veneto: Belluno, Padova, Rovigo, Treviso, Venezia, Vicenza and Verona (Figure 7.9). The questionnaire was divided into three main parts relating, respectively, sociodemographic information, consumption habits, preferences on wine and Passito.

The main goal of the study consisted of analysing how the preferences and consumption habits of Passito vary depending on consumers' characteristics. Applying to Passito survey data either logistic regression or CUB models (Covariates in the mixture of Uniform an shifted Binomial distributions, see Iannario and Piccolo, 2007), Arboretti Giancristofaro et al. (2014) found that gender and residence are the most important covariates, useful in defining segments of consumers, homogeneous from the sociodemographic standpoint, with significant differences in terms of preferences and Passito consumption behaviour.

Accordingly, after clustering the seven provinces of Veneto into three main areas, that is, West, South and North-East (Figure 7.9), a stratified by area

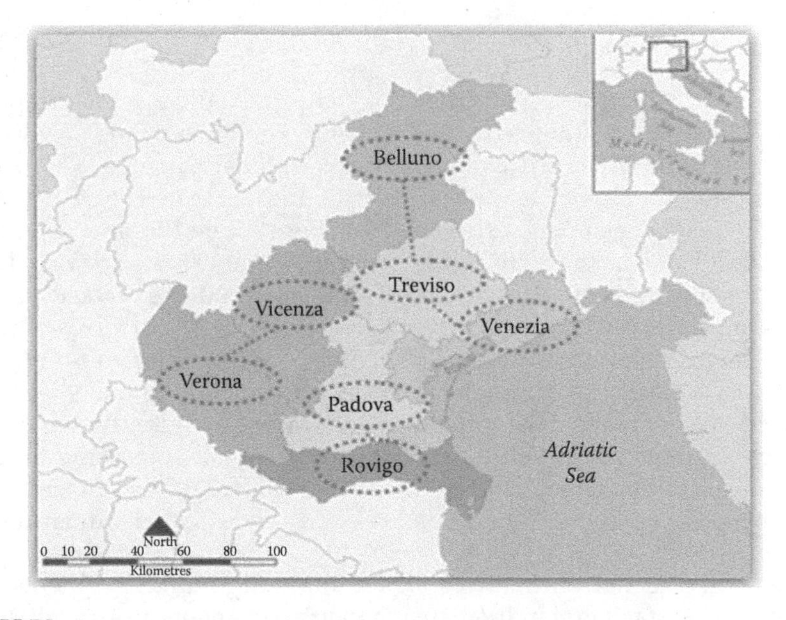

FIGURE 7.9
Clustering of Veneto Provinces into three main areas: (1) West (Verona and Vicenza), (2) South (Padova and Rovigo) and (3) North-East (Belluno, Treviso and Venezia).

TABLE 7.25

Definitions of Items Investigated in Ranking Analysis of Passito Market Segments

Item	Definition
Liking	Level of liking Passito
Aroma	Satisfaction with Passito's aroma
Sweet taste	Satisfaction with Passito's sweet taste
Alcohol content	Satisfaction with Passito's alcohol content
Intensity	Satisfaction with Passito's intensity of taste

of residence ranking analysis was performed to highlight whether the four segments defined as Female–Young, Male–Young, Female–Mature, Male–Mature can be ranked with respect to their overall preference on Passito, where the segmentation by age into the two categories, Young and Mature, was established considering as the cut-off 35 years of age.

In the present work we focussed on ranking analysis of four previously defined market segments from the set of five items on Passito satisfaction as detailed in Table 7.25.

To measure respondents' satisfaction with regard to various aspects of Passito wine, a scale of scores from 1 (not preferred at all) to 7 (very much preferred) was used.

7.5.2 Ranking Analysis

A first look at Passito satisfaction data is provided by radar graphs in Figure 7.10, where score sample means by market segment and area of residence are reported. According to Arboretti Giancristofaro et al. (2014), from a descriptive standpoint it appears that the most satisfied segment seems to be the male consumers, especially for those located in area 1, because the related score means tend to take generally large values. Conversely, the young female segment appears to be the least satisfied one because its score means are generally lower than those of other segments. In terms of which satisfaction dimension is the one that most discriminates the preference toward Passito, the radar graphs seem to suggest that no one single item should be considered as the most valuable feature because the score mean lines by segment are roughly concentric. This is not surprising because, as pointed out by Arboretti Giancristofaro et al. (2014), there is generally a strong positive correlation among the five investigated dimensions of Passito satisfaction.

The main objective of this work was to rank four market segments from customers' satisfaction standpoint using their assessments on several aspects of satisfaction in the consumption of Passito wine. As the questionnaire was submitted to people living in different subareas, a stratified global ranking analysis is advised. We recall that by 'stratified analysis' we mean that

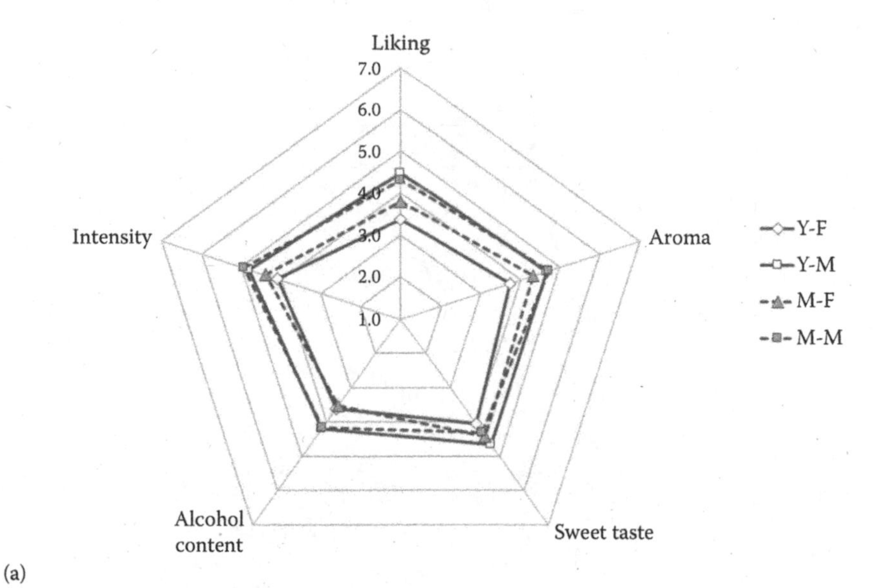

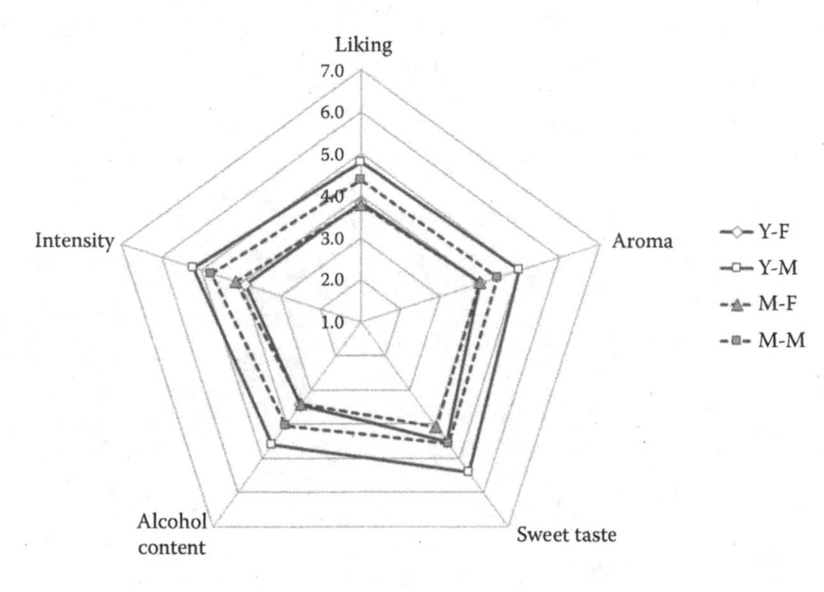

FIGURE 7.10
Radar graphs of score means by area of customer residence. (a) Item-by-item sample score means (all areas). (b) Item-by-item sample score means (area 1). *(Continued)*

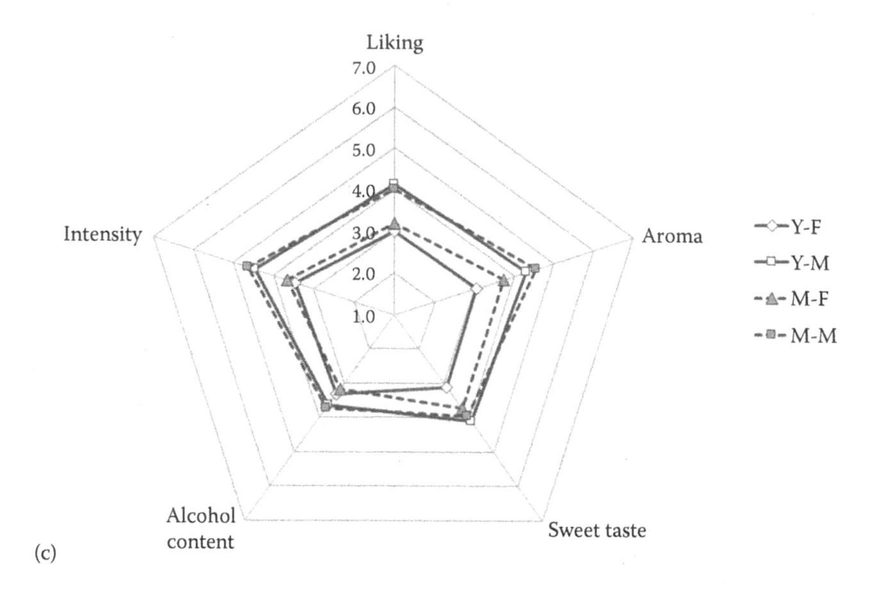

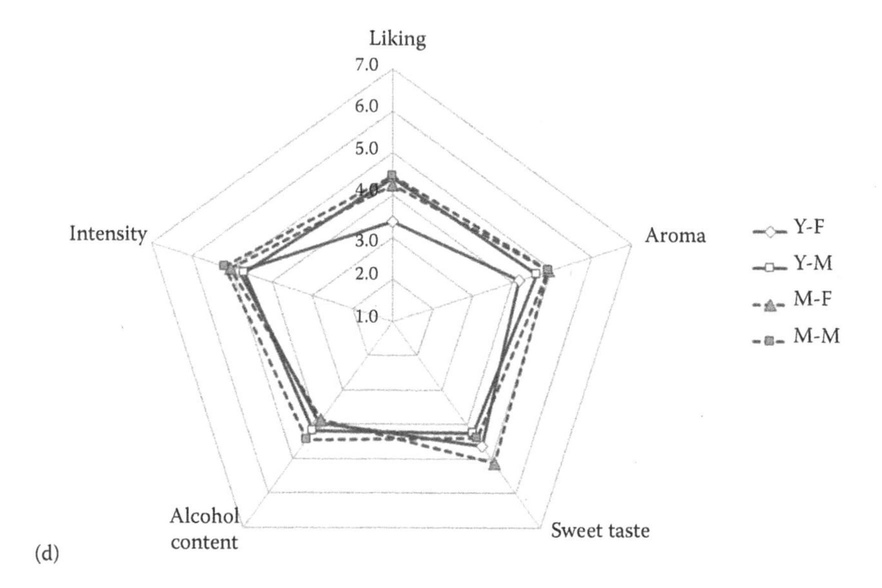

FIGURE 7.10 (CONTINUED)
Radar graphs of score means by area of customer residence. (c) Item-by-item sample score means (area 2). (d) Item-by-item sample score means (area 3).

several separate ranking analyses for each stratum (area of residence) along with an overall ranking analysis should be performed.

The setting of the Passito customers' satisfaction ranking problem (see Chapter 3) can be summarized as follows.

- Type of design: one-way MANOVA with stratification (four independent samples, i.e. four multivariate populations to be ranked and one confounding factor to take into account, i.e. the specific area of customer residence)
- Domain analysis: no
- Type of response variables: ordered categorical (five ordinal 1–7 scores)
- Ranking rule: 'the higher the better'
- Combining function: Fisher
- B (number of permutations: 2000
- Significance α-level: 0.1

Details on multivariate and univariate market segment comparisons via directional permutation tests are presented in Tables 7.26 and 7.27.

Ranking analysis results allow us to easily rank the four market segments under investigation using the customers' judgements on satisfaction items. The most Passito satisfied segment are the young male consumers, while the least involved with Passito appears to be the young female segment. It is also worth noting that the stratified ranking analysis provides several interesting insights into how customers evaluate Passito under different residence areas. In fact, the within-strata estimated rankings by area of residence are not exactly the same, in particular for consumers located in north-east Veneto, all market segments are tied in the first ranking position. Moreover, according to Arboretti Giancristofaro et al. (2014), note that the west area denotes a much more general satisfaction in comparison with the other areas, probably because the most famous and appreciated Passitos of Veneto are produced in Verona (*Recioto di Soave, Recioto di Valpollicella* and *Amarone della Valpolicella*) and Vicenza (*Torcolato di Breganze*).

7.5.3 Conclusions

This study presents the results of a survey on the potential of the Passito wine market considering as the target population consumers living in Veneto, north-east Italy, where the main goal was to identify homogeneous segments of consumers based on preferences and consumption habits. The survey consisted of a questionnaire administrated via the CATI system and involved a sample of around 350 consumers. Multivariate nonparametric ranking methods were applied to compare customer satisfaction toward

TABLE 7.26

Details of Stratified Global Ranking Results on Four Market Segments (α = 10%)

Market Segment	Y-F	Y-M	M-F	M-M
All areas	**4**	**1**	**3**	**1**
West	4	1	2	2
South	4	1	2	2
North-east	1	1	1	1
All Areas				
Y-F		1.00	1.00	1.00
Y-M	**.009**		.180	1.00
M-F	.178	.898		1.00
M-M	**.008**	1.00	.189	
West				
Y-F		1.00	1.00	1.00
Y-M	**.033**		.081	.285
M-F	1.00	.991		1.00
M-M	.526	1.00	.534	
South				
Y-F		.979	1.00	1.00
Y-M	**.033**		.159	1.00
M-F	.721	.977		1.00
M-M	.055	1.00	.153	
North-East				
Y-F		1.00	.862	1.00
Y-M	.378		1.00	1.00
M-F	.241	.514		1.00
M-M	.378	.478	.409	

Passito among four possible market segments defined by combination of gender and age. Ranking results allowed us to rank the segments, showing that young male consumers are the most oriented to be satisfied in the consumption of Passito wine, with particular reference to those who live in the western Veneto.

The ranking results are in agreement with results found in Arboretti Giancristofaro et al. (2014), where it was noted that people living in the western Veneto provinces of Vicenza and Verona show more interest in Passito and drink it more than other consumers from Veneto. Also, men are characterized by a much greater appreciation and knowledge of this wine than women.

TABLE 7.27

Item-by-Item Univariate Directional Permutation p-Values

	All Areas											
	Y-F	Y-M	M-F	M-M	Y-F	Y-M	M-F	M-M	Y-F	Y-M	M-F	M-M
	Liking				*Aroma*				*Sweet Taste*			
Y-F		.999	.979	.999		.999	.978	.999		.973	.937	.749
Y-M	**.000**		**.011**	.294	**.000**		.128	.605	**.026**		.526	.188
M-F	**.021**	.989		.936	**.021**	.871		.891	.063	.473		.142
M-M	**.000**	.705	.063		**.001**	.395	.108		.251	.812	.857	
	Alcohol Content				*Intensity*							
Y-F		.992	.436	.978		.997	.806	.996				
Y-M	**.007**		**.003**	.438	**.003**		.078	.708				
M-F	.564	.997		.980	.194	.922		.957				
M-M	**.021**	.562	**.019**		**.004**	.291	**.043**					
	West											
	Y-F	Y-M	M-F	M-M	Y-F	Y-M	M-F	M-M	Y-F	Y-M	M-F	M-M
	Liking				*Aroma*				*Sweet Taste*			
Y-F		.984	.517	.854		.982	.504	.742		.957	.201	.408
Y-M	**.016**		**.033**	.197	**.017**		**.034**	.129	**.043**		**.001**	.057
M-F	.483	.966		.842	.496	.965		.750	.799	.989		.684
M-M	.145	.803	.158		.258	.870	.249		.592	.943	.316	

(Continued)

TABLE 7.27 (CONTINUED)

Item-by-Item Univariate Directional Permutation p-Values

	Alcohol Content				*Intensity*			
Y-F		.995	.455	.864		.998	.644	.921
Y-M	**.005**		**.009**	.089	**.001**		**.019**	.139
M-F	.545	.990		.904	.355	.980		.843
M-M	.136	.910	.096		.078	.860	.156	

South

	Y-F	Y-M	M-F	M-M	Y-F	Y-M	M-F	M-M	Y-F	Y-M	M-F	M-M
	Liking				*Aroma*				*Sweet Taste*			
Y-F		.997	.679	.976		.998	.907	.996		.990	.883	.952
Y-M	**.003**		**.015**	.434	**.002**		**.034**	.678	**.009**		.124	.269
M-F	.320	.985		.949	.092	.965		.968	.117	.875		.648
M-M	**.024**	.565	.051		**.003**	.321	**.032**		**.048**	.731	.351	
	Alcohol Content				*Intensity*							
Y-F		.851	.369	.783		.987	.608	.976				
Y-M	.149		.073	.597	**.013**		.021	.590				
M-F	.631	.927		.906	.392	.978		.967				
M-M	.216	.403	.093		**.024**	.410	**.033**					

(*Continued*)

TABLE 7.27 (CONTINUED)

Item-by-Item Univariate Directional Permutation p-Values

	North-East											
	Y-F	Y-M	M-F	M-M	Y-F	Y-M	M-F	M-M	Y-F	Y-M	M-F	M-M
	Liking				*Aroma*				*Sweet Taste*			
Y-F		.996	.981	.994		.834	.960	.956		.150	.885	.271
Y-M	**.004**		.404	.514	.165		.835	.825	.850		.985	.657
M-F	**.018**	.596		.601	**.039**	.164		.454	.114	**.014**		**.033**
M-M	**.005**	.486	.398		**.043**	.175	.546		.728	.342	.957	
	Alcohol Content				*Intensity*							
Y-F		.772	.501	.871		.376	.666	.793				
Y-M	.227		.226	.715	.623		.780	.899				
M-F	.499	.773		.887	.333	.220		.640				
M-M	.129	.285	.112		.207	.100	.360					

7.6 Longitudinal Survey to Evaluate Perception of Malodours

The Statistical Group from Department of Management and Engineering of University of Padua, on behalf of the municipalities of Este and Ospedaletto Euganeo (Italy, north-east), carried out a statistical sensory survey on odour perceptions to evaluate the impact on the local population of various odourous sources located in the territory of two towns (Arboretti et al., 2015). Specifically, the objective of this survey was to monitor, from a statistical standpoint, odour perceptions of local citizens to determine the impact of different odour categories on citizens' perception and measure the temporal evolution, also depending on atmospheric and climatic changes. In this way, a map of this phenomenon was provided, taking into account the area involved and the seasonal period. The survey lasted 12 months, starting on 15 February 2010 and ending on 6 February 2011.

7.6.1 Literature Review of Malodour Surveys

Unpleasant odours can cause a serious nuisance in the vicinity of sanitary landfills. Nicolas et al. (2006) proposed the use of the sniffing team method to estimate the odour emission rate from landfill areas. Controlling odours from landfill sites has become an important regulatory issue, requiring accurate and reproducible sampling and measurement. Because the monitoring of the odour annoyance generated by a landfill area is difficult, one of the most representative and most frequently used ways to assess the overall odour level still remains sensory measurement using a panel of judges. Usually, the measurement goal is the determination of the mean odour emission rate from the whole landfill area, expressed in odour unit per second (ouE/s: the 'E' stands for 'European '7', as defined by the European standard EN 13725, 2003; henceforth this abbreviation is used only if that European standard method is applied). Such outcomes can be used for further evaluation of odour concentration percentiles prevailing for typical climatic conditions. Long-term exposure is quantified in terms of frequency of occurrence of hourly averaged concentrations above a certain odour concentration limit.

The determination of the odour emission rate can be based on global field measurements, taking account of the real perception of the odour in the environment in the surroundings of the source. A possible approach is the method of sniffing team observations, which utilise experienced people to evaluate the maximum distance from the source at which the odour is perceived. 'Experienced people' means operators with reliable olfactory performance who always apply the same sniffing procedure. The results of a dozen such measurements allow calculation of the typical odour emission rate with a dispersion model. Sniffing team methods have some advantages over instrumental and olfactometric measurements. The main advantage is that they involve field measurements, by which the global impact of the source is evaluated, allowing consideration of diffuse, surface and less clear sources, such as waste handling or transportation. Furthermore, these methods reflect the actual perceptibility of the odour in the environment. However, the sniffing team observation method also presents many limitations. First, it makes fundamental assumptions: it is valid only if both the meteorological situation and the odour emission do not vary too much during the measurement period. The waste odour around a landfill site is actually emitted as discontinuous puffs, depending on the activities on the landfill tipping face.

Nicolas et al. (2006) discuss the applicability of the sniffing team observation method to estimate the annoyance zone around landfill areas. It is based on 52 measurements made on five different municipal solid waste landfill sites in Wallonia, in the south of Belgium. The causes of the estimation bias are identified and the relative errors are estimated by a sensitivity analysis. The main topic of the paper is the discussion of the applicability of the methodology to diffuse and discontinuous odour sources. The results are supplied only for illustrative purposes. The sniffing method, as applied by the Department of Organic Chemistry at the University of Gent, is described in detail in Van Langenhove and Van Broeck (2001).

7.6.2 Survey Description

As mentioned previously, this study took place in the territories of two towns, that is, Este and Ospedaletto Euganeo, located in the south part of the province of Padova, north-east Italy. In Figure 7.11 the actual boundaries of the area under study are shown with an orthophoto.

The odour sources that have an impact on the territory being studied come from livestock farms and from plants operating locally. Regarding the first sources, many stock raising farms were identified (broilers, ducks, cattle, rabbits, pheasants, guinea fowl, sheep, turkeys, geese and swine), and a couple of farms for weaning. Overall there are 49 livestock farms, 28 in the municipal area of Este and 21 in the municipal area of Ospedaletto Euganeo. Note that the census of the aforementioned farms was obtained from the Veterinary Office of the local National Health Service department. Regarding the second odour sources, the main sources identified are a feed mill, a cement factory and a solid waste treatment landfill.

The current survey consisted of a statistical sensory analysis based on the reporting of odour perceptions by the citizens of the town of Este who took part in the survey are called *sniffers* from now on). The reporting consisted of filling in a questionnaire specifically prepared by the University of Padua and available online at a website, or alternatively, as printed document.

To better present and summarize the survey results, some areas have been joined, so the Este district has been divided into nine areas, six of which belong to the town of Este and three to the town of Ospedaletto Euganeo (see

FIGURE 7.11
Orthophotographic boundaries of Este and Ospedaletto Euganeo (Google Earth).

Table 7.28 and Figure 7.12). This classification in sub-districts was suggested bearing in mind:

- Homogeneity with regard to the issue of odours (similar exposure to odourous sources)
- Morphology of the territory
- Density of the local population

In Table 7.29 the number of sniffers who took part in the survey for each of the nine areas under examination is reported. Notice that each sniffer

TABLE 7.28
Number of the Nine Areas into Which Este and Ospedaletto Euganeo Were Divided

Town	Area No.
Este	1
	2
	3
	4
	5
	6
Ospedaletto Euganeo	7
	8
	9

FIGURE 7.12
Division of Este and Ospedaletto Euganeo into nine areas.

TABLE 7.29
Number of Sniffers in Each of the Nine Areas Examined

Area No.	Count of Sniffers
1	78
2	45
3	48
4	16
5	22
6	12
Total for Este	**221**
7	43
8	6
9	15
Total for Ospedaletto Euganeo	**64**
Overall Total	**285**

represents a household; therefore it consists of at least one person, but normally there would be more than one housemate.

To better study the phenomenon of odour perceptions, to evaluate their temporal trend and to compare reports done at different times in different areas, we defined a weekly synthetic indicator and called it Odour Perception Index (OPI). The formula defining the OPI is

$$\text{OPI}(\text{area} = i, \text{week} = j) = \frac{\sum_{i,j} (\text{duration of reported odors in } i\text{th area during the } j\text{th week})}{\text{number of active sniffers in the } i\text{th area during the } j\text{th week}}.$$

By 'number of active sniffers' we mean the number of sniffers who actually file at least one report (even when no odour is detected). In fact we noticed that, when keeping all other factors fixed, the actual number of truly active sniffers can significantly change during the course of the survey. To better understand the meaning of the indicator, we present some possible values that can be taken by OPI, and the related explanation:

- **OPI = 0** (*minimum value*): There was no report at all during the whole week (the duration sum is 0); therefore, as soon as there is at least one report, IPO has a value greater than zero.
- **OPI = 0.5**: The sum of the duration of all odour reports during a week is equivalent to half a day (12 hours).
- **OPI = 1.0**: The sum of the duration of all odour reports during a week is equivalent to one day (24 hours).
- **OPI = 7** (*maximum value*): All sniffers reported the presence of malodour for an overall duration of 7 days out of 7 (uninterruptedly, for every hour of every day of the week).

In brief, the greater OPI is, the longer the odour perception lasted. Notice that OPI takes into account the number of reports, their durations and the number of sniffers present and truly active in the area under investigation.

7.6.3 Summary Results on Odour Reports

This section contains the main summary results on odour reports recorded in the first trimester of the survey, for the 12 weeks from 15 February 2010 until 9 May 2010. In particular, we elaborated the number of reports in each town, divided by odour taxonomy and potential source of the malodour. Then, to study the phenomenon of odourous perceptions better, to appreciate their temporal evolution and to be able to compare reports done at different times in different areas, we show the values taken by the synthetic indicator OPI. Finally, as an effective and brief tool to summarize odourous perceptions in the span of the whole trimester, we show a series of bubble charts that enable a quick comparison among different areas.

The first synthetic survey result consists of reporting the number of odour perceptions, divided by town and week of accounting (see Table 7.30). To better describe and analyse the temporal trend of the detections, Table 7.30 also indicates duration and average intensity of recorded reports.

The average duration was calculated by assigning a score to each answer concerning the duration of an odourous perception, when the odour is detected continuously. The scale is: 1 = 'less than a minute', 2 = 'a few minutes',

TABLE 7.30

Odour Reports: Number, Average Duration and Intensity, Week by Week in Each Town

Town	Week	No. of Reports	Avg. Duration	Avg. Intensity
Este	15/02/2010–21/02/2010	83	2.96	2.32
	22/02/2010–28/02/2010	54	2.56	2.24
	01/03/2010–07/03/2010	52	3.02	2.27
	08/03/2010–14/03/2010	44	3.02	2.34
	15/03/2010–21/03/2010	74	2.78	2.54
	22/03/2010–28/03/2010	47	2.57	2.45
	29/03/2010–04/04/2010	54	2.69	2.46
	05/04/2010–11/04/2010	63	2.62	2.43
	12/04/2010–18/04/2010	33	2.58	2.64
	19/04/2010–25/04/2010	68	2.55	2.59
	26/04/2010–02/05/2010	29	2.80	2.69
	03/05/2010–09/05/2010	15	2.92	2.93
Este Total		**616**	**2.75**	**2.45**
Ospedaletto Euganeo	15/02/2010–21/02/2010	4	2.00	2.50
	22/02/2010–28/02/2010	5	2.20	3.20
	01/03/2010–07/03/2010	5	2.20	2.20
	08/03/2010–14/03/2010	10	2.10	2.30
	15/03/2010–21/03/2010	28	2.21	2.82
	22/03/2010–28/03/2010	25	2.16	2.84
	29/03/2010–04/04/2010	22	2.09	2.82
	05/04/2010–11/04/2010	40	2.23	2.63
	12/04/2010–18/04/2010	86	2.38	2.67
	19/04/2010–25/04/2010	102	2.29	2.77
	26/04/2010–02/05/2010	76	2.12	2.70
	03/05/2010–09/05/2010	44	2.48	2.89
Ospedaletto Euganeo Total		**447**	**2.26**	**2.73**
Overall Total		**1063**	**2.56**	**2.57**

3 = 'a few hours', 4 = 'half a day', 5 = 'the whole day', 6 = 'the whole night'. We calculated the arithmetic mean of the scores, and the average duration is always between 2 and 3, that is, between a few minutes and a maximum of some hours.

Similarly, we created a scale for the intensity: 1 = 'light', 2 = 'moderate', 3 = 'strong', 4 = 'very strong'. Then we calculated the arithmetic mean, which always presents a value between 2 and 3, so from 'moderate' to 'strong', except for the second week where, in Ospedaletto Euganeo the average intensity was between 'strong' and 'very strong'.

It is important to notice how stable the situation is: in fact, the phenomena observed fall into a precise interval, both in terms of duration and intensity. To take into account the diversity in the number of sniffers per area, we will use the indicator OPI, as described in the previous section. Table 7.31 shows the number of odour reports, divided by type and possible source, and also separately for each town.

By looking at Table 7.31, we can draw two conclusions:

1. For a considerable number of reports, the source is not identified (46% in Este, 32% in Ospedaletto Euganeo); among those reports where the source is identified, the item most commonly indicated is 'municipal waste treatment plant' (41% for Este, and even 69% for Ospedaletto Euganeo).
2. Regarding the type of odour, the most frequent categories are:
   a. 'Harsh, stinging' (29% both for Este and Ospedaletto Euganeo)
   b. 'Putrid, rotten' (20% for Este, 31% for Ospedaletto Euganeo)
   c. 'Animal manure' (18% for Este, 13% for Ospedaletto Euganeo)

It should also be noted that the category 'other smell' was reported for 22% of reports in Este, and 14% in Ospedaletto Euganeo. By crossing the results on potential sources with the types of smell, we can see that the municipal waste plant, reported as the main odour source, is believed to be the cause of a wide range of odour types, even very different ones. In other words, the municipal waste treatment plant is often indicated as the odour source, regardless of the type of smell perceived. In Table 7.32 we report details on OPI values for some of the 12 weeks, either overall or separately for each type of smell.

The evolution of the IPO index for the trimester examined (weekly average values in the trimester) can be plotted as the chart in Figure 7.13.

Studying the trimester report, it is clear that the areas where odour emissions were more intensely perceived are located in the north-east corner of the map examined, with the only exception of the area no. 1. To better understand the phenomenon, we conducted an analysis concerning type, intensity and source of perceptions.

TABLE 7.31

Number of Odour Reports, Divided by Type and Potential Source

		Odour Type								
Town	**Potential Source**	**Animal Related**	**Harsh, Stinging**	**Ammoniacal**	**Hay/ Forage**	**Toasting**	**Animal Manure**	**Putrid, Rotten**	**Other Smell**	**Total**
Este	NOT identified	19	72	6	1	–	70	53	64	285
	Identified	28	107	3	1	6	43	70	73	331
	Poultry stock	11	6	–	–	–	11	9	11	48
	Cattle	3	2	–	–	–	4	1	2	12
	Swine breeding	1	2	–	–	–	2	1	1	7
	Poultry droppings on cropland	–	1	–	–	–	1	1	1	4
	Manure spread on cropland	1	3	–	–	–	1	3	1	9
	Municipal waste treatment plant	6	40	2	–	4	14	46	23	135
	Feed mill	1	–	–	–	–	3	2	2	8
	Cement factory	–	14	–	–	–	–	–	1	15
	Traffic	4	8	1	–	1	–	4	12	30
	Other source	1	31	–	1	1	7	3	19	63
Este Total		47	179	9	2	6	113	123	137	616

(Continued)

TABLE 7.31 (CONTINUED)

Number of Odour Reports, Divided by Type and Potential Source

Town	Potential Source	Odour Type								Total
		Animal Related	Harsh, Stinging	Ammoniacal	Hay/ Forage	Toasting	Animal Manure	Putrid, Rotten	Other Smell	
Ospedaletto Euganeo	NOT identified	14	32	1	–	5	29	41	23	145
	Identified	27	99	6	1	7	27	97	38	302
	Poultry stock	7	3	2	–	1	11	1	2	27
	Cattle	–	1	–	–	–	–	1	–	2
	Swine breeding	6	2	2	–	–	4	5	3	22
	Poultry droppings on cropland	–	2	–	–	–	3	–	2	7
	Manure spread on cropland	1	–	–	–	–	1	–	1	3
	Municipal waste treatment plant	10	86	1	1	1	7	83	18	207
	Feed mill	1	1	–	–	5	1	2	3	13
	Cement factory	1	2	1	–	–	–	2	1	7
	Traffic	–	–	–	–	–	–	1	1	2
	Other source	1	2	–	–	–	–	2	7	12
Ospedaletto Euganeo Total		41	131	7	1	12	56	138	61	447
Overall Total		88	310	16	3	18	169	261	198	1063

TABLE 7.32

Details of OPI Values by Area, Week and Type of Smell

Week_ID	Week	All Smells	Area	Animal Related	Harsh, Stinging	Ammoniacal	Hay/ Forage	Toasting	Animal Manure	Putrid, Rotten	Other Smell
1	15/02/2010–21/02/2010	0.331	1	0.021	0.054	0.000	0.000	0.003	0.012	0.040	0.202
		0.061	2	0.000	0.013	0.004	0.000	0.000	0.004	0.000	0.040
		0.028	3	0.000	0.004	0.000	0.000	0.000	0.017	0.007	0.000
		0.010	4	0.000	0.000	0.000	0.000	0.000	0.010	0.000	0.000
		0.000	5	0.000	0.000	0.000	0.000	0.000	0.000	0.000	0.000
		0.000	6	0.000	0.000	0.000	0.000	0.000	0.000	0.000	0.000
		0.028	7	0.000	0.014	0.000	0.000	0.000	0.000	0.014	0.000
		0.000	8	0.000	0.000	0.000	0.000	0.000	0.000	0.000	0.000
		0.042	9	0.000	0.000	0.000	0.000	0.000	0.000	0.000	0.042
2	22/02/2010–28/02/2010	0.116	1	0.017	0.017	0.000	0.000	0.000	0.005	0.011	0.067
		0.075	2	0.000	0.015	0.001	0.000	0.003	0.005	0.007	0.043
		0.030	3	0.000	0.008	0.000	0.000	0.000	0.000	0.008	0.015
		0.000	4	0.000	0.000	0.000	0.000	0.000	0.000	0.000	0.000
		0.000	5	0.000	0.000	0.000	0.000	0.000	0.000	0.000	0.000
		0.000	6	0.000	0.000	0.000	0.000	0.000	0.000	0.000	0.000
		0.010	7	0.000	0.010	0.000	0.000	0.000	0.000	0.000	0.000
		0.000	8	0.000	0.000	0.000	0.000	0.000	0.000	0.000	0.000
		0.000	9	0.000	0.000	0.000	0.000	0.000	0.000	0.000	0.000
...	...	...	...	...	...	...	...	...	...	...	...

(Continued)

TABLE 7.32 (CONTINUED)

Details of OPI Values by Area, Week and Type of Smell

Week_ID	Week	All Smells	Area	Animal Related	Harsh, Stinging	Ammoniacal	Hay/ Forage	Toasting	Animal Manure	Putrid, Rotten	Other Smell
12	03/05/2010–09/05/2010	0.146	1	0.000	0.083	0.000	0.000	0.000	0.063	0.000	0.000
		0.271	2	0.000	0.000	0.000	0.000	0.000	0.021	0.250	0.000
		0.167	3	0.063	0.010	0.000	0.000	0.000	0.000	0.031	0.063
		0.250	4	0.000	0.125	0.000	0.000	0.000	0.000	0.125	0.000
		0.000	5	0.000	0.000	0.000	0.000	0.000	0.000	0.000	0.000
		0.000	6	0.000	0.000	0.000	0.000	0.000	0.000	0.000	0.000
		0.081	7	0.000	0.039	0.000	0.000	0.005	0.031	0.003	0.003
		0.050	8	0.025	0.000	0.000	0.000	0.000	0.025	0.000	0.000
		0.065	9	0.006	0.006	0.000	0.000	0.000	0.000	0.030	0.024
All weeks	15/02/2010–09/05/2010	0.263	1								
		0.064	2								
		0.151	3								
		0.052	4								
		0.079	5								
		0.023	6								
		0.050	7								
		0.057	8								
		0.150	9								

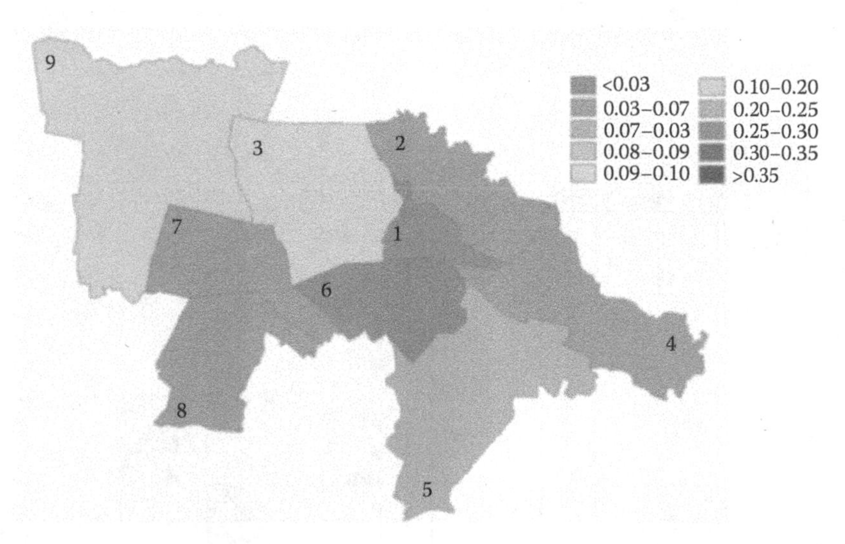

FIGURE 7.13
Cartogram of OPI values for Este and Ospedaletto Euganeo.

As about the most relevant odour types reported by the sniffers, we noticed that the distribution of reports, grouped by odour type, is concentrated mainly on four items: harsh/stinging, putrid/rotten, from animal dejections and all the answers grouped under 'other smell'. In particular, we noticed how the three critical areas, 1, 3 and 9, have a direct relation to the harsh/stinging category.

The phenomenon of odourous perceptions in the towns of Este and Ospedaletto Euganeo, in the trimester studied (from 15 February 2010 to 9 May 2010), appeared to be relevant and widespread. Figure 7.14 summarizes the intensity of the phenomenon in every single week and for each area in which the territory was divided.

The area most affected by the phenomenon is the no. 1, where the peak moments happened in the first, third and fourth weeks, but the situation never went under the minimum warning threshold, represented by the value of 0.10 for the OPI, for the whole period. The phenomenon was also somewhat persistent, although less relevant and homogeneous, in the areas no. 3 (province of Este) and no. 9 (in the province of Ospedaletto Euganeo). Despite the pervasiveness of the phenomenon (only in the area no. 6 were the values of OPI always under 0.10), it is clear that its intensity varies based on the territory. The odour phenomena are not perceived homogeneously in the different areas, in terms of both intensity and time duration.

Generally, odour presence was characterized by a great deal of variability from week to week, possibly due to the different activities done in the stock farms and plants in the territory.

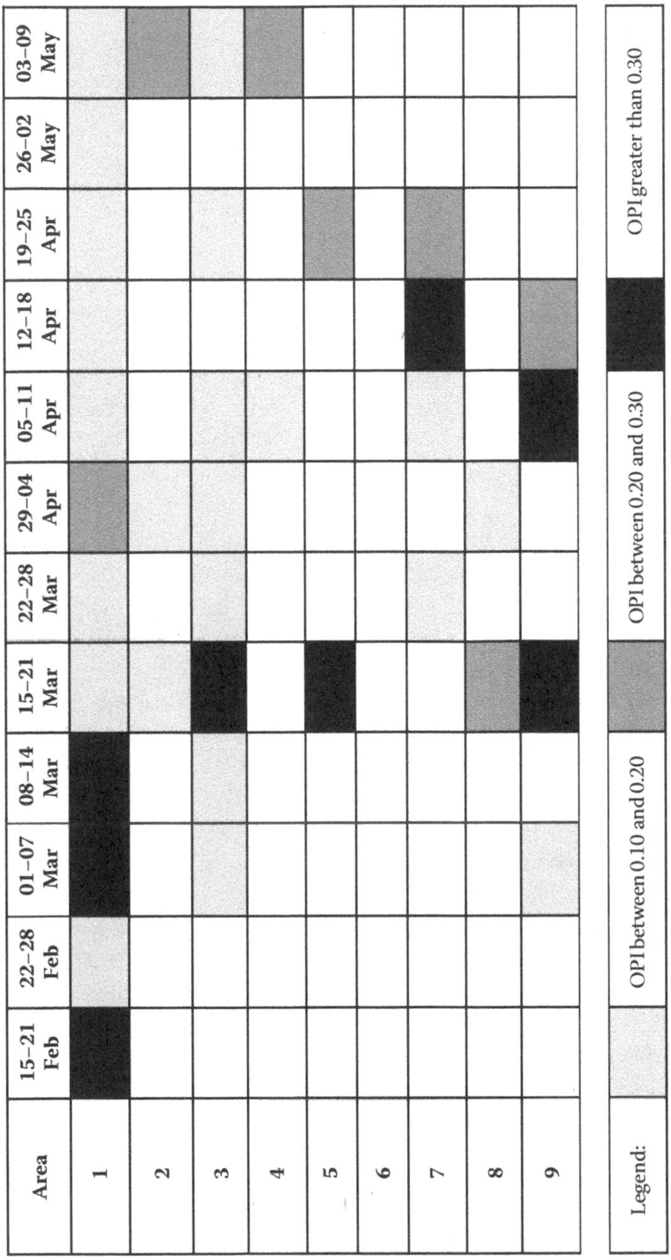

Area	15–21 Feb	22–28 Feb	01–07 Mar	08–14 Mar	15–21 Mar	22–28 Mar	29–04 Apr	05–11 Apr	12–18 Apr	19–25 Apr	26–02 May	03–09 May
1												
2												
3												
4												
5												
6												
7												
8												
9												

Legend: OPI between 0.10 and 0.20; OPI between 0.20 and 0.30; OPI greater than 0.30

FIGURE 7.14
Intensity of odour perceptions, calculated using OPI, by area and week in the first trimester.

Odour types and potential sources identified by the sniffers were numerous, and again variable, depending on the area or the week examined, which suggests that possible corrective measures should not focus on a specific source but rather act in more than one direction.

Still, these technical issues made it difficult to superimpose instrumental measurements on data provided by the sniffers.

Therefore we believe that electronic nose measures should be used as additional data, and not to draw comparisons with what the sniffers perceived, limited to the nose's ability to make out only odour sources perceived during training and specifically for the area being monitored, bearing in mind atmospheric conditions, especially with reference to wind.

In conclusion, in the first trimester of the survey we observed some intensity and diffusion of this phenomenon, but also a great deal of heterogeneity both along the geographical and the temporal dimensions. The lack of homogeneity concerned not only its intensity, duration and number of reports, but also the types of smells perceived and the potential sources identified.

7.6.4 Ranking Analysis

Ranking methods may be quite effective to highlight in which area the odour perceptions have been more frequently and/or intensively perceived. The weekly synthetic indicator OPI by area and type of smell (see Table 7.32) appears to be a very useful dataset because it allows us to compare odour reports done at different times in different areas for different types of odours. In this context, the need to take into account the time effect over the reporting of odour perceptions by sniffers suggests considering as a reference statistical multivariate model a repeated measures design in which the effect of living in a given area on odour perception can be modelled by a discrete or discretized stochastic process for which at most a countable set of data points is observed. In this way an observed OPI curve is nothing more than a set of *repeated measures* in which for each a given curve is observed on a finite or at most a countable number of occasions so that successive responses are dependent. Within this framework we apply the ranking approach for multivariate populations where multivariate directional permutation tests are obtained via nonparametric combination by the so-called *time-to-time* permutation solution we described in Chapter 2, Section 2.3.2.

The setting of the odour perception ranking problem (see Chapter 3) can be summarized as follows.

- Type of design: multivariate repeated measures (MRM design) (nine samples, i.e. nine multivariate populations to be ranked)
- Domain analysis: no
- Type of response variables: continuous (eight continuous responses)

- Ranking rule: 'the higher the better'
- Combining function: Fisher
- B (number of permutations): 5000
- Significance α-level: 0.1

All ranking analysis results are reported in Tables 7.33 and 7.34.

Results allow us to show to rank area 1 and area 3 as the sub-districts where the odour perceptions are stronger. Conversely, area 6 and area 8 are sub-districts where the odour perceptions are weaker.

Table 7.34 displays the set of the univariate permutation directional pairwise p-values, related to the set of the multiple comparison procedures for the problem at hand.

Univariate directional p-values highlight that the most discriminating smells are 'harsh, stinging', 'Animal related', 'Animal manure' and 'Other smell'.

7.6.5 Conclusions

The objective of this work was to monitor, using a suitable Web-based survey and the related reporting questionnaire, the odour perceptions of a group of nonprofessional sniffers. The final goal was to evaluate the temporal evolution of this phenomenon, also depending on atmospheric and climatic changes, and investigate any possible differences among several areas. The survey lasted 12 months and involved around 300 households for a total of more than 500 citizens. We presented details of odour reporting results collected during the first 12 weeks, that is, the first trimester of the survey. Descriptive statistics and ranking analysis provided interesting insights

TABLE 7.33

Global Ranking Results of Odour Perceptions in Nine Areas (α = 10%)

Area	Area 1	Area 2	Area 3	Area 4	Area 5	Area 6	Area 7	Area 8	Area 9
Global ranking	1	3	2	7	3	9	3	8	3
Area 1		.022	.207	.045	.422	.014	.067	.017	1.00
Area 2	1.00		1.00	1.00	1.00	.022	1.00	1.00	1.00
Area 3	1.00	.67		.664	.365	.028	1.00	.034	1.00
Area 4	1.00	1.00	1.00		1.00	1.00	1.00	1.00	1.00
Area 5	1.00	1.00	1.00	1.00		1.00	1.00	1.00	1.00
Area 6	1.00	1.00	1.00	1.00	1.00		1.00	1.00	1.00
Area 7	1.00	1.00	1.00	1.00	1.00	.185		1.00	1.00
Area 8	1.00	1.00	1.00	1.00	1.00	.291	1.00		1.00
Area 9	1.00	1.00	1.00	1.00	.677	.072	1.00	1.00	

TABLE 7.34

Details on Univariate Directional p-Values for Odour Perceptions in Nine Areas

	Animal Related									**Harsh, Stinging**								
	Ar.1	**Ar.2**	**Ar.3**	**Ar.4**	**Ar.5**	**Ar.6**	**Ar.7**	**Ar.8**	**Ar.9**	**Ar.1**	**Ar.2**	**Ar.3**	**Ar.4**	**Ar.5**	**Ar.6**	**Ar.7**	**Ar.8**	**Ar.9**
Ar.1		**.020**	.410	.388	**.016**	**.007**	**.032**	.771	.671		**.000**	**.000**	**.001**	.218	**.000**	**.001**	**.000**	.115
Ar.2	.987		.952	.809	.375	.248	.753	.982	1.00	1.00		.990	.803	.882	.407	.999	.901	.965
Ar.3	.591	.065		.506	.049	**.036**	.187	.833	.753	1.00	**.010**		.413	.521	**.004**	.957	.140	.781
Ar.4	.617	.252	.510		.252	.252	.283	.834	.752	.999	.204	.587		.681	.156	.865	.458	.824
Ar.5	.991	.745	.967	1.00		.495	.877	1.00	1.00	.783	.126	.480	.334		.059	.496	.486	.561
Ar.6	1.00	1.00	1.00	1.00	100		1.00	1.00	1.00	1.00	.610	.997	.875	1.00		1.00	.924	1.00
Ar.7	.971	.278	.815	.749	.183	.060		.961	.955	1.00	**.001**	**.044**	.135	.504	**.001**		**.023**	.672
Ar.8	.231	**.033**	.175	.183	**.033**	**.033**	.047		.627	1.00	.103	.861	.559	.558	.091	.973		.919
Ar.9	.331	**.016**	.250	.255	**.016**	**.016**	.053	.389		.885	**.035**	.219	.183	.444	**.007**	.323	.089	

	Ammoniacal									**Hay/Forage**								
	Ar.1	**Ar.2**	**Ar.3**	**Ar.4**	**Ar.5**	**Ar.6**	**Ar.7**	**Ar.8**	**Ar.9**	**Ar.1**	**Ar.2**	**Ar.3**	**Ar.4**	**Ar.5**	**Ar.6**	**Ar.7**	**Ar.8**	**Ar.9**
Ar.1		**.030**	**.030**	.150	.131	**.030**	**.030**	.200	**.030**		1.00	1.00	1.00	1.00	1.00	1.00	1.00	1.00
Ar.2	.977		.254	.629	.810	.254	.254	.810	.254	1.00		1.00	1.00	1.00	1.00	1.00	1.00	1.00
Ar.3	1.00	1.00		1.00	1.00	1.00	1.00	1.00	1.00	1.00	1.00		1.00	1.00	1.00	1.00	1.00	1.00
Ar.4	.880	.500	.500		.500	.500	.500	.621	.500	1.00	1.00	1.00		1.00	1.00	1.00	1.00	1.00
Ar.5	.894	.251	.251	.629		.251	.251	.744	.251	.500	.500	.500	.500		.500	.500	.500	.751
Ar.6	1.00	1.00	1.00	1.00	1.00		1.00	1.00	1.00	1.00	1.00	1.00	1.00	1.00		1.00	1.00	1.00
Ar.7	1.00	1.00	1.00	1.00	1.00	1.00		1.00	1.00	1.00	1.00	1.00	1.00	1.00	1.00		1.00	1.00
Ar.8	.815	.251	.251	.508	.507	.251	.251		.251	1.00	1.00	1.00	1.00	1.00	1.00	1.00		1.00
Ar.9	1.00	1.00	1.00	1.00	1.00	1.00	1.00	1.00		.500	.500	.500	.500	.500	.500	.500	.500	

(Continued)

TABLE 7.34 (CONTINUED)

Details on Univariate Directional p-Values for Odour Perceptions in Nine Areas

	Toasting									Animal Manure								
	Ar.1	Ar.2	Ar.3	Ar.4	Ar.5	Ar.6	Ar.7	Ar.8	Ar.9	Ar.1	Ar.2	Ar.3	Ar.4	Ar.5	Ar.6	Ar.7	Ar.8	Ar.9
Ar.1		.756	.508	.508	.508	.757	.753	.508	.967		.577	.579	**.044**	.164	**.001**	**.028**	**.002**	.049
Ar.2	.498		.498	.498	.498	.753	.747	.498	.968	.424		.559	**.035**	.141	**.001**	**.043**	**.013**	.058
Ar.3	1.00	1.00		1.00	1.00	1.00	1.00	1.00	1.00	.421	.442		**.044**	.136	**.041**	.080	**.012**	.110
Ar.4	1.00	1.00	1.00		1.00	1.00	1.00	1.00	1.00	.957	.965	958		.560	.204	.408	.195	.445
Ar.5	1.00	1.00	1.00	1.00		1.00	1.00	1.00	1.00	.836	.859	.867	.444		.261	.344	.311	.383
Ar.6	.503	.503	.503	.503	.503		.503	.503	1.00	1.00	1.00	.963	.804	.808		.715	.545	.661
Ar.7	.500	.500	.500	.500	.500	.741		.500	.967	.972	.958	.923	.594	.660	.299		.254	.534
Ar.8	1.00	1.00	1.00	1.00	1.00	1.00	1.00		1.00	.999	.988	.990	.809	.722	.485	.753		.598
Ar.9	.065	.065	.065	.065	.065	.065	.065	.065		.951	.943	.892	.571	.633	.372	.473	.432	
	Putrid, Rotten									Other Smell								
	Ar.1	Ar.2	Ar.3	Ar.4	Ar.5	Ar.6	Ar.7	Ar.8	Ar.9	Ar.1	Ar.2	Ar.3	Ar.4	Ar.5	Ar.6	Ar.7	Ar.8	Ar.9
Ar.1		.514	.813	.521	.242	.287	.735	.196	.892		**.021**	.253	**.040**	**.038**	**.019**	.175	.089	.075
Ar.2	.486		.500	.156	.286	.271	.464	.263	.541	.980		.983	.398	.135	.078	.797	.606	.744
Ar.3	.187	.501		.419	.082	.113	.576	**.024**	.799	.747	**.017**		**.002**	**.004**	**.001**	.231	**.002**	**.021**
Ar.4	.480	.853	.582		.280	.279	.645	.327	.791	.962	.606	.999		.309	.366	.875	.745	.769
Ar.5	.759	.722	.919	.750		.625	.906	.444	.907	.965	.875	.996	.751		.500	.905	.780	.903
Ar.6	.714	.738	.889	.730	.408		.828	.410	.893	.982	.927	.999	.665	.632		.992	.931	1.00
Ar.7	.265	.536	.425	.358	.097	.173		.123	.722	.826	.204	.770	.132	.102	**.015**		.259	.251
Ar.8	.805	.741	.976	.688	.613	.622	.879		.922	.912	.396	.999	.286	.253	.102	.745		.475
Ar.9	.109	.460	.202	.210	.097	.111	.279	.092		.925	.257	.980	.235	.104	**.015**	.751	.534	

on how the odours are perceived and evaluated from the citizens' points of view. These results may suggest to administrators any possible solutions and interventions designed to alleviate the discomfort felt by the possible presence of malodours.

7.7 Evaluation of Customer Satisfaction on Public Transport Services

The main goal of customer satisfaction is the investigation of perceptible assessment of services and products. Regardless of the economic sector in which a public or private product/service is offered/provided, customer satisfaction is worldwide considered one of the most important key factors that must be taken into consideration to effectively manage and monitor the business activities. In general, customer satisfaction provides a leading indicator of user/consumer purchase intentions and loyalty (Farris et al., 2010).

In this work we investigate customer satisfaction on public transport services of university students attending courses at the University of Ferrara, Italy. Using the multivariate ranking methods we demonstrate that the overall global satisfaction on train transportation does not depend on the student place of residence so that when trying to improve the actual and perceived quality of train transportation services there is no need to segment the users with respect to the area of origin of their train transfer to Ferrara.

7.7.1 Introduction

Most transportation service providers collect customer information but customer satisfaction is a specific type of customer information. As pointed out by Cantalupo and Quinn (2002), customer satisfaction is measured by a change in the users' perceptions of the adequacy of service provided according to the mode utilized. Stradling et al. (2007) define satisfaction as involving metrication of both customer perceptions and expectation of service even if in transportation contexts the expectation of performance is not commonly collected.

Focussed segmentation practices emerging in data collection efforts are vital to defining the customer (VKCRC, 2002). However, this segmentation is often determined by transportation agency objectives and directives, not by customer behavior (Kelly and Swindell, 2002). Further studies conducted by Zineldin (2005) and Zheng and Jiaqing (2007) concluded that customers want the best service, whether provided by private or public companies, and by improving quality and responsiveness customers could be satisfied, which would ultimately be helpful for the reputation and profit of the companies. All of the aforementioned work conducted by different researchers can conclude that in the transportation sector, quality of service plays an important role, together with some other moderating variables, which could be stability, capacity and security, as well as feedback from customers, as it helps to enhance the quality of service and provide what customers want.

Satisfaction is basically an intangible concept and varies among individuals and products/services and can depend on covariates that represent characteristics of individuals or objects. The main goal of customer satisfaction analysis often is studying the effect of individual factors on overall satisfaction. In this context, the final purpose of the analysis usually consists of determining and comparing the satisfaction of different groups of customers (market segments), for adapting marketing strategies to market segments' characteristics through targeted communication strategies and product/service differentiation.

As the aspects of a service to be evaluated are less evident than those of a product, customer satisfaction analysis of services presents some specific difficulties. However, the need to assess the quality of services, according to ISO 9000 standards, charters of services and quality management systems, leads many companies to implement procedures for assessing the quality of services and in particular customer satisfaction. This is true also in the public transport sector. In this area some scientific work has tried to develop methodologies for studying the latent aspects of customer satisfaction. Gallo et al. (2009) discuss the application of Rasch Analysis and Simple Components Analysis based on the RV coefficient for studying passengers' satisfaction in the local public transport area. Multidimensionality of customer satisfaction and the different natures of data are considered by Gallo and Ciavolino (2009), who analyse the spatial effects of the territorial dislocation of stations by means of a rating scale model and a spatial structural equation model. Satisfaction in urban public transportation is also studied by Bernini and Lubisco (2005, 2009), who propose an extended dynamic version of the LISREL model to investigate possible changes over time of customer satisfaction, the main factors affecting satisfaction and the effects of customer covariates, with application to the Tram Service in Rimini (Italy). Except for the airline industry, few empirical studies have been dedicated to satisfaction for transportation services. Sumaedi et al. (2012), using structural equation modeling techniques, analysed the relationship between passengers' behavioral intentions and others latent factors affecting them in Jakarta City.

7.7.2 Survey Description

In the spring of 2012 a statistical survey about life and studying conditions of university students was performed in Ferrara, Italy (Bonnini et al., 2013; Bonnini, 2014). A random sample of 747 students was drawn with the stratification method, with gender and type of enrollment (freshman, upperclassman) as stratification variables. The sample size corresponds to 4.2% of the whole population size. Through the CATI method, the students in the sample were asked to answer some questions related to their habits, spare time activities, services used and working and economic conditions. A section of the questionnaire was dedicated to means and services used by the students for travelling from home to University. In particular, among the other considered means of transport, the questionnaire was focussed on the public rail service and on the urban bus service.

For medium-long distances the main mean of transport used by the students for travelling from their home to university, for attending lessons or doing examinations, is the train. In this work our focus is on the commuters, that is, undergraduate students who live outside Ferrara and who do not want or cannot buy or rent a house in Ferrara, and daily or almost daily go to university.

In the sample of the survey on the students of Ferrara, among the undergraduate students the total number of commuters is 136. By considering as segmentation factor the student residence, it is possible to define four groups of commuter students (Figure 7.15): Province of Ferrara – FE (22%), Province of Rovigo – RO (23%), other provinces in Emilia Romagna – EM–ROM (23%), other provinces in Veneto – VEN (33%).

The commuter students who use the railway transfer were asked to provide an opinion about six different aspects of the service quality, by rating them according to a numeric scale from 1 (worst evaluation – maximum dissatisfaction) to 10 (best evaluation – maximum satisfaction). The aspects considered are Costs, On time, Timetable, Comfort, Cleansing, Extension of the railway network. Within this 1–10 rating scale, the value 6 represents the minimum value assumed to correspond to the minimum satisfaction level; that is, all scores below 6 are assumed to indicate a perceived dissatisfaction with the service.

7.7.3 Ranking Analysis

A first look at comfort data is provided by radar graphs in Figure 7.16, where score sample means and top-three-boxes T3B percentages (the percentage of scores greater than 7) by student segmentation, that is, respondents' residence, are reported. From a descriptive standpoint it appears that no one student segment seems to be more or less satisfied, because the related score means and T3B percentages are generally quite similar to each other. As noted by Bonnini (2014), students are mostly unsatisfied about the train transportation service because score means are generally below the value of 6, which is usually assumed to be the minimum acceptable satisfaction level within the considered scale 1–10.

The main goal of this work was to rank the four investigated user segments from students' points of view on using their assessments on customer satisfaction items.

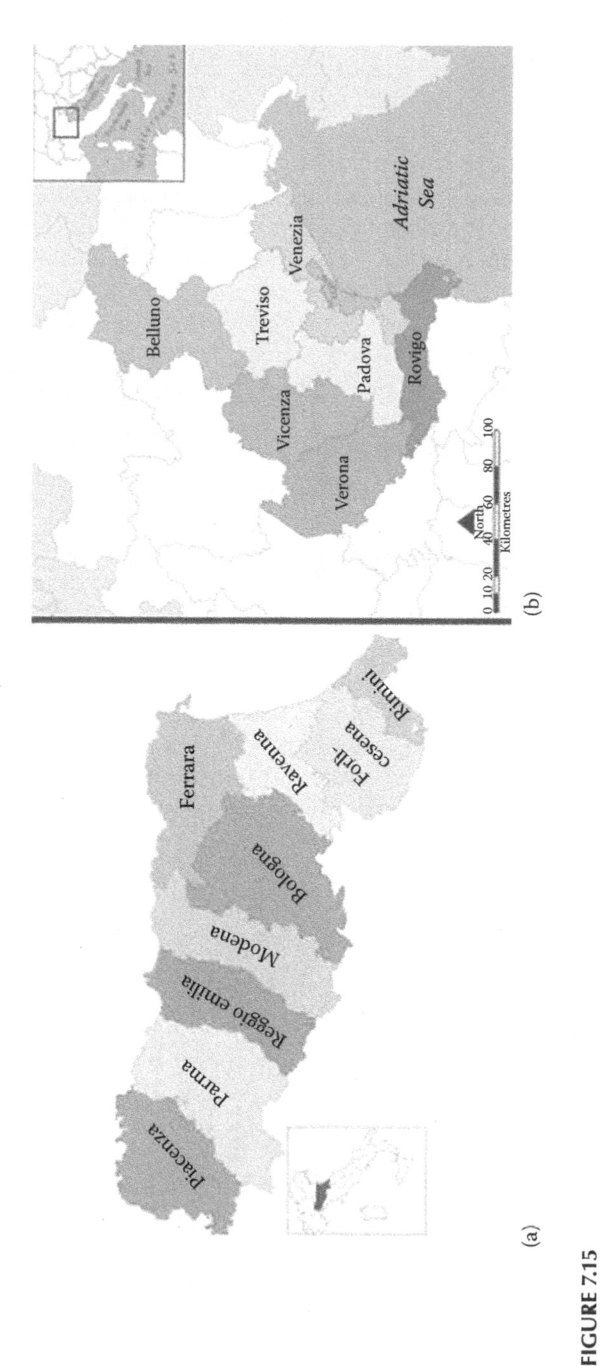

FIGURE 7.15
Provinces of Ferrara and Rovigo and other provinces in Emilia Romagna and Veneto. (a) Ferrara and other provinces of Emilia Romagna. (b) Rovigo and other provinces of Veneto.

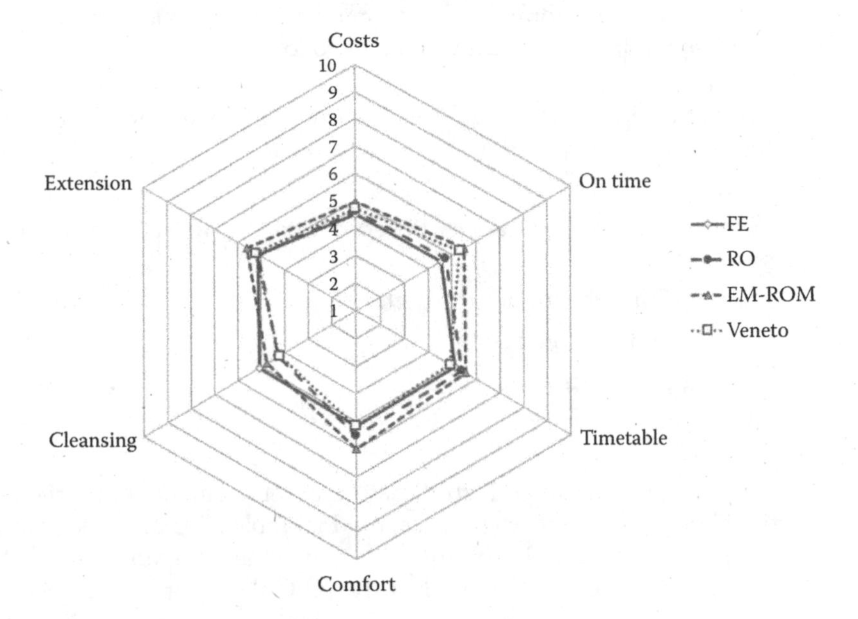

(a)

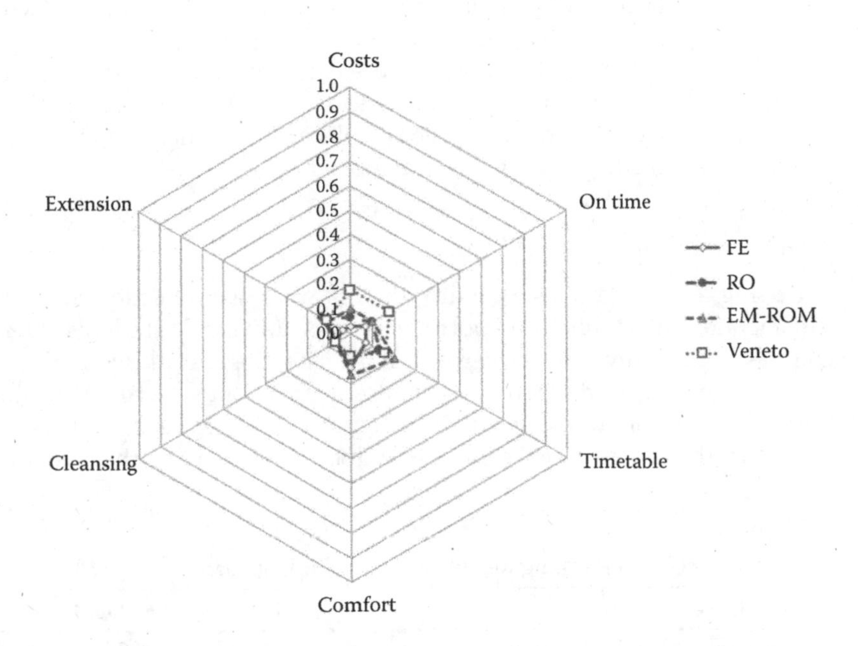

(b)

FIGURE 7.16
Radar graphs of score means and T3B percentages. (a) Item-by-item sample score means. (b) Item-by-item T3B percentages.

The setting of the customer satisfaction on railway services ranking problem (see Chapter 3) can be summarized as follows.

- Type of design: one-way MANOVA (four independent samples, i.e. four multivariate populations to be ranked)
- Domain analysis: no
- Type of response variables: ordered categorical (six ordinal 1–10 scores)
- Ranking rule: 'the higher the better'
- Combining function: Fisher
- B (number of permutations): 2000
- Significance α-level: 0.05

Details on multivariate and univariate market segment comparisons via directional permutation tests are presented in Tables 7.35 and 7.36. All ranking analysis results can be obtained by using the companion book Web-based software (see Chapter 4) within the NPC Web Apps (lstat.gest.unipd.it/npcwebapps/). The reference dataset is labelled Train_satisfaction.csv. To replicate the permutation p-values reported in Tables 7.35 and 7.36 exactly, a seed value equal to 1234 must be set up.

Ranking analysis results allow us to easily rank the four user segments; in particular all segments are tied and ranked in first position. This means that there is no relationship between the place of residence and the perceived customer satisfaction about train transportation service.

7.7.4 Conclusions

On using a suitable questionnaire, in this work we investigated customer satisfaction with public transport services of university students attending courses at the University of Ferrara, Italy. The global ranking analysis performed considering the perceived assessments about several dimensions of the train transportation, such as costs, comfort and cleansing, led us to conclude that the overall global satisfaction does not depend on the student place

TABLE 7.35

Details of Global Ranking Results on Four Student Segments ($\alpha = 5\%$)

Residence	FE	RO	EM-ROM	VEN
Global Ranking	**1**	**1**	**1**	**1**
FE		1.00	1.00	1.00
RO	.951		.844	1.00
EM-ROM	.834	.642		1.00
VEN	1.00	1.00	.847	

TABLE 7.36

Item-by-Item Univariate Directional Permutation *p*-Values

	FE	RO	EM-ROM	VEN	FE	RO	EM-ROM	VEN	FE	RO	EM-ROM	VEN
	Costs				On time				Timetable			
FE		.658	.961	.950		.594	.781	.699		.699	.707	.414
RO	.342		.850	.850	.406		.715	.614	.300		.545	.237
EM-ROM	.038	.149		.447	.219	.284		.387	.292	.455		.171
VEN	.050	.150	.553		.301	.386	.612		.586	.762	.829	
	Comfort				Cleansing				Extension			
FE		.727	.944	.561		.093	.349	.055		.472	.755	.585
RO	.272		.842	.344	.907		.836	.451	.527		.765	.606
EM-ROM	.056	.158		.054	.650	.164		.115	.245	.235		.306
VEN	.438	.655	.946		.945	.549	.885		.414	.393	.693	

of residence. Consequently, when trying to improve the actual and perceived quality of train transportation services there is no need to segment the users with respect to the area of origin of their train transfer to Ferrara. Accordingly, to improve the satisfaction of customers aiming to reach Ferrara by railway, train service providers should consider actions, not user-origin dependent.

References

Alexandris K., Zahariadis P., Tsorbatzoudis C., Grouios G. (2004). An empirical investigation of the relationships among service quality, customer satisfaction and psychological commitment in a health club context. *European Sport Management Quarterly*, 4(1), 36–52.

Arboretti Giancristofaro R., Boatto V., Tempesta T., Vecchiato D. (2014). Statistical modelling for the evaluation of consumers' preferences. In *Communications in Statistics – Simulation and Computation*, doi: 10.1080/03610918.2014.920884.

Arboretti Giancristofaro R., Bonnini S., Corain L., Vidotto D. (2015). Environmental odor perception: Testing procedure for spatial comparisons between the heterogeneities of types and possible sources of odor with application to the odor perceptions in the area of Este (Italy). *Environmetrics*, in press.

Arboretti Giancristofaro R., Pesarin F., Salmaso L. (2007). Nonparametric approaches for multivariate testing with mixed variables and for ranking on ordered categorical variables with an application to the evaluation of PhD programs. In S. Sawilowsky (ed.), *Real Data Analysis*. Quantitative Methods in Education and the Behavioral Sciences: Issues, Research and Teaching, Ronald C. Serlin, Series Editor. Charlotte, NC: Information Age Publishing, pp. 355–385.

Bernini C., Lubisco A. (2005). A new extension of dynamic simplex model for the public transport customer satisfaction. *Statistica*, LXV(4), 367–386.

Bernini C., Lubisco A. (2009). Modelling dynamic customer satisfaction in urban public transportation. *Global and Local Economic Review*, 13(2), 87–107.

Bolzan M. (2010). L'identità professionale del personale dirigente degli enti locali del Veneto: Competenze, fabbisogno formativo in relazione al territorio e missione dell'ente. In Bolzan M. (ed.), *Competenze e processi formativi per i dirigenti degli enti locali*. Atti a cura di Cleup, pp. 13–54.

Bolzan M., Boccuzzo G. (2011). Skills, formative needs and criticalities of municipal managers in Italy. In *Improving the Quality of Public Services: A Multinational Conference on Public Management International Conference*, Moscow 27–29 June, Panel 12: Padova, Italy: Understanding Contemporary Public Administration.

Bolzan M., Corain L., De Giuli V., Salmaso L. (2014). An evaluation of performance of Territorial Services Center (TSC) by a nonparametric combination ranking method. The IQuEL Italian project. In D. Vicari, A. Okada, G. Ragozini, C. Weihs (eds.), *JCS-CLADAG12: Analysis and Modeling of Complex Data in Behavioural and Social Sciences*, pp. 47–54. Studies in Classification, Data Analysis, and Knowledge Organization. Berlin: Springer, Physica-Verlag.

Bonnini S. (2014). Multivariate approach for comparative evaluations of customer satisfaction with application to transport services. *Communications in Statistics: Simulation and Computation*, doi: 10.1080/03610918.2014.941685.

Bonnini S., Nanetti D., Polastri R. (2013). *Indagine sulle condizioni di vita e di studio degli studenti dell'Università di Ferrara e impatto sul territorio*, UnifePress, Ferrara.

Cantalupo J., Quinn M. (2002). Using customer survey data to monitor the progress of Delaware's statewide long-range transportation plan, In *Seventh TRB Conference on the Application of Transportation Planning Methods*. Transportation Research Board of the National Academies, Washington, DC.

Carrozzo E., Cichi I., Corain L., Salmaso L. (2012). Permutation-based control charts for ordered categorical response variables with application to monitoring of customer satisfaction. In *Quaderni di Statistica*, Special Issue International Conference on "Methods and models for latent variables", Naples, 17–19 May 2012, 14, pp. 65–68.

Chalip L. (2001). Sport and tourism: Capitalising on the linkage. In D. Kluka, G. Schilling (eds.), *The Business of Sport*. Oxford: Meyer & Meyer, pp. 77–88.

Crilley G., Murray D., Howat G., March H., Adamson D. (2002). Measuring performance in operational management and customer service quality: A survey of financial and non-financial metrics from the Australian golf industry. *Journal of Leisure Property*, 2, 369–380.

Dolnicar S., Leisch F. (2003). Winter tourist segments in Austria: Identifying stable vacation styles using bagged clustering techniques. *Journal of Travel Research*, 41(3), 281–292.

Farris P. W., Bendle N. T., Pfeifer P. E., Reibstein D. J. (2010). *Marketing Metrics: The Definitive Guide to Measuring Marketing Performance.* Upper Saddle River, NJ: Pearson Education.

Franch M., Martini U., Tommasini D. (2003). Mass-ski tourism and environmental exploitation in the Dolomites: Some considerations regarding the Tourist Development Model. In E. Christou, M. Sigala (eds.), *International Scientific Conference on Sustainable Tourism Development and the Environment*, 2–5 October 2003, Chios Island, Greece, p. 12.

Gallo M., Ciavolino E. (2009). Multivariate statistical approaches for customer satisfaction into the transportation sector. *Global and Local Economic Review*, 13(2), 55–70.

Gallo M., D'Ambra A., Camminatiello I. (2009). The evaluation of passenger satisfaction in the local public transport: A strategy for data analysis. *Global and Local Economic Review*, 13(2), 71–86.

Gibson H. J. (2002). Sport tourism at a crossroad? Considerations for the future. In S. Gammon, J. Kurtzman (eds.), *Sport Tourism: Principles and Practice*. Eastbourne: LSA, pp. 111–128.

Gibson H. J. (2003). Sport tourism. In J. B. Parks, J. Quarterman (eds.), *Contemporary Sport Management*. Champaign, IL: Human Kinetics.

Greenwell T. C., Fink J. S., Pastore D. L. (2002a). Assessing the influence of the physical sports facility on customer satisfaction within the context of the service experience. *Sport Management Review*, 5(2), 129–148.

Greenwell T. C., Fink J. S., Pastore D. L. (2002b). Perceptions of the service experience: Using demographic and psychographic variables to identify customer segments. *Sport Marketing Quarterly*, 11, 233–241.

Hall C. M. (1992). Review, adventure, sport and health tourism. In B. Weiler, C. M. Hall (eds.), *Special Interest Tourism*. London: Belhaven Press, pp. 186–210.

Iannario M., Piccolo D. (2007). A new statistical model for the analysis of customer satisfaction, *Quality Technology and Quantitative Management*, 7(2), 149–168.

ISO 9001 (2008). *Quality Management Systems: Requirements*. http://www.iso.org.

Jackson G. A. M., Weed M. E. (2003). The sport–tourism interrelationship. In B. Houlihan (ed.), *Sport and Society*. London: SAGE, pp. 235–251.

Kelley S. W., Turley L. W. (2001). Consumer perceptions of service quality attributes at sport events. *Journal of Business Research*, 54, 161–166.

Kelly J. M., Swindell D. (2002). Service quality variation across urban space: First steps toward a model of citizen satisfaction, *Journal of Urban Affairs*, 24(3), 271–288.

Ko Y. J., Pastore D. (2004). Current issues and conceptualizations of service quality in the recreation sport industry. *Sport Marketing Quarterly*, 13, 159–167.

Kouthouris C., Alexandris K. (2005). Can service quality predict customer satisfaction and behavioral intentions in the sport tourism industry? An application of the SERVQUAL model in an outdoor setting. *Journal of Sport & Tourism*, 10, 101–111.

Manning R. E. (2007). *Parks and Carrying Capacity*. Washington, DC: Island Press.

Matzler K., Füller J., Renzl B., Herting S., Späth S. (2008). Customer satisfaction with alpine ski areas: The moderating effects of personal, situational, and product factors. *Journal of Travel Research*, 46, 403–413.

Matzler K., Pechlaner H., Hattenberger G. (2004). *Lifestyle-Typologies and Market Segmentation: The Case of Alpine Skiing Tourism*. Bolzano, Italy: European Academy.

Matzler K., Siller H. (2004). Linking travel motivations with perceptions of destinations: The case of youth travelers in alpine summer and winter tourism. *Tourism Review*, 58(4), 6–11.

McDonald M. A., Sutton W. A., Milne G. R. (1995). TEAMQUAL measuring service quality in professional team sports. *Sport Marketing Quarterly*, 4(2), 9–15.

MIUR – Ministero dell'Istruzione, dell'Università e della Ricerca (Ministry of Education, University and Research) (2003). Lo stato della ricerca nelle Università (The state of research in Italian Universities). In *Università obiettivo valutazione, 1 Analisi e commenti, Atenei*, Rome, Italy: MIUR, pp. 89–113.

Nicolas J., Craffe F., Romain A. C. (2006). Estimation of odor emission rate from landfill areas using the sniffing team method. *Waste Management*, 26, 1259–1269.

Pechlaner H., Tschurtschenthaler P. (2003). Tourism policy, tourism organisations and change management in alpine regions and destinations: A European perspective. *Current Issues in Tourism*, 6(6), 508–539.

Petrick J. F., Backman S. J. (2002a). An examination of the determinants of golf travelers' satisfaction. *Journal of Travel Research*, 40, 3, 252–258.

Petrick J. F., Backman S. J. (2002b). An examination of the construct of perceived value for the prediction of golf travelers' intentions to revisit. *Journal of Travel Research*. 41, 1, 38–45.

Petrick J. F., Backman S. J. (2002c). An examination of golf travelers' satisfaction, perceived value, loyalty and intentions to revisit. *Tourism Analysis*, 6(3/4), 223–237.

Rapporto eGov Italia 2010, Milano. Internet access and use of ICT in enterprises in 2011. http://epp.eurostat.ec.europa.eu/185/2011. (Accessed 13 December 2011).

Richards G. (1996). Skilled consumption and UK ski holidays. *Tourism Management*, 17(1), 25–34.

Shonk D. J., Chelladurai P. (2008). Service quality, satisfaction, and intent to return in event sport tourism. *Journal of Sport Management*, 22, 587–602.

Smith C., Jenner P. (1990). Activity holidays in Europe. *Travel and Tourism Analyst*, 5, 58–78.

Stradling S., Anable J., Carreno J. (2007). Performance, importance and user disgruntlement: A six-step method for measuring satisfaction with travel modes. *Transportation Research Part A*, 41, 98–106.

Sumaedi S., Bakti G. M. Y., Yarmen M. (2012). The empirical study of public transport passengers' behavioral intentions: The roles of service quality, perceived sacrifice, perceived value, and satisfaction (Case study: Paratransit passengers in Jakarta, Indonesia). *International Journal for Traffic and Transport Engineering*, 2(1), 83–97.

Van Langenhove H., Van Broeck G. (2001). Applicability of sniffing team observations: Experience of field measurements. *Water Science and Technology*, 44, 65–70.

VKCRC – Virchow Krause & Co, Chamberlain Research Consultants (2002). Develop a mechanism to measure customer satisfaction with products and services of the department, SPR-0092-0207. Wisconsin Department of Transportation.

Wakefield K. L., Blodgett J. G., Sloan H. J. (1996). Measurement and management of the sportscape. *Journal of Sport Management*, 10, 15–31.

Weed M. (2009). Progress in sports tourism research? A meta-review and exploration of futures. *Tourism Management*, 30, 615–628.

Weed M. E. (2006). Sports tourism research 2000–2004: A systematic review of knowledge and a meta evaluation of method. *Journal of Sport & Tourism*, 11(1), 5–30.

Weed M. E., Bull C. J. (2004). *Sports Tourism: Participants, Policy & Providers*. Oxford: Elsevier.

Weiermair K., Fuchs M. (1999). Measuring tourist judgment on service quality. *Annals of Tourism Research*, 26(4), 1004–1021.

Williams P., Fidgeon P. R. (2000). Addressing participation constraint: A case study of potential skiers. *Tourism Management*, 21(3), 379–393.

Yoon Y., Uysal M. (2005). An examination of the effects of motivation and satisfaction on destination loyalty: A structural model. *Tourism Management*, 26, 45–56.

Zheng L., Jiaqing W. (2007). Summary of the application effect of bus rapid transit at Beijing South-Centre Corridor of China. *Journal of Transportation Systems Engineering and Information Technology*, 7(4), 137–142.

Zineldin M. (2005). Quality and customer relationship management as competitive strategy in the Swedish industry. *TQM Magazine*, 17(4), 329–344.

8

Cytological and Histological Analysis by Image Processing: Ranking Approach

The present chapter aims to apply innovative multivariate ranking-based statistical methods to biomedical problems concerned with image processing and shape analysis. Shapes and images are special kinds of data that show a functional structure and have implementation issues of major importance mainly in biomedical fields. Despite the existence and relevance of many statistical methods and approaches presented in the specialized literature, we believe there is still room for development in refining multivariate approaches and algorithms to achieve significant improvements in decision-making effectiveness and readiness of the results obtained. In this connection, this chapter aims to contribute to research progress both in terms of multivariate analysis and of inferential results, introducing the multivariate ranking approach in image and shape analysis.

8.1 Ranking of Estradiol Exposures on Intracellular Calcium Concentration in Bovine Brain–Derived Endothelial Cells

Oetrogens diversely affect various physiological processes by genomic or nongenomic mechanisms, in both excitable and nonexcitable cells. In addition to the trophic effects of oestrogens in promoting cell growth and differentiation, recent experimental evidence highlights their involvement in the regulation of intracellular Ca^{2+} homeostasis (Nilsen and Diaz Brinton, 2003). Suman et al. (2012) tested estradiol (E_2) effects on intracellular Ca^{2+} homeostasis by varying the exposure time to the hormone (8, 24 and 48 h), where calcium measurements were performed with genetically encoded Ca^{2+} probes (Cameleons) targeted to the main subcellular compartments involved in intracellular Ca^{2+} homeostasis (cytosol, endoplasmic reticulum, mitochondria). Suman et al. (2012) proved that mitochondrial Ca^{2+} uptake significantly decreased after 48 h of exposure to E_2, whereas cytosolic and endoplasmic reticulum responses were unaffected. Taking again into consideration the experimental data of Suman et al. (2012), in this work we aim to rank, for each specific intracellular organelle, the E_2 exposure times with respect to the intracellular calcium concentration.

8.1.1 Introduction

The concentration of calcium [Ca^{2+}] in a cell is managed by specific intracellular organelles, such as the endoplasmic reticulum (ER) and mitochondria. In the lumen of ER, the [Ca^{2+}] depends mainly on the balanced activity of the inositol triphosphate (IP3) receptor (IP3R) and of the sarco-endoplasmic reticulum Ca^{2+} ATPase (SERCA), which drives the ion from the cytoplasm into the organelle (Berridge et al., 2003). The storage of Ca^{2+} in the ER is regulated by Ca^{2+}-binding proteins that act as buffers. Stimulation of cells by agonists coupled to IP3 production causes Ca^{2+} release into the cytosol through the IP3R. Subsequently, the released Ca^{2+} can exit from the cell by the calcium extrusion systems of the plasma membrane, can be stored back into the ER or can enter into the mitochondrial matrix through the mitochondrial Ca^{2+} uniporter, whose molecular identity has recently been defined (De Stefani et al., 2011). The Ca^{2+} is extruded by mitochondria through cation exchangers, namely, the Na^+/Ca^{2+} exchanger (NCX) and H^+/Ca^{2+} exchanger and through the transient opening of the permeability transition pore (PTP). The opening of this inner mitochondrial membrane channel leads to the permeabilization of the inner membrane, a phenomenon known as mitochondrial permeability transition (MPT). The MPT causes the release of cations, the loss of mitochondrial transmembrane potential and eventually outer membrane damage (Rasola et al., 2010). E_2 controls numerous cellular processes, including cell growth and differentiation, by modulating Ca^{2+} homeostasis (McCarthy, 2008). E_2 can influence Ca^{2+} handling by two different molecular mechanisms: a genomic action ('slow mechanism'), that is, the binding of E_2 to either of the oestrogen receptor isoforms resulting in the regulation of gene transcription and a nongenomic action ('rapid mechanism') generally initiated at the plasma membrane by E_2 and promoting the activation of signal transduction pathways (Vasudevan and Pfaff, 2008). Several studies have highlighted the capability of E_2 to induce or repress the expression of a variety of Ca^{2+} channels. Gu et al. (2001) have reported that E_2 increases the expression of two subunits of the L-type voltage-dependent Ca^{2+} channels (L-VOCCs). Another study by Ritchie (2008) has demonstrated an increased density of L-VOCCs after E_2 treatment. To improve our understanding of the effect of E_2 on intracellular Ca^{2+} homeostasis, we used an in vitro model based on an immortalized endothelial cell line derived from foetal bovine cerebellum. The effects of oestrogens have been investigated mainly in excitable cells. However, recent studies have demonstrated that these steroids also act on nonexcitable cells, including endothelial cells and dermal fibroblasts (Arnal et al., 2010; Salazar-Colocho et al., 2008). Recent findings suggest that the endothelium and vascular smooth muscle are controlled by circulating steroid hormones and by steroids synthesized locally. Increasing evidence indicates that E_2 exerts its protective effects through mitochondrial mechanisms in cerebral endothelial cells (Guo et al., 2010). Other experimental data show that E_2 has significant effects on cell migration in human skin fibroblasts (Stevenson et al., 2009). To evaluate whether intracellular Ca^{2+} handling is significantly affected by E_2 treatment, we performed experiments on endothelial cells treated with various concentrations of E_2 (1, 10 and 100 nM) for various incubation times (8, 24 and 48 h). We monitored Ca^{2+} oscillations in the main intracellular compartments involved in Ca^{2+} handling, namely, the cytosol, ER and mitochondria, by means of genetically encoded, specifically targeted Ca^{2+} probes called Cameleons (Palmer et al., 2006).

8.1.2 Materials and Methods

8.1.2.1 Cell Cultures

Primary cell cultures were derived from foetal bovine cerebellum. Male bovine foetuses at 4 months were obtained at nearby commercial abattoirs on accidental slaughtering of pregnant cows. Animals were treated according to the European Community Council Directive (86/609/EEC) concerning animal welfare during the commercial slaughtering process and were constantly monitored under mandatory official veterinary medical care. Age was determined by crown–rump length based on commonly accepted reference tables (McGeady et al., 2006). The use of bovine foetal cells is a reliable tool for studying the molecular mechanisms of Ca^{2+} regulation and thus follows the joint declaration of the American, European and Japanese Societies for Neuroscience that endorses replacement of laboratory animals whenever a valuable scientific alternative model is available. Primary cell culture was performed following an established laboratory protocol (Peruffo et al., 2004). To obtain a stable cell line, cells were immortalized following the protocol by Takenouchi et al. (2007). At day 2 in culture, cells were transfected with pSV3neo plasmid (LGC Promochem, Teddington, UK) by using the cationic lipid Lipofectamine 2000 (Invitrogen, Carlsbad, CA, USA). Subsequently, a selection medium was added and resistant cells were selected with the antibiotic G418 (400 μg/ml; Gibco, Life Techologies BRL).

8.1.2.2 Cell Line Characterization

Staining analyses were performed following an established laboratory protocol (Peruffo et al., 2008). The immortalized cell line was characterized immunocytochemically with antibodies against neural and endothelial markers: anti-neurofilament-200 (Sigma–Aldrich, dilution 1:40), anti-glial fibrillary acidic protein (GFAP; DakoCytomation, Glostrup, Denmark, dilution 1:200), anti-integrin αM (Santa Cruz Biotechnology, Santa Cruz, CA, USA, dilution 1:200), anti-vimentin (Sigma–Aldrich, dilution 1:100), anti-von Willebrand factor (vWF; Sigma–Aldrich, dilution 1:250) and anti-endothelial NO synthase (e-NOS; Abcam, dilution 1:250). Cells were then incubated with secondary antibodies against IgG tetramethyl rhodamine-isothiocyanate (DakoCytomation, dilution 1:100) and IgG-fluorescein isothiocynate (Santa

Cruz Biotechnology, dilution 1:100). Immunostained cultures were observed under a Leika TCS confocal microscope. Negative controls were performed by substituting primary antibodies with bovine serum albumin in phosphate-buffered saline.

8.1.2.3 Fluorescence-Imaging Experiments

Endothelial-like cells were grown in DMEM-F12 containing 10% foetal calf serum, L-glutamine (2 mM), penicillin (30 μg/ml), streptomycin (50 μg/ml) and G418 (400 μg/ml), at 37°C in a humidified atmosphere containing 5% CO_2. Cells were transfected with genetically encoded Ca^{2+} probes (Cameleons) targeted to the main subcellular compartments involved in intracellular Ca^{2+} homeostasis. The cDNA used for transfection coded for the Ca^{2+} probes D3cpv (cytosol), ERD1 (ER) and 4mtD3cpv (mitochondria; kindly provided by Prof. R.Y. Tsien, University of San Diego, USA; Palmer et al., 2004, 2006). Transfection with Cameleons was performed at 70–80% cell confluence by using the Ca^{2+}-phosphate technique in the presence of 2 μg of DNA. Experiments were conducted in an open-topped chamber thermostatically controlled at 37°C on a fluorescence microscope system (Nikon TE2000-E stage) fitted with an immersion oil objective (CFl Plan Apochromat TIRF 60X, N.A. 1.45). Cells were maintained in an extracellular physiological solution containing 135 mM NaCl, 5 mM KCl, 0.4 mM KH_2PO_4, 1 mM $MgCl_2$, 5.5 mM glucose, 20 mM HEPES (pH 7.4 at 37°C) and were challenged either with ATP or with cyclopiazonic acid (CPA). FRET recordings were obtained through a Beamsplitter (OES, Padua; emission filters HQ 470/30M for CFP [cyano fluorescent protein] and HQ 535/30M for yellow fluorescent protein [YFP]) and a dichroic mirror 505 LD. Images were collected by a back-thinned charge-coupled device camera (MicroMax 512 BFT, Princeton Instruments, USA) with a 300-ms exposure time. Figure 8.1 provides an example of fluorescence images of an endothelial-like cell transiently transfected with the cDNA coding for the cytosolic (Figure 8.1a), endoplasmic reticulum (Figure 8.1b) and mitochondrial (Figure 8.1c) Ca^{2+}, where the bar represents 10 μm.

The ratio values (R) were obtained as follows: $R = (F_{cpv} - F_{bgcpv})/(F_{cfp} - F_{bgcfp})$, where F_{cpv}, F_{cfp} are the fluorescence emission of the two fluorophores

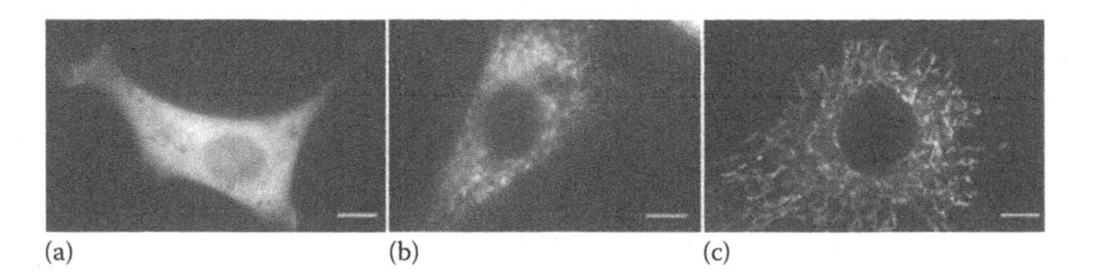

FIGURE 8.1
Example of fluorescence images by type of endothelial-like cell. (a) Cytosolic endothelial-like cell. (b) ER endothelial-like cell. (c) Mitochondrial endothelial-like cell.

and F_{bgcpv}, F_{bgcfp} represent background values for cpV (circularly permuted Venus) and CFP, respectively (Palmer et al. 2006). The data were normalized to $\Delta R/R_0$ values, where R_0 is the fluorescence emission ratio at time 0 (i.e., the basal Ca^{2+} level before ATP stimulation) and ΔR is the increase in fluorescence emission ratio at any point. Analysis of the data was performed with NIS-elements AR software (Nikon Instruments, Japan).

Analysis of these data was performed with HCImage software (Hamamatsu, Japan). Figure 8.2 provides an example of kinetics of fluorescence ratio (R) changes, that is, yellow fluorescent protein (YFP) fluorescence emission/cyano fluorescent protein (CFP) fluorescence emission.

Observed profiles of kinetics of R, that is, fluorescence ratio changes, were summarized into several derived variables: Peak, Peak Time, Rate, Total Time and Area. These five parameters are numerically calculated from each observed profile and represent the traditional aspects to take into consideration when performing kinetics analysis. Note that R profiles can be viewed as samples drawn from population of curves or trajectories (Ferraty and Vieu, 2006). To make inference on the possible E_2 time exposure effect, R profiles were then converted into a set of suitable so-called *derived variables* intended to reduce and summarize the information of intracellular calcium concentration. In general, examples of derived variables are the Principal Components, the wavelet or Fourier coefficients and more generally all the basis function approximations suitable to compress information from response profiles (Corain et al., 2014). Sometimes, as in this case, these transformations are used when they have a precise physical interpretation with respect to the specific inferential problem at hand.

8.1.2.4 Exposure of Immortalized Cell Line to E_2

All experiments were performed 48 h after transfection with the cDNA coding for Cameleons. In the cells transfected with Ca^{2+} probes D3cpv (cytosol) and ERD1 (ER), E_2 (Sigma–Aldrich, Milan, Italy) was added to the culture medium for 8, 24, or 48 h at a final concentration of 10 nM. In the cells transfected with the Ca^{2+} probe 4mtD3cpv (mitochondria), E_2 was added to the culture medium for 8 and 24 h at a final concentration of 10 nM and for 48 h at a final concentration of 1, 10 and 100 nM. To achieve 48-h cell exposure, E_2 was added at the time of transfection. For 24-h exposure, E_2 was added 24 h after transfection. Finally, to achieve 8-h exposure, E_2 was added 8 h before the fluorescence-imaging experiment (40 h after transfection).

8.1.3 Results

A first look at experimental data is provided by radar graphs in Figure 8.3, where sample means by endothelial-like type of cell are reported. It appears that both endothelial-like cells of cytosol and mitochondria may be affected in intracellular calcium concentration by the treatment effects whereas

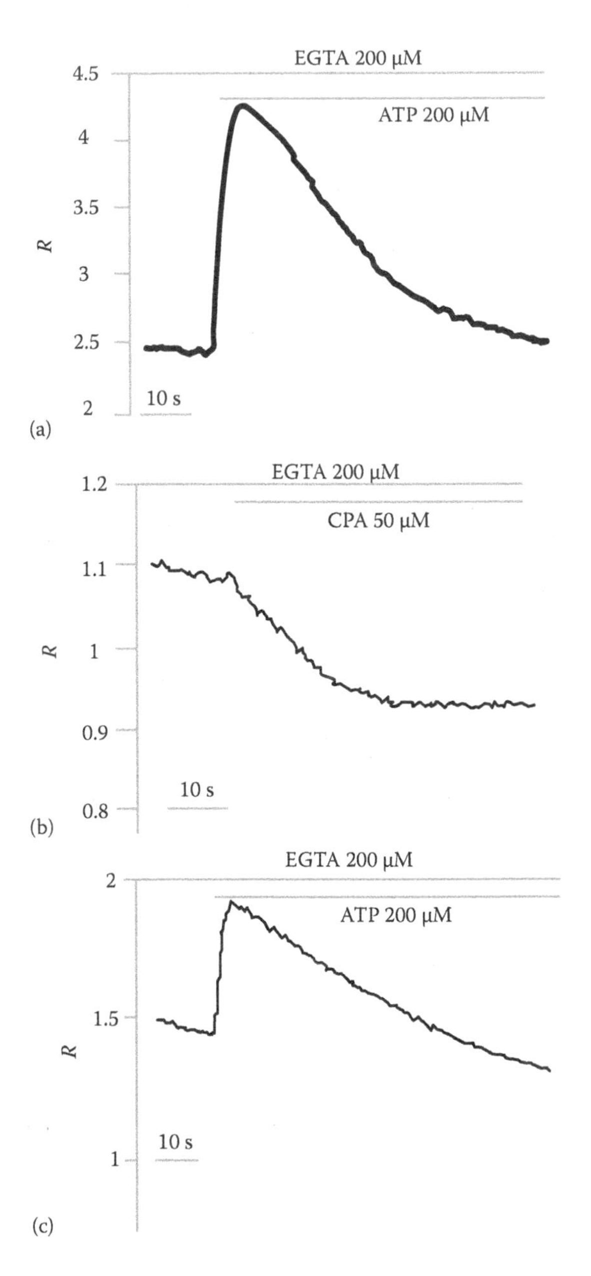

FIGURE 8.2
Example of kinetics of fluorescence ratio (R) changes by type of endothelial-like cell. (a) Cytosol of endothelial-like cell. (b) ER of endothelial-like cell. (c) Mitochondria of endothelial-like cell.

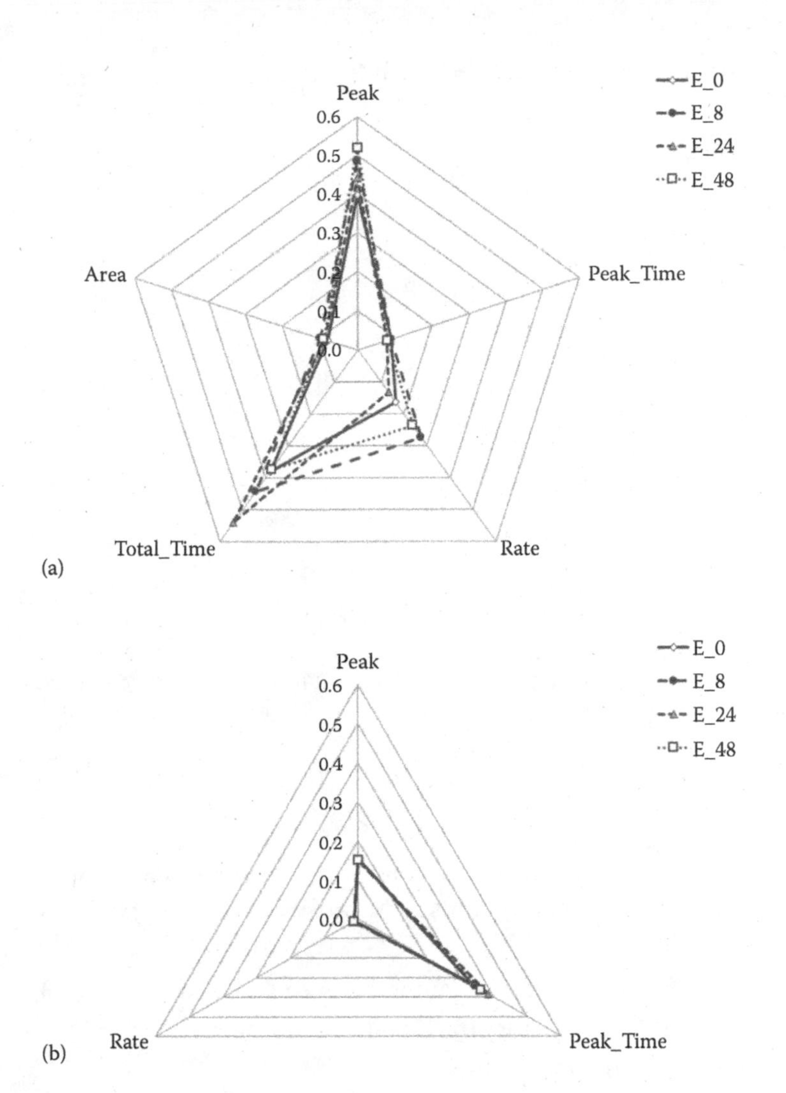

FIGURE 8.3
Radar graphs of kinetic curve aspects by endothelial-like type of cell. (a) Cytosol of endothelial-like cell sample means. (b) ER of endothelial-like cell sample means. *(Continued)*

ER cells seem not to be affected. Moreover, as pointed out by Suman et al. (2012), to discriminate among exposure times, the most relevant kinetic curve aspects seem to be related to the curve peak.

The main objective of the proposed statistical analysis is to rank the four investigated exposure times using experimental data from several aspects of

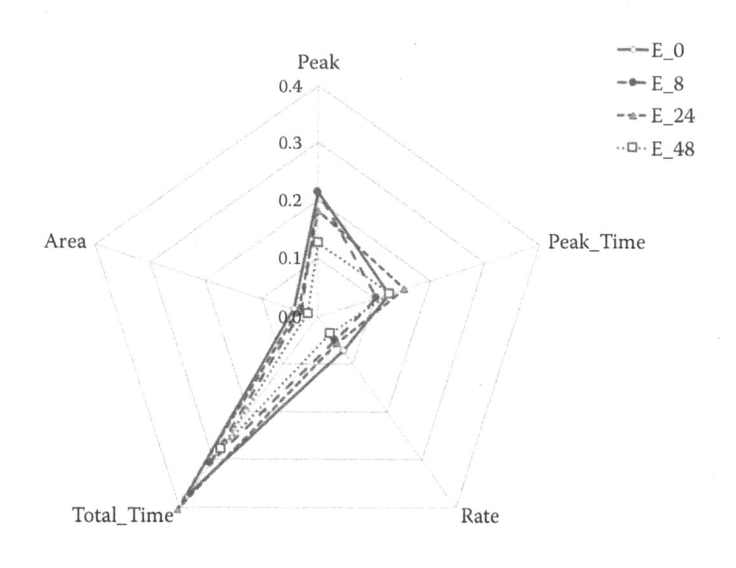

(c)

FIGURE 8.3 (CONTINUED)
Radar graphs of kinetic curve aspects by endothelial-like type of cell. (c) Mitochondria of endothelial-like cell sample means.

the related Ca^{2+} expression kinetic curves. The setting of the exposure times ranking problem (see Chapter 3) can be summarized as follows.

- Type of design: one-way MANOVA (four independent samples, i.e., four multivariate populations to be ranked)
- Domain analysis: no
- Type of response variables: numeric (five numerical responses)
- Ranking rule: 'the higher the better'
- Combining function: Fisher
- B (number of permutations): 2000

Details on multivariate and univariate exposure time comparisons via directional permutation tests are presented in Tables 8.1 and 8.2.

All ranking analysis results can be obtained by using the companion book Web-based software (see Chapter 4) within the NPC Web Apps (lstat.gest.unipd.it/npcwebapps/). The reference datasets are labelled Ca_concentration_cytosol.csv, Ca_concentration_ER.csv, and Ca_concentration_mitochondria.csv. To exactly replicate the permutation p-values reported in Tables 8.1 and 8.2 exactly, a seed value equal to 1234 must be set up.

In agreement with the findings in Suman et al. (2012), ranking analysis results allow us to conclude that for either endothelial-like cells of cytosol and ER there is no E_2 exposure effect at all, because all exposure times are

TABLE 8.1

Details of Global Ranking Results for Exposure Times by Endothelial-Like Cells ($\alpha = 5\%$)

	Cytosol			
Exposure Time	0 h	8 h	24 h	48 h
Global Ranking	**1**	**1**	**1**	**1**
0-h		.985	1.00	1.00
8-h	1.00		1.00	1.00
24-h	.888	1.00		1.00
48-h	.891	.900	1.00	
	Endoplasmic Reticulum			
Exposure Time	0 h	8 h	24 h	48 h
Global Ranking	**1**	**1**	**1**	**1**
0-h		.605	1.00	1.00
8-h	1.00		1.00	1.00
24-h	1.00	1.00		1.00
48-h	1.00	.567	1.00	
	Mitochondria			
Exposure Time	0 h	8 h	24 h	48 h
Global Ranking	**1**	**1**	**1**	**1**
0-h		.853	.764	**.012**
8-h	1.00		.717	**.018**
24-h	1.00	1.00		**.015**
48-h	1.00	1.00	1.00	

tied and equally ranked in the first position. Conversely, ranking results on endothelial-like cells of mitochondria allow us to rank the four considered exposure times using the related experimental measurements on Ca^{2+} expression kinetic curves. Specifically, the less Ca^{2+} expressed exposure time is 48 h, while all the other times are tied in the first ranking position. This finding confirms that there is evidence supporting the conclusion that after 48 h of E_2 exposure a global multivariate reduction effect occurs in Ca^{2+} stored in endothelial-like cells of mitochondria.

8.1.4 Discussion and Conclusions

In the present work, we investigated the 'slow' (genomic) action exerted by E_2 on the Ca^{2+} handling of endothelial-like cells. Our data show a significant global decrease in mitochondrial Ca^{2+} uptake in cells incubated with E_2 for 48 h, whereas both cytosolic and ER responses to E_2 are unaffected. We further studied the effect on mitochondrial Ca^{2+} handling, as it is today largely accepted that these organelles play a fundamental role in intracellular Ca^{2+} homeostasis, possibly modulate cytosolic Ca^{2+} oscillations and act

TABLE 8.2

Kinetic Curve Aspect-by-Aspect Univariate Directional Permutation p-Values

	Cytosol											
	0 h	8 h	24 h	48 h	0 h	8 h	24 h	48 h	0 h	8 h	24 h	48 h
	Peak				Peak_Time				Rate			
0 h		.735	.683	.853		.504	.339	.326		.854	.333	.802
8 h	.266		.345	.603	.500		.355	.328	.146		.076	.356
24 h	.318	.656		.774	.662	.648		.460	.668	.925		.888
48 h	.148	.397	.227		.679	.678	.542		.198	.644	.112	
	Total_Time				Area							
0 h		.786	.924	.469		.754	.640	.663				
8 h	.216		.785	.194	.246		.333	.373				
24 h	.077	.216		.069	.360	.667		.541				
48 h	.531	.808	.931		.338	.627	.460					
	Endoplasmic Reticulum											
	0 h	8 h	24 h	48 h	0 h	8 h	24 h	48 h	0 h	8 h	24 h	48 h
	Peak				Peak_Time				Rate			
0 h		.464	.497	.473		.558	.730	.599		.690	.332	.520
8 h	.541		.467	.493	.447		.751	.562	.315		.242	.354
24 h	.506	.540		.505	.277	.253		.427	.671	.762		.620
48 h	.530	.510	.498		.404	.442	.582		.483	.649	.384	

(*Continued*)

TABLE 8.2 (CONTINUED)

Kinetic Curve Aspect-by-Aspect Univariate Directional Permutation p-Values

	Mitochondria											
	0 h	**8 h**	**24 h**	**48 h**	**0 h**	**8 h**	**24 h**	**48 h**	**0 h**	**8 h**	**24 h**	**48 h**
	Peak				**Peak_Time**				**Rate**			
0 h		.493	.220	**.004**		.649	.764	.039		.193	.227	**.023**
8 h	.508		.288	**.019**	.351		.614	**.003**	.808		.619	.205
24 h	.781	.712		**.022**	.237	.387		**.016**	.773	.382		.063
48 h	.997	.982	.978		.962	.998	.984		.978	.795	.938	
	Total Time				**Area**							
0 h		.196	.597	**.010**		.158	.236	**.011**				
8 h	.804		.791	.057	.843		.661	.084				
24 h	.404	.209		**.017**	.765	.340		.048				
48 h	.991	.944	.984		.990	.917	.953					

as Ca^{2+} buffers (Berridge et al., 2003; Rizzuto et al., 1993). Our experiments performed with a probe specifically targeted to mitochondria have revealed a significant decrease in the Ca^{2+} peak when cells are treated with E_2 (10 nM) for 48 h.

Our results are in agreement with the data obtained by Moreira et al. (2011) demonstrating a decreased capacity of heart and liver mitochondria to accumulate Ca^{2+}, because of a higher susceptibility to PTP opening. Moreover, Moro et al. (2010) have found that oestrogen deficiency inhibits the MPTP in isolated murine mitochondria and supplementation of mice with E_2 restores this process. In conclusion, our findings strongly suggest that E_2 is involved in the modulation of PTP. Possibly, E_2 genomic action is associated with the expression of proteins that modulate PTP opening. E_2 has been suggested to affect the expression of numerous mitochondria-interacting proteins, regulators or components of PTP, such as the antiapoptotic protein Bcl-2 (Fan et al., 2008; Moro et al., 2010; Zhi et al., 2007) and adenine nucleotide translocator (Moro et al., 2010; Too et al., 1999). Whether E_2 modulation on PTP opening has a bearing, in our experimental model, on the activation of pro-apoptotic pathways remains to be thoroughly ascertained. Further experiments are required to determine whether and the manner in which the regulation of PTP opening by oestrogens affects MPTP.

8.2 Ranking of Trophic Effects from Oestrogen Exposure of Primary Cell Cultures from Foetal Bovine Cerebellum

During brain development, growth, differentiation and circuit formation of the neural cells are under the influence of neuroactive steroids. It is widely known that neurosteroids are modulators of cell morphology and motility in different cellular lines and the consequence of estradiol-induced neurotrophism in early developmental stages is filopodial outgrowth (de Lacalle, 2006; Sanchez and Simoncini, 2010). In mammals, the brains of males and females show differences in the volume of specific hypothalamic nuclei, in the density of neurons, in the complexity of dendritic arborizations and even in the expression of neuropeptides (de Lacalle, 2006). Most of the published data discuss the key role that oestrogens play, through the regulating effects of oestrogen receptors (ERs) and the enzyme aromatase 450, in the areas of the central nervous system (CNS) that are principally managed by neurosteroids, such as the hypothalamus (the main sexually dimorphic region of the CNS), the hippocampus and the cerebral cortex involved in cognitive functions and memory (de Lacalle, 2006; Montelli et al., 2012; Peruffo et al., 2011).

To increase insight into the effect of E_2 on differentiation of neural growing in vitro, the effects of oestrogens on neuritic development were evaluated with or without 100 nM 17β-estradiol added to the medium components.

Cerebellar neurons from sexually segregate bovine foetuses were cultured following an established laboratory protocol (Peruffo et al., 2008). This study focussed on understanding the trophic actions of E_2 on the growth of neurons in primary cultures obtained from foetal bovine cerebellum. Taking again into consideration the experimental data of Montelli et al. (2016), in this work we aim at ranking several primary cell culture populations with respect to the possible trophic effect induced by oestrogen exposure.

8.2.1 Introduction

In the CNS neurosteroids can affect several functions to regulate the growth and differentiation of cells during development till adulthood (Mellon and Vaudry, 2001; Tsutsui, 2008). Neurosteroids influence higher cognitive functions, pain mechanisms, fine motor skills, neuroprotection and neuroregeneration as well as the formation of sexually dimorphic neural circuits (Chowen et al., 2000).

The effects of neurosteroids are mediated by steroid hormones receptors which can act both as nuclear transcription factors, that regulate gene expression, as well as modulators of rapid cytoplasmic signal transduction pathways activated at the plasma membrane (McCarthy, 2008). At the basis of brain plasticity is the ability of neurons to remodel their mutual connections. When a cell remodels its shape, the actin cytoskeleton is reorganized, and protrusive membrane structures, such as filopodia, are formed at the leading edge: these structures generate the locomotive force in migrating cells and are critical in the generation of cell–cell interconnections.

It is widely known that neurosteroids are modulators of cell morphology and motility in different cellular lines and the consequence of E_2-induced neurotrophism in early developmental stages is filopodial outgrowth (de Lacalle, 2006; Sanchez and Simoncini, 2010). Depending on the brain region, E_2 can either increase or decrease both the degree of dendritic branching as well as the density of dendritic spines; moreover E_2 continues to act on mature neurons regulating their structural plasticity at mature synapses (Sanchez and Simoncini, 2010).

In mammals, the brains of males and females show differences in the volume of specific brain nuclei, in the density of neurons, in the complexity of dendritic arborizations and even in the expression of neuropeptides (de Lacalle, 2006). The volumetric differences of these regions and the microscopic differences in the morphometry of individual cells both represent the biological underpinnings for behaviour in adulthood.

Most of the published data discuss the key role that oestrogens play, through the regulating effects of ERs and P450Arom, in the areas of the CNS that are characterized by the main anatomical sexual differences, the hypothalamus (the main sexually dimorphic region of the CNS), the hippocampus and the cerebral cortex involved in cognitive functions and memory (de Lacalle, 2006; Peruffo et al., 2011). It is difficult to explore via in vivo research the wide range of effects associated with brain differentiation in large mammals with long gestation periods. We propose the in vitro model based on primary cerebellar culture because it represents a good system to study a steroid effect during brain growth and differentiation (Peruffo et al., 2008).

An in vitro experimental model represents a simplified system that can be easily controlled and yield reproducible results. Because of all the features mentioned previously, we think that the in vitro model that we propose may represent a helpful alternative for studying the role that oestrogens exert on neuronal morphology and the sexual differentiation of the CNS during early embryonic development.

Most of the data concerning the cellular and molecular mechanisms that regulate brain growth and differentiation through steroid actions derives from experimental data obtained in rodents, a species with a short gestation. In the present study we use the bovine as an alternative model to rodents. In comparative neuroscience, the bovine could be an alternative animal model useful for studying the role of oestrogen in the development of foetal neural structures. Bovine tissues are easily obtained at commercial slaughterhouses and allow considerable saving of experimental animal lives. This species has a large encephalon and the duration of gestation is long (9 months). The bovine brain is one of the largest among mammals and highly convoluted, a fact that may be of interest when comparisons are made with the human.

In this work, to clarify the action of E_2 on foetal bovine cerebellum, we analysed the trophic actions of E_2 on cellular growth and dendritic development in neurons by in vitro studies using foetal bovine foetuses. We monitored cerebellar culture from both male and female bovine foetuses, pointing out the differences that took place as a result of E_2 treatment.

Data were collected by measuring morphometric parameters in neurons. Our results suggest that E_2 produces a strong trophic effect on neurons, highlighting that effects of E_2 on neurons from male cerebella were stronger than effects of E_2 on neurons from female foetuses. The present study moreover illustrates the potential use of tissue cultures for analysis of morphological change in the CNS during development of a long gestation mammalian species.

8.2.2 Materials and Methods

At least 10 randomly selected fields, each one containing readily identifiable neurons, were photographed using a Leika TCS confocal microscope equipped with a laser beam system (Figure 8.4).

All images were acquired at the same magnification with oil immersion and saved with a pixel distribution of 1024 × 1024. Z stacks were taken with 30 steps every 0.12 μm. Morphological analysis was manually performed utilizing the ImageJ software package using the line segment tool and the scaling option. Measurements are expressed in micrometres.

On a total of 829 identified neuron, in our analysis we measured the following morphological endpoints (see Figure 8.5): the whole area and the

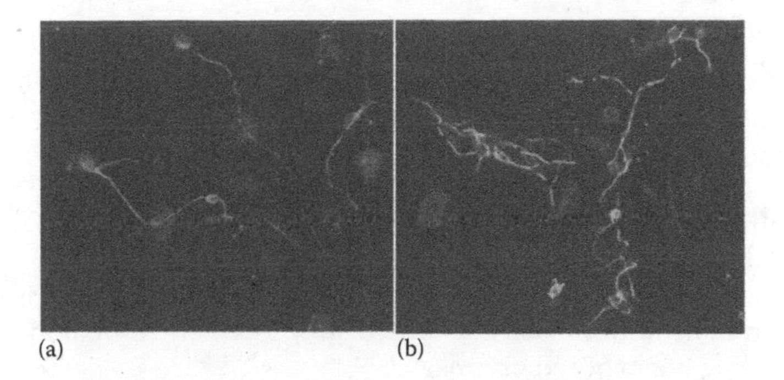

FIGURE 8.4
An example of one digital photo of an E_2 exposed and not-E_2 exposed cell culture. (a) Sample image of one not-E_2 exposed primary cell culture. (b) Sample image of one E_2 exposed primary cell culture.

whole perimeter of neuronal and glial cells (somata; represented by the black ellipsis in Figure 8.5); the possible presence of cells with primary, secondary branches, and with third- and fourth-order branches; the total number of each order of dendritic branches per cell of any given order (labeled in Figure 8.5 with the number 1, 2, 3 and 4); the total branch length (i.e., the sum of all dendritic segments) per hierarchic order per neuron and the total branch diameter (i.e., the sum of all dendritic diameters) per order per neuron.

Note that branch number, length and diameter can be observed only when the neurons have at least one branch, so that when there are no branches the value is missing, which is a typical situation when data dependent on the treatment effect are missing.

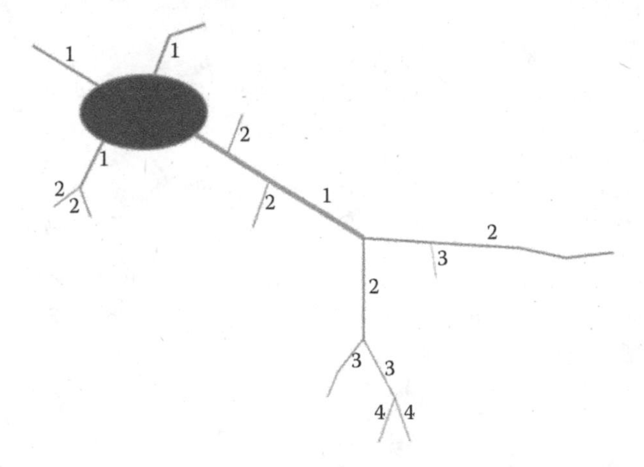

FIGURE 8.5
Simplified representation of the considered morphological endpoints where numbers represent the order of dendritic branches.

8.2.3 Results

What we expect is that both male and female E_2-exposed primary cell cultures should show a trophic effect to oestrogen exposure in such a way that all morphological endpoints should increase in size and number. Table 8.3 reports some descriptive statistics on the four considered groups, that is, male control (MC), male E_2-exposed (ME_2), female control (FC) and female E_2-exposed (FE_2).

From Table 8.3, it appears that somata and branch measurements seem to increase in male and female E_2-exposed groups when compared with the corresponding nonexposed groups.

To formally compare the morphological endpoints of E_2-exposed versus control primary cells we performed a stratified by gender multivariate two-sample permutation test analysis (Pesarin and Salmaso, 2010), where the multivariate combined permutation tests were calculated using the Fisher combination function and the univariate permutation tests in Table 8.4 have been adjusted by multiplicity using the Simes–Hommel method (Westfall

TABLE 8.3

Descriptive Statistics on Morphological Endpoints by Treatment Group (Sample Mean and Std. Deviation, in Parentheses, or Percentage)

Endpoint	FC (N = 249)	MC (N = 229)	FE_2 (N = 213)	ME_2 (N = 168)
Somata area (μm²)	1182 (423)	1070 (315)	1372 (600)	1121 (379)
Somata perimeter (μm)	149 (35)	141 (32)	161 (43)	141 (32)
Presence of primary branches	100.0%	100.0%	100.0%	100.0%
Presence of second-order branches	51.0%	27.5%	76.5%	64.3%
Presence of third-order branches	10.4%	2.2%	29.1%	11.3%
Presence of fourth-order branches	0.8%	0.0%	6.6%	0.6%
Number of primary branches	2.1 (1.0)	1.9 (0.8)	2.6 (1.2)	2.0 (0.9)
Number of second-order branches	2.2 (1.1)	2.1 (0.7)	3.0 (1.6)	2.0 (0.8)
Number of third-order branches	2.2 (1.3)	1.4 (0.5)	2.4 (1.0)	2.1 (0.7)
Number of fourth-order branches	2.0 (0.7)	–	2.6 (1.2)	2.0 (–)
Primary branches length (μm)	253.8 (163.6)	217.5 (119.1)	297.9 (181.3)	239.7 (143.1)
Second-order branches length (μm)	167.0 (125.7)	140.4 (113.8)	251.9 (200.3)	125.6 (101.3)
Third-order branches length (μm)	90.0 (53.0)	125.5 (121.8)	135.8 (100.3)	101.4 (96.0)
Fourth-order branches length (μm)	52.1 (38.1)	–	146.9 (153.4)	56.1 (–)
Primary branches diameter (μm)	7.4 (3.7)	7.1 (3.0)	11.0 (5.2)	8.1 (4.2)
Second-order branches diameter (μm)	6.3 (3.3)	6.6 (3.1)	10.2 (5.7)	7.4 (3.8)
Third-order branches diameter (μm)	5.8 (4.0)	4.4 (2.6)	7.8 (3.5)	7.3 (3.1)
Fourth-order branches diameter (μm)	4.3 (0.3)	–	7.6 (4.0)	6.4 (–)

et al., 2011). Note that we added also some intermediate combinations with respect to several suitable endpoint domains to highlight any possible global significance towards sets of similar endpoints.

Permutation p-values of Table 8.4 should be interpreted from bottom to top: considering a significance α-level equal to 5%, first of all we start from the final combined test suggesting that there is a strong evidence of a global overall difference in the morphometric cell structure of E_2-exposed cells when compared with the control cells (p-value = 0.0005). This multivariate global difference is found on both gender strata because there is a significant global difference in the within-gender combined tests separately either for male (p-value = 0.0045) and female (p-value = 0.0013) cells. When considering

TABLE 8.4

Stratified Two-Sample Permutation p-Values for Comparing E_2-Exposed vs. Control Primary Cells: Univariate Permutation Tests, Combined Tests by Domain and by Gender and Final Overall Combined Permutation Test (Values in Boldface Highlight 5% Significant p-Values)

	Gender	
Morphological Endpoint	M	F
Somata area	0.3752	**0.0090**
Somata perimeter	1.0000	**0.0045**
Somata size	0.3752	**0.0045**
Presence of primary branches	1.0000	1.0000
Presence of second-order branches	**0.0090**	**0.0045**
Presence of third-order branches	**0.0045**	**0.0030**
Presence of fourth-order branches	0.7319	**0.0022**
Branches presence	**0.0045**	**0.0022**
Primary branches length	0.2789	**0.0098**
Second-order branches length	0.6909	**0.0020**
Third-order branches length	0.9798	**0.0405**
Fourth-order branches length	1.0000	0.5283
Branches length	0.2789	**0.0020**
Primary branches diameter	**0.0120**	**0.0015**
Second-order branches diameter	0.3463	**0.0013**
Third-order branches diameter	0.3103	**0.0387**
Fourth-order branches diameter	1.0000	0.3624
Branches diameter	**0.0120**	**0.0013**
Within-sex combined test	**0.0045**	**0.0013**
Final combined test	**0.0005**	

the combination by domain, some different results are found in the male and female stratum; in particular it appears that male cells show a smaller set of domains where the difference do exist (i.e., somata size and branch length). In general, because the number of univariate significant p-values is greater for female than for male cells, it seems that female E_2-exposed cells show a much stronger trophic effect.

A multivariate ranking analysis is then advised to find a possible global ranking of the trophic effect for the four classes: female control (FC), male control (MC), male E_2-exposed (ME_2) and female E_2-exposed (FE_2). Owing to the large amount of missing data (see Table 8.3), we leave out from the ranking analysis the third- and fourth-order branches data. Results of ranking analysis are displayed in Tables 8.5 and 8.6, where the univariate and multivariate directional permutation p-values (calculated with 2000 CMC permutations) have been considered and where we set up the ranking problem as follows.

- As test statistics, we used the difference of sample means in case of missing values ${}_{DM\text{-}MV}T$ for the endpoints related to branch number, length and diameter, while for the remaining endpoints we used the difference of sample means ${}_{DM}T$.
- As combining function, we applied the Fisher's combining function.
- As correction for multiplicity we applied the Shaffer (1986) method.

8.2.4 Discussion and Conclusions

From ranking analysis results ($\alpha = 5\%$), it is worth noting that, as expected, both male and female E_2-exposed primary cell cultures show a significant trophic effect when compared with the corresponding control group. Interestingly, the control cells from female samples have larger morphological endpoints than the control cells from male samples; moreover, the E_2-exposed male cells and the female control cells do not show globally different morphological features because they both rank in the second position. These are interesting results that suggest new insights on the relevance of the sexual differentiation process during brain development. Specifically, the ability of

TABLE 8.5

Details of Global Ranking Results for the Four Gender-Exposed Groups ($\alpha = 5\%$)

Treatment	FC	MC	FE_2	ME_2
Global Ranking	**2**	**4**	**1**	**2**
FC		**.000**	1.00	.060
MC	1.00		1.00	.933
FE_2	**.000**	**.001**		**.001**
ME_2	.378	**.009**	1.00	

TABLE 8.6

Morphometric Endpoint-by-Endpoint Univariate Directional Permutation p-Values

	FC	MC	FE_2	ME_2	FC	MC	FE_2	ME_2	FC	MC	FE_2	ME_2
	Somata Area				**Somata Perimeter**				**Presence of Prim. Branches**			
FC		**.000**	1.00	.066		**.002**	.999	**.010**		1.00	1.00	1.00
MC	.999		1.00	.923	.998		1.00	.595	1.00		1.00	1.00
FE_2	**.000**	**.000**		**.000**	**.001**	**.000**		**.000**	1.00	1.00		1.00
ME_2	.936	.076	1.00		.993	.402	1.00		1.00	1.00	1.00	
	Presence of 2nd Ord. Branches				**Number of Prim. Branches**				**Number of 2nd Ord. Branches**			
FC		**.000**	1.00	.997		**.021**	1.00	.278		.138	1.00	.048
MC	1.00		1.00	1.00	.983		1.00	.902	.858		1.00	.347
FE_2	**.000**	**.000**		**.006**	**.000**	**.000**		**.000**	**.000**	**.000**		.000
ME_2	**.004**	**.000**	.998		.759	.118	1.00		.951	.660	1.00	
	Length of Prim. Branches				**Length of 2nd Ord. Branches**				**Diameter of Prim. Branches**			
FC		**.003**	.997	.181		.079	1.00	**.003**		.127	1.00	.965
MC	.997		1.00	.954	.926		1.00	.193	.865		1.00	.998
FE_2	**.003**	**.000**		**.006**	**.000**	**.000**		**.000**	**.000**	**.000**		**.000**
ME_2	.821	.043	1.00		.997	.806	1.00		.035	**.002**	1.00	
	Diameter of 2nd Ord. Branches											
FC		.727	1.00	.991								
MC	.266		1.00	.922								
FE_2	**.000**	**.000**		**.000**								
ME_2	**.009**	.073	1.00									

E_2 to activate trophic effects in the male cerebellar cells could be due to the important role of E_2 in inducing masculinization in the male brain during foetal development of the CNS while, on the contrary, female differentiation is the default (Peruffo et al., 2011). Moreover, in the last decade several publications have shown structural and physiological differences in the cerebellum between male and female, that is, volumetric differences in the grey matter (Fan et al., 2008), and dimorphic expression profiles of specific proteins (Abel et al., 2011). Furthermore, recently it has been noted that there are sex differences in the incidence of all the CNS disorders, including neurodevelopmental and neurodegenerative disorders such as Parkinson's and Alzheimer's diseases, autism and schizophrenia (Andreasen and Black, 2001; Keller et al., 2003). For this reason the investigation of morphological differences in the brain between males and females is important for designing novel hormone-based therapeutic agents that hopefully will have optimal effectiveness.

8.3 Ranking of Storing Conditions in the Tanning Industry to Preserve Bovine Skin

The goal of this study was to investigate the possible relation between storing conditions used in the tanning industry to properly preserve bovine skin and tissue degeneration with respect of histological dermal components. As input data we used true-colour images of histologically stained skin sections collected thanks to an open-source software named hisTOOLogy (http://www.centropiaggio.unipi.it/content/histoology-download). We used bovine skin samples evaluated at different periods and storing conditions (salting and refrigeration) after the skinning of the animals at the slaughterhouse. In a histological analysis performed using an optical microscope Montelli et al. (2015) noted a progressive reduction in time of the number of cellular nuclei, as well as collagen and acidic polysaccharide content in the dermal tissue, while the elastic fibre networks maintained their organization independently of the preservation method and time. Qualitative histological data performed by expert grader assessments were elaborated (Montelli et al., 2015) using suitable nonparametric permutation tests for ordered categorical responses within the randomized complete block design (Arboretti Giancristofaro et al., 2012). Results allowed a deeper analysis of the dermal components, revealing also a reduction of collagen content in salted hides and the progressive reduction of cellular nuclei and acidic polysaccharides, although with different dynamics of degradation.

Taking again into consideration the quantitative histological image–based data of Montelli et al. (2015), in this work we aimed to rank the storing conditions used in the tanning industry with respect to their ability to preserve from the tissue degeneration of bovine skin histological dermal components.

8.3.1 Introduction

The dermis is the layer of skin between the epidermis and the hypodermis. It consists of connective tissue divided into two parts: the superficial area called papillary region and a deep area called reticular dermis (Watt and Fujiwara, 2011). The structural components of the dermis are collagen, elastic fibers, extrafibrillar matrix and different types of cells (fibroblasts, immune cells, sensory and glandular cells). These dermal components provide a combination of flexibility and tensile strength to the skin (Bernstein and Uitto, 1996; Egbert et al., 2014; Varani et al., 2000).

Histological samples of the skin are widely used in pathology and dermatology, that is, to diagnose skin cancers and other diseases, and in forensic medicine to establish the time of death (Sadimin and Foran, 2012). Animal skin is an important byproduct of the slaughtering process, and leather industries are interested in postmortem alterations of the skin components, as scientific knowledge of dermal modifications can influence the quality of tanned skin, improve the production process and reduce waste and pollution.

Traditional histological analysis is based on expert grade observations of the morphology and distribution of cells and specific structural components of the tissue. However, even the most accurate qualitative investigation may lead to misjudgements, misinterpretations and subjective observer errors. Scientific analyses should be reproducible, free from potential involuntary human bias and allow comparison of results.

Image analysis is an emerging discipline in the biomedical sciences, and digital histopathological slides are widely used to diagnose the presence and progression of cancer (Pantanowitz et al., 2013). In particular, histological sections can be used for specific measurements coupling a digital acquisition with a standardized staining procedure. After image digital scanning, data-processing algorithms can be implemented to derive useful quantitative information for histological analysis. Bearing this in mind, we developed a computer-based method that enables the quantitative determination of dermal structural components by using scanned sections of bovine dermal skin, which we previously stained using specific and standardized histological procedures. The automatic measurement of dermal components yields data expressed as percentage of stained surface and thus allows comparison and relative statistical analysis of specimens sampled at different times and/or preserved in diverse ways.

To investigate the tissue degeneration process, we used bovine skin specimens sampled at different times and/or preserved in a diverse manner. In fact the bovine skin is a well-known model, easy to collect and allows a wide range of histological studies, such as the degradation of samples. Sammarco (1978) reported that monitoring the slaughtering process, the decay of fresh dermal tissue properties starts immediately after the machine-operated removal of the skin from the animal. However, only a few studies were performed with the aim to evaluate the decay of the skin structure or to quantify the changes in skin structure and composition (Baroni et al., 2012; Bhattacharyya and Thomas, 2004; Fantasia et al., 2013; Uitto et al., 1989). Most of the studies reported in the literature describe human cutaneous aging during life or histological changes of the human skin in the postmortem period (Bardale et al., 2012; Baumann, 2007).

With the emerging need for reliable data for the quantitative evaluation of specific histological parameters of interest, we here aim to integrate image analysis and statistical knowledge to develop a method for the analysis of histological samples. Using bovine dermal skin as a standard reference, we demonstrate that the method here proposed highlights histological differences that cannot be detected with standard observations.

8.3.2 Materials and Methods

8.3.2.1 Animal and Specimen Collection

The bovine hides were obtained from commercial abattoirs. Animals were treated according to the European Community Council directive concerning animal welfare during the commercial slaughtering process (86/609/EEC) and were constantly monitored under mandatory official veterinary medical care. Skin samples were taken from a series of adult male bovine (total number 8) from the hips of each animal. A detailed sampling plan was organized, as shown in Table 8.7, where we labelled the different sampling times as

- T0: Four independent samples were collected from hides just after the skinning; these are referred to as *fresh hide control samples.*
- T1: Five hours after bovine death, four independent samples were stored for 11 days in a refrigerator at 4°C; these are referred to as *refrigerated hides.*
- T2: Five hours after bovine death, four independent samples were stored in salt for 19 days; these are referred to as *short time salted hides.*
- T3: Five hours after bovine death, four independent samples were stored in salt for 84 days; these are referred to as *long time salted hides.*

8.3.2.2 Histological Staining

Skin samples 10 cm^2 wide and 1 cm thick were collected from the hips of the animals using a scalpel blade, and fixed for 24 h in 4% formaldehyde. Each sample was reduced to fragments of 1 cm^2, dehydrated and embedded in paraffin. From each sample, transverse sections of 4 μm thickness were obtained using a Leica R M2035 microtome. Slides were stained with conventional laboratory methods: (1) haematoxylin and eosin (H&E) for identification of general structure; (2) Masson's trichrome technique and (3) orcein staining, to evaluate degeneration of collagen and elastic fibers and (4) periodic acid-Schiff–positive technique (Alcian Blue – PAS) to detect neutral and

TABLE 8.7
Sampling Time and Plan Organized to Collect Bovine Hides

Sampling Time	Conditions	Description
T0	Fresh hides	Four fresh hide controls (collected a few minutes after the skinning)
T1	Refrigerated hides	Four refrigerated hides (conserved at 4°C 5 h after the bovine death, sampled 11 days after the skinning)
T2	Short time salted hides	Four short time salted hides (conserved in salt 5 h after the bovine death, sampled 19 days after the skinning)
T3	Long time salted hides	Four long time salted hides (conserved in salt 5 h after bovine death sampled 84 days after the skinning)

acidic polysaccharides. All methods were performed according to Bonucci (1981). The sections were then dehydrated and cover-slipped with balsam for light microscopy, and subsequently visualized using both 10× and 20× optical zooms with an Olympus DP70 RGB camera controlled by Cell Software (Olympus), interfacing a microscope Olympus Vanox AHBT3. Figure 8.6 displays several sample images by stained methods, representing slides of fresh (T0), refrigerated (T1), short time salted (T2) and long time salted hides (T3).

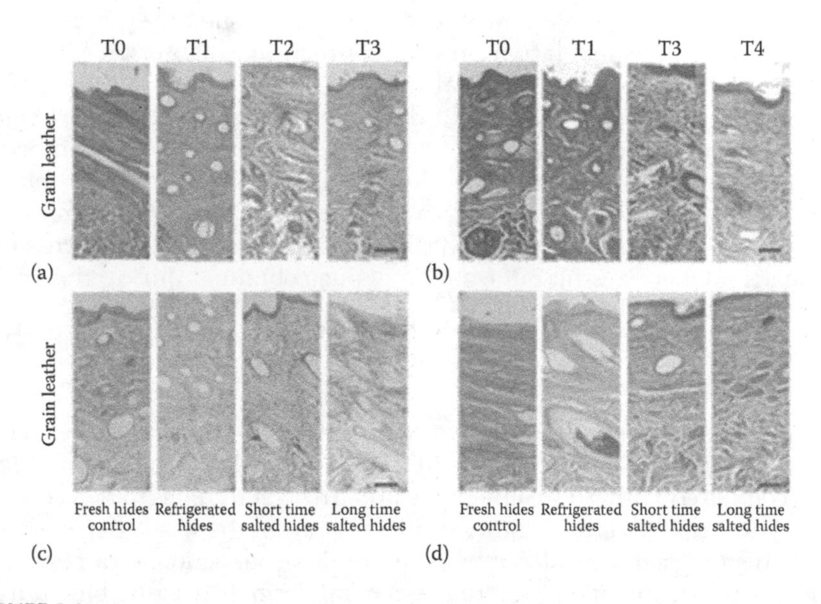

FIGURE 8.6
Sample images by treatment condition and staining method: (a) Hematoxylin and eosin (H&E). (b) Masson's trichrome technique. (c) Orcein staining. (d) Alcian Blue – PAS (periodic acid-Schiff–positive technique) staining. Bars represent 500 μm.

8.3.2.2.1 hisTOOLogy Software Analysis

Images were visualized using an optical microscope (Olympus Vanox AHBT3, Italy) equipped with a RGB camera (Olympus DP70, Italy) on slides stained as previously described. For each histological section, $n = 5$ images were visualized and analyzed, using the purposely developed image analysis tool named hisTOOLogy (it can be requested at a.tirella@centropiaggio.unipi.it). This software was developed to handle and process optical microscopy images representing histological stained slices. In particular, hisTOOLogy returns the amount of colour per unit area of the most common histological dyes (H&E, Alcian Blue–PAS, orcein and Masson's trichrome). Separately for two dermal layers, that is papillary and reticular, three different sections were evaluated for each one of the four treatment conditions T0,…,T3 and five independent images were considered. Thus, for each staining we obtained a 40 × 6 dataset where each single value was calculated as the average of three sections. For example, for the H&E staining, the six columns are labelled as Hemat_10×, Hemat_20×, Eos_10×, Eos_20×, White_10×, White_20×, that is, they represent the percentage relative amount of haematoxylin (deep blue-purple), eosin (pink) and white, measured using both 10× and 20× optical zooms.

HisTOOLogy is an open-source software, developed in a MATLAB® environment, performing both image processing and analysis. A user-friendly graphical user interface (GUI) allows a fast and easy track of analyses performed by nonexpert users. Specifically, hisTOOLogy returns a value representing the amount of 'colour' per tissue area. This value is obtained after a segmentation process, defined via RGB or CYMK channel thresholds, for each histological staining (e.g., Masson, H&E). The GUI allows (1) loading of an image; (2) definition of the pixel size; (3) selection of the stain used; (4) identification of the regions carrying information, separating them from the background; (5) selection of the optimal threshold for each color in the stain (enabled only for expert users) and (6) acquisition of the values. Finally, hisTOOLogy automatically creates a datamatrix, where all the information/values obtained during the analysis are stored.

8.3.3 Results

In this section we present the multivariate ranking analysis results related to the H&E staining. The main goal is to rank the T1, T2, T3 storage conditions when compared with T0 treatment (i.e., the fresh skin hides, our gold standard) with respect to their ability to preserve from the tissue degeneration of skin histological dermal components. In this connection, as a reminder, H&E colours are intended to stain the general dermal structure blue-purple and pink, that is, nuclei of cells (and a few other objects, such as keratohyalin granules and calcified material) and eosinophilic structures. In this way, lower blue-purple and pink percentages than the gold standard T0 indicate that the general dermal structure has been damaged. In a complementary way, higher white percentages than the gold standard T0 suggest tissue degeneration as well.

A first look at experimental data is provided by radar graphs in Figure 8.7, where sample means of colour percentages by dermal layer are reported. It

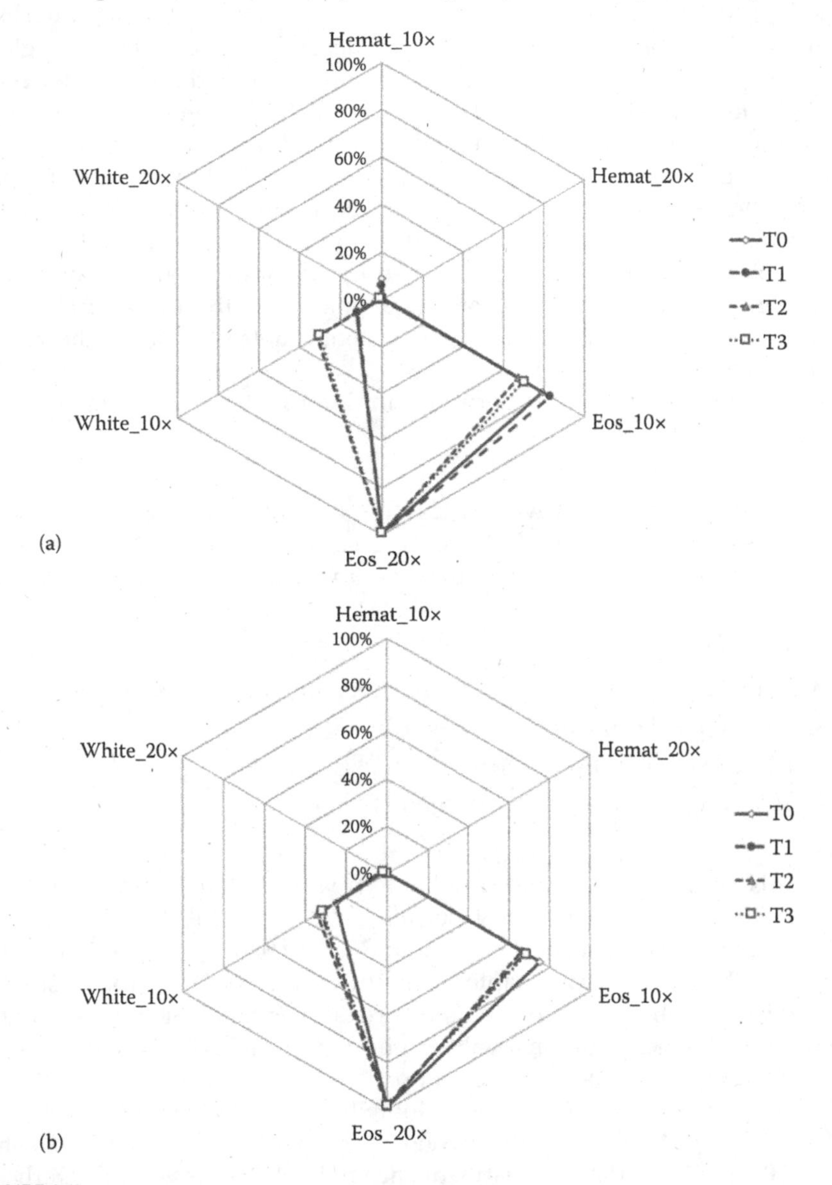

FIGURE 8.7
Radar graphs of colour percentages by layer and treatment. (a) Colour percentage sample means measured in the papillary layer. (b) Colour percentage sample means measured in the reticular layer.

appears that eosin stain percentage and white may be seemingly affected by the storage conditions, especially in the papillary layer.

The main objective of the proposed ranking analysis is to rank the four investigated treatments using experimental data from H&E staining colours that can be viewed as proxies for the possible tissue degeneration of skin histological dermal components. For this goal, because the staining H&E colours was measured in two different dermal layers, a stratified global ranking analysis is advised. We recall that by 'stratified analysis' we mean that several separated ranking analyses for each stratum (layer) along with an overall ranking analysis should be performed.

It is worth noting that, to keep consistency with the blue-purple and pink colors, the white color percentage must be previously transformed via a reciprocal function so that the rule 'the higher the better' holds for all the three colours. Accordingly, a higher estimated ranking position of a given treatment, possibly equal or close to the gold standard T0, has to be interpreted as showing that the related treatment is able to preserve the general dermal structure.

The setting of the storage conditions ranking problem (see Chapter 3) can be summarized as follows.

- Type of design: one-way MANOVA with stratification (four independent samples, i.e., four multivariate populations to be ranked, and one confounding factor to take into account, i.e., the specific skin layer)
- Domain analysis: no
- Type of response variables: numeric (six numerical responses)
- Ranking rule: 'the higher the better'
- Combining function: Fisher
- *B* (number of permutations): 2000

Details on multivariate and univariate treatment comparisons and ranking via directional permutation tests are presented in Tables 8.8 and 8.9.

All ranking analysis results can be obtained by using the companion book Web-based software (see Chapter 4) within the NPC Web Apps (lstat.gest.unipd.it/npcwebapps/). The reference dataset is labelled Dataset_HE.csv. To replicate the permutation p-values reported in Tables 8.8 and 8.9 exactly, a seed value equal to 1234 must be set up.

In agreement with findings in Montelli et al. (2015), ranking analysis results allow us to conclude that, when ranked with respect to the gold standard T0, all T1, T2 and T3 treatments are not able to preserve properly the general dermal structure but the best storing condition, that is, the one that damages the dermal structure less and makes it more similar to fresh skin, is T1 (conservation in refrigerator). In fact, while as expected T0 is ranked in

TABLE 8.8
Details of Global Ranking Results by Dermal Layer (α = 5%)

	All Layers			
Treatment	**T0**	**T1**	**T2**	**T3**
Global Ranking	**1**	**2**	**3**	**4**
T0		**.003**	**.003**	**.002**
T1	1.00		**.005**	**.003**
T2	1.00	1.00		**.002**
T3	1.00	1.00	1.00	
	External Papillary Layer			
Treatment	**T0**	**T1**	**T2**	**T3**
Global Ranking	**1**	**2**	**3**	**4**
T0		**.005**	**.018**	**.009**
T1	1.00		**.008**	**.010**
T2	1.00	1.00		**.010**
T3	1.00	1.00	1.00	
	Internal Reticular Layer			
Treatment	**T0**	**T1**	**T2**	**T3**
Global Ranking	**1**	**2**	**2**	**2**
T0		**.015**	**.018**	**.009**
T1	1.00		.034	.189
T2	1.00	.069		.243
T3	1.00	.577	1.00	

the first position, all the other treatments are ranked in a worst ranking position; in particular T3 (long time salted hides) appears to be the worst storage condition as it is always ranked as the least preserving treatment. It is also worth noting that the stratified ranking analysis provides several interesting insights into how the storage conditions differently affect the dermal structure of distinct dermal layers. In fact, the within- layer estimated rankings are similar but not exactly the same; specifically if the goal is to better preserve the dermal structure of reticular layer, which is from the standpoint of the tanning industry the less valuable part of the tissue, then there is no difference between all three storage conditions because T1, T2 and T3 are tied in the second position. Conversely, this is not true when the goal is to preserve the dermal structure of the papillary layer, for which T1 and T2 appear better than T3.

8.3.4 Discussion and Conclusions

In this study we applied the multivariate ranking analysis for the comparison of several different storage conditions used in the tanning industry with

TABLE 8.9

H&E Colour-by-Colour Univariate Directional Permutation p-Values

						All Layers						
	T0	**T1**	**T2**	**T3**	**T0**	**T1**	**T2**	**T3**	**T0**	**T1**	**T2**	**T3**
		Hemat_ 10×				**Hemat_20×**				**Eos_10×**		
T0		**.021**	**.001**	**.001**		**.001**	**.001**	**.001**		.177	**.001**	**.001**
T1	.979		**.001**	**.001**	1.00		**.033**	**.001**	.824		**.003**	**.004**
T2	1.00	1.00		**.004**	1.00	.974		**.002**	1.00	.998		.900
T3	1.00	1.00	.997		1.00	1.00	1.00		1.00	.997	.100	
		Eos_20×				**1/White_ 10×**				**1/White_ 20×**		
T0		.046	.669	**.010**		.214	**.001**	**.001**		**.001**	**.001**	**.001**
T1	.955		.979	.339	.787		**.002**	**.003**	1.00		.070	**.005**
T2	.333	**.022**		**.001**	1.00	.999		.741	1.00	.930		**.001**
T3	.992	.66	1.00		1.00	.998	.260		1.00	.996	1.00	
						External Papillary Layer						
	T0	**T1**	**T2**	**T3**	**T0**	**T1**	**T2**	**T3**	**T0**	**T1**	**T2**	**T3**
		Hemat_10×				**Hemat_20×**				**Eos_10×**		
T0		**.013**	**.003**	**.003**		**.005**	**.003**	**.003**		.909	**.006**	**.003**
T1	.992		**.004**	**.004**	1.00		.032	**.004**	.096		**.004**	**.004**
T2	1.00	1.00		**.004**	1.00	.972		**.004**	.998	1.00		.879
T3	1.00	1.00	1.00		1.00	1.00	1.00		1.00	1.00	.124	

(Continued)

TABLE 8.9 (CONTINUED)

H&E Colour-by-Colour Univariate Directional Permutation p-Values

					External Papillary Layer							
	T0	**T1**	**T2**	**T3**	**T0**	**T1**	**T2**	**T3**	**T0**	**T1**	**T2**	**T3**
	Eos_20×				**1/White_10×**				**1/White_20×**			
T0		.990	.701	**.018**		.395	**.003**	**.003**		**.011**	**.003**	**.003**
T1	.015		**.013**	**.004**	.610		**.004**	**.004**	.994		**.011**	**.004**
T2	.310	.991		**.004**	1.00	1.00		.629	1.00	.993		**.004**
T3	.985	1.00	1.00		1.00	1.00	.375		1.00	1.00	1.00	
					Internal Reticular Layer							
	T0	**T1**	**T2**	**T3**	**T0**	**T1**	**T2**	**T3**	**T0**	**T1**	**T2**	**T3**
	Hemat_10×				**Hemat_20×**				**Eos_10×**			
T0		.934	**.003**	**.003**		**.005**	**.003**	**.003**		**.008**	**.022**	**.015**
T1	.071		**.004**	**.004**	1.00		**.010**	.948	.997		.356	.700
T2	1.00	1.00		.984	1.00	1.00		1.00	.981	.648		.740
T3	1.00	1.00	.076		1.00	.943	1.00		.988	.303	.263	
	Eos_20×				**1/White_10×**				**1/White_20×**			
T0		**.009**	.612	.050		**.008**	**.022**	**.015**		**.009**	.443	.064
T1	.996		1.00	.881	.997		.408	.689	.996		1.00	.881
T2	.396	**.004**		.032	.981	.596		.707	.560	**.004**		.029
T3	.953	.122	.971		.988	.314	.297		.939	.123	.975	

respect to their capability to preserve the histological components of dermal layer. Histological analyses are routinely used in the medical and biological fields, including forensic medicine and pathology, to describe structures of tissues (Bacci et al., 2011; Paterson et al., 2006). Evaluations rely mainly on the direct microscopic examination of tissue sections and expert evaluation based on the personal experience of the observer. However, this judgement is subjective and could lead to high variability (Andrion et al., 1995; Ismail et al., 1989). To overcome the limitations of the subjective-based histological analyses, in recent decades routine histological analyses have been coupled with different computational approaches (Kakasheva-Mazhenkovska et al., 2011; Kobayashi et al., 1999; Macenko et al., 2009; Miedema et al., 2012; Miot and Brianezi, 2010). Digital image acquisition and analysis coupled with standardized histological staining procedures allow high-quality data collection, more precise investigations, and comparable results that are not operator dependent.

Many previous studies were dedicated to the investigation of the dermal layer (e.g., cutaneous aging mechanisms and effects caused by solar radiations Fisher et al., 1996; Nishimori et al., 2001), while in forensic medicine there are few reports concerning the evaluation of postmortem histological changes in human skin (Bardale et al., 2012; Kovarik et al., 2005; Zherebtsov et al., 1977). The main topic debated in these studies was the determination of the postmortem interval (PMI), an important issue because of its possible significance in legal investigations. Moreover, the molecular complexity of the dermal layer has been studied with the aim of quantifying the proportion of different isoforms of collagen in the skin (Naylor et al., 2011).

The ranking analysis results on H&E staining data allow us to confirm the findings in Montelli et al. (2015): a dynamic of degradation of the general dermal structure was noticed, pointing out a specific ranking of the rate of degradation among different storage conditions and by distinct dermal layers (papillary vs. reticular). The proposed histological protocol, image processing and statistical ranking analysis could be extended to many different research areas, including pathology and forensic medicine. This methodology could have an important impact also on quality control processes and monitoring required by the leather industries, which are the main customers of the tanning industry, where histological analysis of the skin is crucial to improve manufacturing processing of leather goods.

Acknowledgement

The authors wish to thank the University of Padova (CPDA131533/13) for providing the financial support for this research.

References

Abel J. M., Witt D. M., Rissman E. F. (2011). Sex differences in the cerebellum and frontal cortex: Roles of estrogen receptor alpha and sex chromosome genes. *Neuroendocrinology*, 93(4), 230–240.

Andreasen N. C., Black D. W. (2001). *Introductory Textbook of Psychiatry*, Vol. 3. Washington, DC: American Psychiatric Publishing.

Andrion A., Magnani C., Betta P. G., Donna A., Mollo F., Scelsi M., Bernardi P., Botta M., Terracini B. (1995). Malignant mesothelioma of the pleura: Interobserver variability. *Journal of Clinical Pathology*, 48, 856–860.

Arboretti Giancristofaro R., Corain L., Ragazzi S. (2012). A comparison among combination-based permutation statistics for randomized complete block design. *Communications in Statistics – Simulation and Computation*, 41(7), 964–979.

Arnal J. F., Fontaine C., Billon-Galés A., Favre J., Laurell H., Lenfant F., Gourdy P. (2010). Estrogen receptors and endothelium. *Arteriosclerosis, Thrombosis, and Vascular Biology*, 30, 1506–1512.

Bacci S., DeFraia B., Romagnoli P., Bonelli A. (2011). Advantage of affinity histochemistry combined with histology to investigate death causes: Indications from sample cases. *Journal of Forensic Sciences*, 56, 1620–1625.

Bardale R. V., Tumram N. K., Dixit P. G., Deshmukh A. Y. (2012). Evaluation of histologic changes of the skin in postmortem period. *American Journal of Forensic Medicine and Pathology*, 33, 357–361.

Baroni E. R., Biondo-Simões M. de L., Auersvald A., Auersvald L. A., MontemorNetto M. R., Ortolan M. C., Kohler J. N. (2012). Influence of aging on the quality of the skin of white women: The role of collagen. *Acta Cirurgica Brasileira*, 27, 736–740.

Baumann L. (2007). Skin ageing and its treatment. *Journal of Pathology*, 211, 241–251.

Bernstein E. F., Uitto J. (1996). The effect of photodamage on dermal extracellular matrix. *Clinics in Dermatology*, 14, 143–151.

Berridge M. J., Bootman M. D., Roderick H. L. (2003). Calcium signalling: Dynamics, homeostasis and remodelling. *Nature Reviews Molecular Cell Biology*, 4, 517–529.

Bhattacharyya T. K., Thomas J. R. (2004). Histomorphologic changes in aging skin: Observations in the CBA mouse model. *Archives of Facial Plastic Surgery*, 6, 21–25.

Bonucci E. (1981). *Manuale di istochimica*. Rome: Lombardo.

Chowen J. A., Azcoitia I., Cardona-Gomez G. P., Garcia-Segura L. M. (2000). Sex steroids and the brain: Lessons from animal studies. *Journal of Pediatric Endocrinology and Metabolism*, 13(8), 1045–1066.

Corain L., Melas V. B., Salmaso L., Pepelyshev A. (2014). New insights on permutation approach for hypothesis testing on functional data. *Advances in Data Analysis and Classification*, 8, 339–356, 2014.

de Lacalle S. (2006). Estrogen effects on neuronal morphology, *Endocrine*, 29(2), 185–190.

De Stefani D., Raffaello A., Teardo E., Szabò I., Rizzuto R. (2011). A forty-kilodalton protein of the inner membrane is the mitochondrial calcium uniporter. *Nature*, 476, 336–340.

Egbert M., Ruetze M., Sattler M., Wenck H., Gallinat S., Lucius R., Weise J. M. (2014). The matricellular protein periostin contributes to proper collagen function and is downregulated during skin aging. *Journal of Dermatological Science*, 73, 40–48.

Fan L., Pandey S. C., Cohen R. S. (2008). Estrogen affects levels of Bcl-2 protein and mRNA in medial amygdala of ovariectomized rats. *Journal of Neuroscience Research*, 86, 3655–3664.

Fantasia J., Lin C. B., Wiwi C., Kaur S., Hu Y. P., Zhang J., Southall M. D. (2013). Differential levels of elastin fibers and TGF-β signaling in the skin of Caucasians and African Americans. *Journal of Dermatological Science*, 70, 159–165.

Ferraty F., Vieu P. (2006). *Nonparametric Functional Data Analysis*. Heidelberg: Springer-Verlag.

Fisher G. J., Datta S. C., Talwar H. S., Wang Z. Q., Varani J., Kang S., Voorhees J. J. (1996). Molecular basis of sun-induced premature skin ageing and retinoid antagonism. *Nature*, 379, 335–339.

Gu Y., Preston M. R., Magnay J., El Haj A. J., Publicover S. J. (2001). Hormonally-regulated expression of voltage-operated Ca^{2+} channels in osteocytic (MLO-Y4) cells. *Biochemical and Biophysical Research Communications*, 282, 536–542.

Guo J., Krause D. N., Horne J., Weiss J. H., Li X., Duckles S. P. (2010). Estrogen-receptor-mediated protection of cerebral endothelial cell viability and mitochondrial function after ischemic insult in vitro. *Journal of Cerebral Blood Flow & Metabolism*, 3, 545–554.

Ismail S. M., Colclough A. B., Dinnen J. S., Eakins D., Evans D. M., Gradwell E., O'Sullivan J. P., Summerell J. M., Newcombe R. G. (1989). Observer variation in histopathological diagnosis and grading of cervical intraepithelial neoplasia. *British Medical Journal*, 298, 707–710.

Kakasheva-Mazhenkovska L., Milenkova L., Kostovska N., Gjokik G. (2011). Histomorphometrical characteristics of human skin from capillitium in subjects of different age. *Prilozi*, 32, 105–118.

Keller A., Castellanos F. X., Vaituzis A. C., Jeffries N. O., Giedd J. N., Rapoport J. L. (2003). Progressive loss of cerebellar volume in childhood-onset schizophrenia. *American Journal of Psychiatry*, 160, 128–133.

Kobayashi A., Takehana K., Eerdunchaolu, Iwasa K., Abe M., Yamaguchi M. (1999). Morphometric analysis of collagen: A comparative study in cow and pig skins. *Anatomia, Histologia, Embryologia*, 28, 235–238.

Kovarik C., Stewart D., Cockerell C. (2005). Gross and histologic postmortem changes of the skin. *American Journal of Forensic Medicine and Pathology*, 26, 305–308.

Macenko M., Niethammer M., Marron J. S., Borland D., Woosley J. T., Guan X., Schmitt C., Thomas N. E. (2009). A method for normalizing histology slides for quantitative analysis. *Biomedical Imaging: From Nano to Macro, 2009. ISBI '09. IEEE International Symposium on*, 1107–1110.

McCarthy M. M. (2008). Estradiol and the developing brain. *Physiological Reviews*, 88, 91–134.

McGeady T. A., Quinn P. J., FitzPatrick E. S., Ryan M. T. (2006). *Veterinary Embryology*. Oxford: Blackwell.

Mellon S. H., Vaudry H. (2001). Biosynthesis of neurosteroids and regulation of their synthesis (review). *International Review of Neurobiology*, 46, 33–78.

Miedema J., Marron J., Niethammer M., Borland D., Woosley J., Coposky J., Wei S., Reisner H., Thomas N. E. (2012). Image and statistical analysis of melanocytic histology. *Histopathology*, 61, 436–444.

Miot H. A., Brianezi G. (2010). Morphometric analysis of dermal collagen by color clusters segmentation. *Anais Brasileiros de Dermatologia*, 85, 361–364.

Montelli S., Peruffo A., Zambenedetti P., Rossipal E., Giacomello M., Zatta P., Cozzi B. (2012). Expression of aromatase P450(AROM) in the human fetal and early postnatal cerebral cortex. *Brain Research*, 1475, 11–18.

Montelli S., Suman M., Corain L., Cozzi B., Peruffo A. (2016). Sexually diergic trophic effects of estradiol exposure on developing bovine cerebellar granule cells. *Neuroendocrinology*, in press.

Montelli S., Corain L., Cozzi B., Peruffo A. (2015). Histological analysis of the skin dermal components in bovine hides stored under different conditions. *Journal of the American Leather Chemists Association*, 110, 54–61.

Moreira P. I., Custódio J. B., Nunes E., Oliveira P. J., Moreno A., Seiça R., Oliveira C. R., Santos M. S. (2011). Mitochondria from distinct tissues are differently affected by 17β-estradiol and tamoxifen. *Journal of Steroid Biochemistry and Molecular Biology*, 123, 8–16.

Moro L., Arbini A. A., Hsieh J. T., Ford J., Simpson E. R., Hajibeigi A., Oz O. K. (2010). Aromatase deficiency inhibits the permeability transition in mouse liver mitochondria. *Endocrinology*, 151, 1643–1652.

Naylor E. C., Watson R. E., Sherratt M. J. (2011). Molecular aspects of skin ageing. *Maturitas*, 69, 249–256.

Nilsen J., Diaz Brinton R. (2003). Mechanism of estrogen-mediated neuroprotection: Regulation of mitochondrial calcium and Bcl-2 expression. *Proceedings of the National Academy of Sciences of the USA*, 100, 2842–2847.

Nishimori Y., Edwards C., Pearse A., Matsumoto K., Kawai M., Marks R. (2001). Degenerative alterations of dermal collagen fiber bundles in photodamaged human skin and UV-irradiated hairless mouse skin: Possible effect on decreasing skin mechanical properties and appearance of wrinkles. *Journal of Investigative Dermatology*, 117, 1458–1463.

Palmer A., Jin C., Reed J. C., Tsien R. Y. (2004). Bcl2-mediated alterations in endoplasmic reticulum Ca^{2+} analyzed with an improved genetically encoded fluorescent sensor. *Proceedings of the National Academy of Sciences of the USA*, 101, 17404–17409.

Palmer A. E., Giacomello M., Kortemme T., Hires S. A., Lev-Ram V., Baker D., Tsien R. Y. (2006). Ca^{2+} indicators based on computationally redesigned calmodulin-peptide pairs. *Chemistry & Biology*, 13, 521–530.

Pantanowitz L., Valenstein P. N., Evans A. J., Kaplan K. J., Pfeifer J. D., Wilbur D. C., Collins L. C., Colgan T. J. (2013). Review of the current state of whole slide imaging in pathology. *Journal of Pathology Information*, 2, 36.

Paterson S. K., Jensen C. G., Vintiner S. K., McGlashan S. R. (2006). Immunohistochemical staining as a potential method for the identification of vaginal epithelial cells in forensic casework. *Journal of Forensic Sciences*, 51, 1138–1143.

Peruffo A., Massimino M. L., Ballarin C., Carmignoto G., Rota A., Cozzi B. (2004). Primary cultures from fetal bovine brain. *NeuroReport*, 15, 1719–1722.

Peruffo A., Buson G., Cozzi B., Ballarin C. (2008). Primary cell cultures from fetal bovine hypothalamus and cerebral cortex: A reliable model to study P450Arom and alpha and beta estrogen receptors in vitro. *Neuroscience Letters*, 434(1), 83–87.

Peruffo A., Giacomello M., Montelli S., Corain L., Cozzi B. (2011). Expression and localization of aromatase P450AROM, estrogen receptor-α and estrogen receptor-β in the developing fetal bovine frontal cortex. *General and Comparative Endocrinology*, 172(2), 211–217.

Pesarin F., Salmaso L. (2010). *Permutation Tests for Complex Data: Theory, Applications and Software.* Wiley Series in Probability and Statistics. Chichester, UK: John Wiley & Sons.

Rasola A., Sciacovelli M., Pantic B., Bernardi P. (2010). Signal transduction to the permeability transition pore. *FEBS Letters*, 584, 1989–1996.

Ritchie A. K. (2008). Estrogen increases low voltage-activated calcium current density in GH3 anterior pituitary cells. *Endocrinology*, 132, 1621–1629.

Rizzuto R., Brini M., Murgia M., Pozzan T. (1993). Microdomains with high Ca^{2+} close to IP3-sensitive channels that are sensed by neighboring mitochondria. *Science*, 262, 744–747.

Sadimin E. T., Foran D. J. (2012). Pathology imaging informatics for clinical practice and investigative and translational research. *North American Journal of Medical Sciences*, 5, 103–109.

Salazar-Colocho P., Del Río J., Frechilla D. (2008). Involvement of the vascular wall in regenerative processes after CA1 ischemic neuronal death. *International Journal of Developmental Neuroscience*, 26, 541–550.

Sammarco U. (1978). *Tecnologia Conciaria*. Milan, IT: Editma-Rescaldina.

Sanchez A. M., Simoncini T. (2010). Extra-nuclear signaling of ERalpha to the actin cytoskeleton in the central nervous system, *Steroids*, 75, 528–532.

Shaffer J. P. (1986). Modified sequentially rejective multiple test procedure. *Journal of the American Statistical Association*, 81, 826–831.

Stevenson S., Sharpe T. D., Thronton J. (2009). Effects of oestrogen agonists on human dermal fibroblasts in an in vitro wounding assay. *Experimental Dermatology*, 18, 988–990.

Suman M., Giacomello M., Corain L., Ballarin C., Montelli S., Cozzi B., Peruffo A. (2012). Estradiol effects on intracellular Ca^{2+} homeostasis in bovine brain-derived endothelial cells. *Cell and Tissue Research*, 350(1), 309–318.

Takenouchi T., Iwamaru Y., Sato M., Yokoyama T., Shinagawa M., Kitani H. (2007). Establishment and characterization of SV40 large T antigen-immortalized cell lines derived from fetal bovine brain tissues after prolonged cryopreservation. *Cell Biology International*, 31, 57–64.

Too C. K., Giles A., Wilkinson M. (1999). Estrogen stimulates expression of adenine nucleotide translocator ANT1 messenger RNA in female rat hearts. *Molecular and Cellular Endocrinology*, 150, 161–167.

Tsutsui K. (2008). Neurosteroids in the Purkinje cell: Biosynthesis, mode of action and functional significance. *Molecular Neurobiology*, 37, 116–125.

Uitto J., Olsen D. R., Fazio M. J. (1989). Extracellular matrix of the skin: 50 years of progress. *Journal of Investigative Dermatology*, 92, 61S–77S.

Varani, J., Warner R. L., Gharaee-Kermani M., Phan S. H., Kang S., Chung J. H., Wang Z. Q., Datta S. C., Fisher G. J., Voorhees J. J. (2000). Vitamin A antagonizes decreased cell growth and elevated collagen-degrading matrix metalloproteinases and stimulates collagen accumulation in naturally aged human skin. *Journal of Investigative Dermatology*, 114, 480–486.

Vasudevan N., Pfaff D. W. (2008). Non-genomic actions of estrogens and their interaction with genomic actions in the brain. *Frontiers in Neuroendocrinology*, 29, 238–257.

Watt F. M., Fujiwara H. (2011). Cell-extracellular matrix interactions in normal and diseased skin. *Cold Spring Harbor Perspectives in Biology*, 3, 1–14.

Westfall P. H., Tobias R. D., Rom D., Wolfinger R. D. (2011). *Multiple Comparisons and Multiple Tests Using SAS*, 2nd ed. Cary, NC: SAS Institute.

Zherebtsov L. D., Vasilevskiĭ V. K., Bremzen S. A. (1977). Postmortem changes in skin color. *Arkhiv Patologii/Archives of Pathology*, 39, 52–58.

Zhi X., Honda K., Sumi T., Yasui T., Nobeyama H., Yoshida H., Ishiko O. (2007). Estradiol-17beta regulates vascular endothelial growth factor and Bcl-2 expression in HHUA cells. *International Journal of Oncology*, 31, 1333–1338.

Index

Page numbers followed by f and t indicate figures and tables, respectively.

D

E

F

N

S